Current Topics in Membranes, Volume 65

Current Topics in Membranes, Volume 65

Claudins

Current Topics in Membranes, Volume 65

Claudins

Edited by

Alan S. L. Yu
Division of Nephrology, Department of Medicine and
Department of Physiology and Biophysics
University of Southern California Keck School of Medicine
Los Angeles, California, USA

ELSEVIER

AMSTERDAM • BOSTON • HEIDELBERG • LONDON
NEW YORK • OXFORD • PARIS • SAN DIEGO
SAN FRANCISCO • SINGAPORE • SYDNEY • TOKYO
Academic Press is an imprint of Elsevier

Academic Press is an imprint of Elsevier
30 Corporate Drive, Suite 400, Burlington, MA 01803, USA
525 B Street, Suite 1900, San Diego, CA 92101-4495, USA

First edition 2010

ISBN: 978-0-12-381039-7
ISSN: 1063-5823

For information on all Academic Press publications
visit our website at elsevierdirect.com

Printed and bound by CPI Group (UK) Ltd, Croydon, CR0 4YY

Transferred to Digital Print 2011

To my mother, Patsy, and my late father, P.K., for their teaching and values

To my wife, Jodie, and son, Daniel, for their support, understanding and inspiration

Contents

Contributors

Numbers in parentheses indicate the pages on which the authors' contributions begin.

Salah Amasheh (195) Institute of Clinical Physiology, Campus Benjamin Franklin, Charité, Berlin, Germany

Roland Bücker (195) Department of General Medicine, Campus Benjamin Franklin, Charité, Berlin, Germany

Ingolf E. Blasig (97) Leibniz Institute for Molecular Pharmacology (FMP), Berlin-Buch, Germany

Zea Borok (177) Department of Medicine; Department of Biochemistry and Molecular Biology, Will Rogers Institute Pulmonary Research Center, Keck School of Medicine, University of Southern California, Los Angeles, California, USA

Edward D. Crandall (177) Department of Medicine; Department of Pathology, Keck School of Medicine; Department of Chemical Engineering and Materials Science, Viterbi School of Engineering, University of Southern California, Los Angeles, California, USA

Christopher Davis (273) Institute for Biomedical Research, University of Birmingham, Birmingham, United Kingdom

Jérôme Devaux (229) Département Signalisation Neuronale, CRN2M, UMR, CNRS, Université de la Méditerranée—Université Paul Cézanne, IFR Jean Roche, Marseille, France

Per Flodby (177) Department of Medicine, Will Rogers Institute Pulmonary Research Center, Keck School of Medicine, University of Southern California, Los Angeles, California, USA

Michael Fromm (39) Institute of Clinical Physiology, Campus Benjamin Franklin, Charité, Berlin, Germany

Mikio Furuse (1) Division of Cell Biology, Department of Physiology and Cell Biology, Kobe University Graduate School of Medicine, Kobe, Japan

Bozena Fykkolodziej (229) Center for Molecular Medicine and Genetics, Wayne State University School of Medicine, Detroit, Michigan, USA

Dorothee Günzel (39) Institute of Clinical Physiology, Campus Benjamin Franklin, Charité, Berlin, Germany

Erika Garay (113) Department of Physiology, Biophysics and Neuroscience, Center for Research and Advanced Studies (Cinvestav), México D.F., México

Lorenza González-Mariscal (113) Department of Physiology, Biophysics and Neuroscience, Center for Research and Advanced Studies (Cinvestav), México D.F., México

Alexander Gow (229) Center for Molecular Medicine and Genetics; Carman and Ann Adams Department of Pediatrics; Department of Neurology, Wayne State University School of Medicine, Detroit, Michigan, USA

Helen J. Harris (273) Institute for Biomedical Research, University of Birmingham, Birmingham, United Kingdom

Reiner F. Haseloff (97) Leibniz Institute for Molecular Pharmacology (FMP), Berlin-Buch, Germany

Jianghui Hou (151) Department of Internal Medicine—Renal Division, Washington University School of Medicine, St. Louis, Missouri, USA

Kwang-Jin Kim (177) Department of Medicine; Department of Physiology and Biophysics, Keck School of Medicine; Department of Pharmacology and Pharmaceutical Sciences, School of Pharmacy; Department of Biomedical Engineering, Viterbi School of Engineering, University of Southern California, Los Angeles, California, USA

Martin Konrad (151) Department of Pediatrics, University of Münster, Münster, Germany

Susanne M. Krug (39) Institute of Clinical Physiology, Campus Benjamin Franklin, Charité, Berlin, Germany

Jane A. McKeating (273) Institute for Biomedical Research, University of Birmingham, Birmingham, United Kingdom

Patrice J. Morin (293) Laboratory of Cellular and Molecular Biology, National Institute on Aging, NIH, Biomedical Research Center;

Department of Pathology, Johns Hopkins Medical Institutions, Baltimore, Maryland, USA

Jörg Piontek (97) Leibniz Institute for Molecular Pharmacology (FMP), Berlin-Buch, Germany

Miguel Quirós (113) Department of Physiology, Biophysics and Neuroscience, Center for Research and Advanced Studies (Cinvestav), México D.F., México

Rita Rosenthal (39) Institute of Clinical Physiology, Campus Benjamin Franklin, Charité, Berlin, Germany

Eveline E. Schneeberger (21) Molecular Pathology Unit, Department of Pathology, Massachusetts General Hospital, Charlestown, Massachusetts, USA

Jörg-Dieter Schulzke (195) Department of General Medicine; Department of Gastroenterology, Campus Benjamin Franklin, Charité, Berlin, Germany

Michael Schumann (195) Department of Gastroenterology, Campus Benjamin Franklin, Charité, Berlin, Germany

Tammy-Claire Troy (255) Regenerative Medicine Program, Sprott Centre for Stem Cell Research, Ottawa Hospital Research Institute, Ottawa, ON, Canada

Kursad Turksen (255) Regenerative Medicine Program, Sprott Centre for Stem Cell Research, Ottawa Hospital Research Institute; Department of Cellular and Molecular Medicine, Faculty of Medicine, University of Ottawa; Department of Medicine, Division of Endocrinology; Department of Medicine, Division of Dermatology, Ottawa Hospital, Ottawa, ON, Canada

Blanca L. Valle (293) Laboratory of Cellular and Molecular Biology, National Institute on Aging, NIH, Biomedical Research Center, Baltimore, Maryland, USA

Alan S. L. Yu (79) Division of Nephrology, Department of Medicine and Department of Physiology and Biophysics, University of Southern California Keck School of Medicine, Los Angeles, California, USA

Preface

Alan S. L. Yu

Tight junctions are essential for epithelial tissues to form barriers between body compartments, and also to selectively allow passive transport of water and solutes as needed. The late Shoichiro Tsukita, along with his wife and scientific collaborator, Sachiko Tsukita, played a pivotal role in developing this field. Of his many discoveries, the one that probably has had the greatest scientific impact, as he himself acknowledged, was the identification of claudins, which was published in 1998. We now know that claudins, uniquely among tight junction proteins, can polymerize into tight junction strands, mediate intercellular adhesion, constitute the paracellular barrier in epithelia and endothelia, and form pores that allow selective transport of small ions through that barrier. Research into the biology of claudins, and their role in the function of epithelialized organs, has increased at a phenomenal rate, and so this book is intended to survey the current state of knowledge in this field and reflect on future directions. This book is targeted not solely to afficionados in this field, for whom this will hopefully serve as a valuable reference, but more generally to scientists and trainees in epithelial biology who might wish to gain a broad perspective on what is known about the role of claudins in epithelial function.

This book is loosely divided into five sections. The introductory chapter is written by Mikio Furuse, who spearheaded the discovery of claudins in Tsukita's laboratory. Chapters 2 and 3 focus on morphological and biophysical methods that are currently being used to study claudin function. The third section, from Chapters 4 to 6, is devoted to the general biology of claudins, including their structure and physiological function, the characteristics of their intermolecular interactions, and what is known about how they are regulated. The physiological role of claudins in the barrier and transport functions of various epithelialized organs, namely the kidney, lung, intestine, nervous system, and skin, are covered in Chapters 7–11. The final two Chapters (12 and 13) highlight novel roles of claudins that may be of particular clinical importance: their role as receptors for hepatitis C virus, and their role in cancer biology. A common theme that emerges from all of these chapters is that claudins likely play important roles in many human diseases, both genetic and acquired, and that we have only begun to scratch the surface in uncovering such roles.

I am deeply indebted to our distinguished panel of authors, who took time out from their busy schedules and endless grant-writing in order to write these chapters and meet a tight production deadline. I wish to thank Sid Simon and Dale Benos for inviting me to edit this book, and the highly professional staff at Elsevier for all their expert assistance, which made my job so much easier. I am fortunate to have had the support and mentorship of many wonderful teachers, colleagues, and collaborators over the years, particularly Barry Brenner, who first ignited my passion for scientific endeavour, and Jim Anderson, who has continued to encourage me in my studies of the tight junction. Finally, this book is a tribute to the memory of Shoichiro Tsukita, a brilliant scientist without whom this field would not exist, and who remains a shining source of inspiration to all of us in this field.

Previous Volumes in Series

Current Topics in Membranes and Transport

Volume 23 Genes and Membranes: Transport Proteins and Receptors* (1985)
Edited by Edward A. Adelberg and Carolyn W. Slayman

Volume 24 Membrane Protein Biosynthesis and Turnover (1985)
Edited by Philip A. Knauf and John S. Cook

Volume 25 Regulation of Calcium Transport across Muscle Membranes (1985)
Edited by Adil E. Shamoo

Volume 26 Na^+–H^+ Exchange, Intracellular pH, and Cell Function* (1986)
Edited by Peter S. Aronson and Walter F. Boron

Volume 27 The Role of Membranes in Cell Growth and Differentiation (1986)
Edited by Lazaro J. Mandel and Dale J. Benos

Volume 28 Potassium Transport: Physiology and Pathophysiology* (1987)
Edited by Gerhard Giebisch

Volume 29 Membrane Structure and Function (1987)
Edited by Richard D. Klausner, Christoph Kempf, and Jos van Renswoude

Volume 30 Cell Volume Control: Fundamental and Comparative Aspects in Animal Cells (1987)
Edited by R. Gilles, Arnost Kleinzeller, and L. Bolis

Volume 31 Molecular Neurobiology: Endocrine Approaches (1987)
Edited by Jerome F. Strauss, III, and Donald W. Pfaff

**Part of the series from the Yale Department of Cellular and Molecular Physiology.*

Volume 32 Membrane Fusion in Fertilization, Cellular Transport, and Viral Infection (1988)
Edited by Nejat Düzgünes and Felix Bronner

Volume 33 Molecular Biology of Ionic Channels* (1988)
Edited by William S. Agnew, Toni Claudio, and Frederick J. Sigworth

Volume 34 Cellular and Molecular Biology of Sodium Transport* (1989)
Edited by Stanley G. Schultz

Volume 35 Mechanisms of Leukocyte Activation (1990)
Edited by Sergio Grinstein and Ori D. Rotstein

Volume 36 Protein–Membrane Interactions* (1990)
Edited by Toni Claudio

Volume 37 Channels and Noise in Epithelial Tissues (1990)
Edited by Sandy I. Helman and Willy Van Driessche

Current Topics in Membranes

Volume 38 Ordering the Membrane Cytoskeleton Trilayer* (1991)
Edited by Mark S. Mooseker and Jon S. Morrow

Volume 39 Developmental Biology of Membrane Transport Systems (1991)
Edited by Dale J. Benos

Volume 40 Cell Lipids (1994)
Edited by Dick Hoekstra

Volume 41 Cell Biology and Membrane Transport Processes* (1994)
Edited by Michael Caplan

Volume 42 Chloride Channels (1994)
Edited by William B. Guggino

Volume 43 Membrane Protein–Cytoskeleton Interactions (1996)
Edited by W. James Nelson

Volume 44 Lipid Polymorphism and Membrane Properties (1997)
Edited by Richard Epand

Volume 45 The Eye's Aqueous Humor: From Secretion to Glaucoma (1998)
Edited by Mortimer M. Civan

Volume 46 Potassium Ion Channels: Molecular Structure, Function, and Diseases (1999)
Edited by Yoshihisa Kurachi, Lily Yeh Jan, and Michel Lazdunski

Volume 47 Amiloride-Sensitive Sodium Channels: Physiology and Functional Diversity (1999)
Edited by Dale J. Benos

Volume 48 Membrane Permeability: 100 Years since Ernest Overton (1999)
Edited by David W. Deamer, Arnost Kleinzeller, and Douglas M. Fambrough

Volume 49 Gap Junctions: Molecular Basis of Cell Communication in Health and Disease
Edited by Camillo Peracchia

Volume 50 Gastrointestinal Transport: Molecular Physiology
Edited by Kim E. Barrett and Mark Donowitz

Volume 51 Aquaporins
Edited by Stefan Hohmann, Søren Nielsen and Peter Agre

Volume 52 Peptide–Lipid Interactions
Edited by Sidney A. Simon and Thomas J. McIntosh

Volume 53 Calcium-Activated Chloride Channels
Edited by Catherine Mary Fuller

Volume 54 Extracellular Nucleotides and Nucleosides: Release, Receptors, and Physiological and Pathophysiological Effects
Edited by Erik M. Schwiebert

Volume 55 Chemokines, Chemokine Receptors, and Disease
Edited by Lisa M. Schwiebert

Volume 56 Basement Membranes: Cell and Molecular Biology
Edited by Nicholas A. Kefalides and Jacques P. Borel

Volume 57 The Nociceptive Membrane
Edited by Uhtaek Oh

Volume 58 Mechanosensitive Ion Channels, Part A
Edited by Owen P. Hamill

Volume 59 Mechanosensitive Ion Channels, Part B
Edited by Owen P. Hamill

CHAPTER 1

Introduction: Claudins, Tight Junctions, and the Paracellular Barrier☆

Mikio Furuse
Division of Cell Biology, Department of Physiology and Cell Biology,
Kobe University Graduate School of Medicine, Kobe, Japan

I. OVERVIEW

Identification and characterization of claudin family membrane proteins over the last decade has presented the current concept for the molecular basis of tight junction (TJ) functions. Claudins are the major constituent of the core structure of TJs designated TJ strands. In most epithelia, TJ strands consist of mosaics of different claudin subtypes. Since each claudin subtype shows its unique characteristics and tissue expression pattern, the

☆This paper is dedicated to the memory of Shoichiro Tsukita (1953-2005).

Current Topics in Membranes, Volume 65 1063-5823/10 $35.00

DOI: 10.1016/S1063-5823(10)65001-6

combination and proportion of claudin subtypes determines the barrier/channel property of the paracellular pathway in each cell type. The analyses of claudin-deficient mice and hereditary diseases with claudin gene mutations have demonstrated crucial roles of TJs *in vivo*. Live imaging of claudins fused with fluorescent proteins has clarified the dynamic nature of TJs.

II. INTRODUCTION

Tight junctions (TJs) contribute to epithelial barrier function by restricting the diffusion of solutes through the intercellular space of epithelial and endothelial cellular sheets with size and charge selectivity. Since Farquhar and Palade discovered TJs as an element of the junctional complex in electron microscopic analyses of a variety of mammalian epithelial cells in 1963 (Farquhar & Palade, 1963), the morphology of TJs and their crucial roles in epithelial barrier function have attracted the attention of many cell biologists and physiologists. Despite intense attempts to clarify the molecular nature of the core structure of TJs, the essential integral membrane components of TJs were not identified until 1998, when Shoichiro Tsukita and coworkers finally discovered claudins (Furuse, Fujita, Hiiragi, Fujimoto, & Tsukita, 1998). The identification of claudins has opened a way to analyze the barrier property of TJs using molecular biological approaches. Indeed, the data accumulated during the past decade have greatly enhanced our knowledge of the architecture and characteristics of TJs at the molecular level, and several comprehensive reviews have recently been published (Angelow, Ahlstrom, & Yu, 2008; Van Itallie & Anderson, 2006). In this chapter, the aspects of how claudins were identified and how they constitute the TJ barrier are summarized.

III. HISTORY OF THE IDENTIFICATION OF CLAUDINS

A. *Identification of TJ-Associated Proteins by Immunological Approaches*

On ultrathin section electron microscopy, TJs are visualized as a series of focal contacts between the plasma membranes of adjacent cells (Farquhar & Palade, 1963). On freeze-fracture electron microscopy, TJs appear as a belt-like network of intramembranous particle strands (TJ strands) (Staehelin, 1973). These observations led to the following proposed three-dimensional structure of TJs: each TJ strand associates laterally with another TJ strand in the apposing membrane of an adjacent cell to form a "paired" TJ strand, in which the intercellular space is obliterated to work as a diffusion barrier. However, the nature of TJ strands remained controversial during the 1970s and 1980s. TJ strands were once regarded to be proteins (Staehelin, 1973), but a new idea that TJ strands were composed of inverted cylindrical micelles of lipids was subsequently proposed (Pinto da Silva & Kachar, 1982). In 1984, Stevenson and Goodenough reexamined the molecular nature of TJ strands using a TJ-enriched plasma membrane fraction isolated from mouse hepatocytes and demonstrated that they were resistant to solubilization with deoxycholic acid (Stevenson & Goodenough, 1984), thus suggesting that TJ strands are detergent-resistant. Using this TJ-enriched membrane fraction as an antigen, they generated monoclonal antibodies and carried out localization-based screening. Finally, they identified ZO-1, which is the first molecular component of TJs (Stevenson, Siliciano, Mooseker, & Goodenough, 1986). Although the biochemical characterization of ZO-1 revealed that it is not an adhesion molecule which generates TJ strands but a cytoplasmic component, this work proceeded to subsequent studies that identified other TJ-associated molecules, including cingulin (Citi, Sabanay, Jakes, Geiger, & Kendrick-Jones, 1988), 7H4 antigen (Zhong et al., 1993), and symplekin (Keon, Schafer, Kuhn, Grund, & Franke, 1996), by immunological approaches. Furthermore, ZO-2 (Gumbiner, Lowenkopf, & Apatira, 1991) and ZO-3 (Haskins, Gu, Wittchen, Hibbard, & Stevenson, 1998) were identified as TJ-associated cytoplasmic proteins and binding partners of ZO-1.

On the other hand, more direct attempts to identify the cell surface proteins that are responsible for TJ formation by functional assays were performed. Gumbiner and Simons generated monoclonal antibodies against MDCK cells, and screened these antibodies in terms of their inhibitory effects on the generation of transepithelial electric resistance (TER), which reflects the barrier function of TJs, during cell–cell junction formation. They obtained an inhibitory antibody, and further analyses revealed that this antibody recognized E-cadherin of MDCK cells (Gumbiner & Simons, 1986), implying that cadherin-mediated cell adhesion is a prerequisite for TJ formation. However, the molecular components of TJs were not found using this approach.

The first integral membrane protein of TJs was identified by an immunological approach. Tsukita and colleagues generated rat monoclonal antibodies against the adherens junction and TJ-enriched plasma membrane fraction isolated from the chick liver, which might have had higher antigenicity for immunized rodents than the fraction isolated from the rodent liver. After localization-based screening of the monoclonal antibodies, an integral membrane protein with four transmembrane domains that was localized at TJs was obtained and designated occludin (Furuse et al., 1993). For some years thereafter, occludin was expected to be the key molecule of TJs. Indeed, this notion was supported by several studies in which the overexpression of occludin or its deleted forms and the addition of a peptide corresponding to the extracellular domain of occludin modified TJ function in epithelial cells (Balda et al., 1996; Bamforth, Kniesel, Wolburg, Engelhardt, & Risau, 1999; McCarthy et al., 1996; Wong & Gumbiner, 1997). On the other hand, the diversity of the barrier properties of TJs, such as their conductance and charge selectivity, suggested biochemical diversity of the extracellular part of TJs, which could not be explained by the idea that occludin is the only constituent of the TJ barrier. Moreover, it was subsequently reported that the TJs of Sertoli cells in the guinea pig and human testis did not contain occludin (Moroi et al., 1998), suggesting that there are certain occludin subtypes or other TJ-associated integral membrane proteins that differ from occludin. Finally, it was demonstrated that occludin-null epithelial cells generated from occludin gene-knockout embryonic stem cells still had TJs with a normal appearance (Saitou et al., 1998, 2000), indicating the existence of unknown membrane components of TJs other than occludin.

B. Identification of the Claudin Family

The fact that occludin-deficient epithelial cells possess normal-looking TJs sent Tsukita and his colleagues right back where they started. To identify the major constituent of TJ strands, they first attempted to look for occludin-binding proteins based on the simple assumption that the TJ adhesion molecules may associate with occludin within the plasma membrane. However, neither yeast two-hybrid screening nor coprecipitation experiments using occludin probes detected any occludin-binding proteins. Consequently, they finally went back to the TJ-enriched membrane fraction isolated from the chick liver (hereafter designated "the junctional membrane fraction"), from which occludin was originally identified (Furuse et al., 1993). Detailed analyses of this fraction revealed that occludin could be detected as a protein band by SDS-PAGE, but the fraction had many contaminating components, such as adherens junctions, gap junctions, and membrane-undercoating cytoskeletons, all of which were connected to the TJ membranes. Therefore, the junctional membrane fraction was treated with sonication, which was sufficiently effective to mechanically separate the TJs from other membranous and cytoskeletal components without disruption of the membrane contacts of TJs, and subjected to sucrose density gradient ultracentrifugation. SDS-PAGE analyses of the resulting fractions indicated that occludin was highly concentrated in several fractions, in which another single protein band showed the identical distribution to that of occludin. Amino acid sequencing of this protein band determined two chicken peptide sequences, which both had homologous sequences among the mouse nucleotides in the database. Subsequent cDNA cloning clarified two related proteins with four transmembrane domains. When these proteins were exogenously expressed with epitope tags, they were found to be localized at TJs in cultured epithelial cells. Therefore, they were designated claudin-1 and claudin-2 from the Latin word "claudere," which means "to close," as components of the TJ barrier (Furuse, Fujita, et al., 1998). Furthermore, the overexpression of claudin-1 and claudin-2 in TJ-negative fibroblasts led to the reconstitution of TJ strands at cell–cell contact planes (Furuse, Sasaki, Fujimoto, & Tsukita, 1998), demonstrating that claudins have TJ-forming activity.

At that time, several related proteins had already been identified and analyzed as molecules whose implications in TJs were unknown. Oligodendrocyte-specific protein (OSP), which had been identified as one of the major proteins of oligodendrocytes (Bronstein, Micevych, & Chen, 1997), was later designated claudin-11. The cellular receptor for *Clostridium perfringens* enterotoxin (CPE) (Katahira, Inoue, Horiguchi, Matsuda, & Sugimoto, 1997) was revealed to be claudin-4.

IV. STRUCTURE OF CLAUDINS

Claudins comprise a multigene family consisting of ~24 members in humans and mice (Morita, Furuse, Fujimoto, & Tsukita, 1999; Wilcox et al., 2001). Claudins are integral membrane proteins with molecular masses of 22–27 kDa that contain four membrane-spanning regions, a short N-terminal cytoplasmic domain, two extracellular loops, one intracellular loop, and a C-terminal cytoplasmic domain. The first extracellular loop of ~50 amino acids has a common motif, GLW(2 aa)C(8–10 aa)C. This domain also contains charged amino acids, whose number and distribution depend on the claudin subtypes, and determines the charge selectivity and conductance in paracellular transport. However, its three-dimensional structure has not yet been solved. The second extracellular loop contains ~20 amino acids and is hypothesized to fold into a helix-turn-helix and form dimers (Piontek et al., 2008). At its end, the cytoplasmic C-terminal domain contains a PDZ domain-binding motif that binds to PDZ domain-containing plaque proteins such as ZO-1, ZO-2, ZO-3 (Itoh et al., 1999), MUPP-1 (Hamazaki, Itoh, Sasaki, Furuse, & Tsukita, 2002), and PATJ (Roh, Liu, Laurinec, & Margolis, 2002). The C-terminal cytoplasmic tail influences the stability of claudins and their targeting to TJs (Muller et al., 2003; Ruffer & Gerke, 2004; Van Itallie, Colegio, & Anderson, 2004). Serine or threonine residues in the C-terminal cytoplasmic regions of several claudins are phosphorylated (Aono & Hirai, 2008; D'Souza, Agarwal, & Morin, 2005; Ikari et al., 2006; Ishizaki et al., 2003). This modification seems to regulate the localization of claudins and the barrier property of TJs, but the molecular mechanisms remain unknown. Claudins have membrane proximal CxxC motifs in the intracellular loop and the juxtamembrane region of the C-terminal cytoplasmic domain. Both of these regions in claudin-14 are palmitoylated (Van Itallie, Gambling, Carson, & Anderson, 2005), and this modification of claudin-14 is required for its efficient localization at TJs, but not the stability of the claudin-14 protein or the assembly of TJ strands. The cysteine motifs are conserved in most claudin subtypes.

V. TJ FORMATION BY THE CLAUDIN FAMILY

Claudins constitute the structural core of TJs. When claudins are overexpressed in mouse L fibroblasts lacking endogenous claudins, the exogenous claudins concentrate into cell–cell contacts and generate TJ strands with the induction of cell-adhesive activity (Furuse, Sasaki, et al., 1998; Kubota et al., 1999). Conversely, addition of the C-terminal half of CPE, which binds to the second extracellular loop of some claudin subtypes (Fujita et al., 2000), results in the removal of these claudins, accompanied by a reduction in the barrier function of TJs (Sonoda et al., 1999). Furthermore, claudin-11-based TJ strands in Sertoli cells (Morita, Sasaki, Fujimoto, Furuse, & Tsukita, 1999) are lost in claudin-11-deficient mice (Gow et al., 1999). These loss-of-function studies confirm that claudins are key structural and functional components of TJs.

The reconstitution of TJ-like structures in L fibroblasts by exogenous claudin expression has revealed several basic characteristics of claudins within TJ strands. Claudins assemble into cell–cell contact planes from both of the adjacent cells, and this concentration of claudins at the cell border is never observed in the absence of adjacent cells (Furuse, Sasaki, et al., 1998). Therefore, the interactions of claudins from adjacent membranes appear to induce fibril-like polymerization of claudin-based TJ strands within the plasma membrane. Claudins cannot form TJ strands in mouse EpH4 mammary epithelial cells lacking ZO-1 and ZO-2, TJ-associated plaque proteins that bind to claudins (Umeda et al., 2006), suggesting that ZO-1 and ZO-2 support claudin polymerization in the plasma membrane in epithelial cells. However, claudins whose C-terminal domains containing the PDZ domain-binding motif were deleted or masked with epitope tags still reconstituted TJ strands in L cells (Furuse, Sasaki, et al., 1998), indicating that claudins have the ability to polymerize into TJ strands autonomously without interacting with cytoplasmic components in this system. This discrepancy in TJ- strand formation between fibroblasts and epithelial cells has not yet been solved, although it is of interest to understand the detailed mechanism for claudin polymerization within the lipid bilayer.

When nonfixed or quick-frozen samples are observed by freeze-fracture replica electron microscopy, TJ strands are visualized as chains of intramembrane particles (Staehelin, 1973). It is reasonable to think that each particle comprises the unit structure of a TJ strand. This unit may be a claudin oligomer rather than a monomer, similar to the case for a connexon, which is the basic unit of a gap junction and is composed of a connexin hexamer. Although there is no direct evidence to support this idea, overexpressed human claudin-4 in insect cells exhibits oligomers up to hexamers when solubilized with perfluoro-octanoic acid (POA), followed by POA-PAGE (Mitic, Unger, & Anderson, 2003).

VI. COMPLEX STRUCTURE OF TJ STRANDS FORMED BY MULTIPLE CLAUDIN SUBTYPES

The existence of more than 20 subtypes of claudins is of interest when considering their subtype-dependent functions and expression patterns. In most epithelial cell types, different claudin subtypes are coexpressed, and the combinations and proportions of the claudin subtypes depend on the cell types, thus creating the biochemical diversity of TJs. Cell biological analyses indicated that TJ strands are generally comprised of the mosaics of multiple claudin subtypes. Cocultures of L cells expressing claudin-1, claudin-2, or claudin-3 revealed that the combinations of claudin-1/-3 and claudin-2/-3, but not claudin-1/-2, could form heterotypic TJ strands (Furuse, Sasaki, & Tsukita, 1999). In similar experiments using HeLa cells, the combinations of claudin-1/-3 and claudin-3/-5 showed heterophilic assembly, but claudin-4/-5, claudin-1/-4, and claudin-3/-4 did not, although the amino acid sequences of claudin-3 and -4 are very similar (Daugherty, Ward, Smith, Ritzenthaler, & Koval, 2007). These results indicate that claudin subtypes can form heterotypic as well as homotypic TJs, but that the compatibility depends on the pair of subtypes involved. Coexpression experiments of two claudin subtypes among mouse claudin-1, claudin-2, and claudin-3 in L cells revealed that the different claudin subtypes could copolymerize into single TJ strands in a heteromeric manner (Furuse et al., 1999) and that claudin-3 and -4 also formed heteromeric TJ strands (Daugherty et al., 2007). On the other hand, different claudin subtypes may be separated into different TJ strands in the same cell surface. In cell–cell contacts between outer hair cells and Dieter cells in the inner ear, claudin-14 is concentrated in the apical TJ strands, while claudin-6 and -9 are localized in the lateral TJ strands (Nunes et al., 2006), suggesting that the heteromeric compatibility depends on the combination. In addition, some claudins may not form TJ strands by themselves. Claudin-16 and -19 are expressed in the thick ascending limb (TAL) of Henle in the nephron of the kidney and play crucial roles in the reabsorption of divalent cations such as Mg^{2+} and Ca^{2+} in this segment (Angelow, El-Husseini, Kanzawa, & Yu, 2007; Konrad et al., 2006; Simon et al., 1999). Overexpressed claudin-16 cannot form TJ strands, at least in L fibroblasts, but can be incorporated into claudin-19-derived TJ strands, probably by a heteromeric interaction with claudin-19 (Hou et al., 2008). Furthermore, this interaction is required for the assembly of these claudins into TJ strands *in vivo*. In claudin-19-knockdown mice, the assembly of claudin-16 into the TJ strands in the TAL of Henle is hampered, while the accumulation of claudin-19 into TJ strands is remarkably reduced in claudin-16-knockdown mice (Hou et al., 2009).

VII. FUNCTIONAL DIVERSITY OF THE BARRIER PROPERTY OF CLAUDIN-BASED TJS

Physiological studies on epithelial transport have shown that the barrier property of TJs varies among cell types. The TER, which reflects the barrier strength of TJs to ions, and the ratio of Na^+ permeability to Cl^- permeability of the paracellular pathway vary among different types of epithelia (Powell, 1981). Intensive examinations during the past decade have clarified that each claudin subtype has its own unique characteristics and that the combination of the claudin subtypes determines the barrier properties of TJs. Analyses involving the overexpression or RNAi-mediated suppression of claudins in epithelial cell lines, followed by electrophysiological measurements of the TER and diffusion potential, have categorized the individual claudins into barrier and channel/pore-forming subtypes with charge discrimination (Table I). It should be noted, however, that the results depend on the background of the cell lines used (Van Itallie, Fanning, & Anderson, 2003). For example, in MDCK II cells with cation selectivity, claudin-4 expression decreases Na^+ permeability while claudin-2 expression has little effect. In contrast, in LLC-PK1 cells with anion selectivity, claudin-4 expression has no effect while claudin-2 expression increases Na^+ permeability (Van Itallie et al., 2003). The background expression of endogenous claudins in these cell lines seems to influence the results of these experiments. Another concern of this type of studies is that the overexpression or removal of a certain claudin may influence the expression level of other claudin subtypes (Yu, Enck, Lencer, & Schneeberger, 2003).

Although the three-dimensional structures of claudin-based barriers and pores are totally unknown, accumulating evidence has shown that the charge selectivity of the paracellular pathway is predominantly attributed to the structure of the first extracellular loop of claudins. The generation of chimeric molecules of claudin-4 and -2 clearly revealed this aspect (Colegio, Van

TABLE I
Predicted Functions of Claudin Subtypes to Ion Permeation Evaluated in Overexpression or Knockdown in Cultured Epithelial Cells

Claudin	Predicted function	Reference
Clandin-1	Ion barrier	Inai, Kobayashi, and Shibata (1999), McCarthy et al. (1996)
Claudin-2	Cation channel	Amasheh et al. (2002), Hou, Gomes, Paul, and Goodenough (2006), Van Itallie et al. (2003), Yu et al. (2009)
Claudin-4	Cation barrier	Van Itallie, Rahner, and Anderson (2001)
	Cation barrier or anion channel	Hou et al. (2006)
Claudin-5	Cation barrier	Wen, Watry, Marcondes, and Fox (2004)
Claudin-6	Anion barrier	Sas, Hu, Moe, and Baum (2008)
Claudin-7	Cation channel, anion barrier	Alexandre, Lu, and Chen (2005)
	Cation barrier or anion channel	Hou et al. (2006)
Claudin-8	Cation barrier	Yu et al. (2003)
Claudin-9	Anion barrier	Sas et al. (2008)
Claudin-10a	Anion channel	Van Itallie et al. (2006)
Claudin-10b	Cation channel	Van Itallie et al. (2006)
Claudin-11	Cation barrier	Van Itallie et al. (2003)
Claudin-14	Cation barrier	Ben-Yosef et al. (2003)
Claudin-15	Cation channel	Van Itallie et al. (2003)
Claudin-16	Cation channel	Hou, Paul, and Goodenough (2005)
Claudin-19	Cation barrier	Angelow et al. (2007)

Itallie, Rahner, & Anderson, 2003). Furthermore, claudin-15 mutants in which negatively charged amino acids in the first extracellular loop were replaced with positively charged amino acids reversed the charge selectivity when introduced into MDCK II cells (Colegio, Van Itallie, McCrea, Rahner, & Anderson, 2002). Subsequent studies demonstrated that point mutations in the charged amino acids in the first extracellular loop affected the charge selectivity (Alexandre, Jeansonne, Renegar, Tatum, & Chen, 2007; Yu et al., 2009).

The channel or pore property of claudins has been well investigated from the viewpoint of the molecular mechanism behind paracellular transport. As a representative of a pore-type claudin, claudin-2 has been analyzed in detail. Claudin-2 provides high conductance with cation selectivity to TJs

(Amasheh et al., 2002; Furuse, Furuse, Sasaki, & Tsukita, 2001; Van Itallie et al., 2003; Yu et al., 2009). In tracer permeation assays using polyethylene glycol as a neutral probe, the overexpression of claudin-2 in MDCK cells increased the pore diameters to ~8 Å in the cellular sheet, suggesting that claudin-2 can form pores of ~8 Å in diameter (Van Itallie et al., 2008). Diffusion potential measurements of permeability to organic ions indicated that the pores formed by claudin-2 were ~7.5 Å in diameter (Yu et al., 2009). Furthermore, site-directed mutagenesis studies clarified that aspartate-65 within the first extracellular loop is responsible for the cation selectivity as well as the high conductance of claudin-2 (Yu et al., 2009). When aspartate-65 was replaced with cysteine, the resulting claudin-2 mutant lost its charge and size selectivity, and formed dimers via intermolecular disulfide bonds, suggesting that claudin-2 pores are multimeric (Angelow & Yu, 2009).

These studies have established the current concept that the combination of claudin subtypes, which all have unique characteristics, determines the barrier/channel property of TJs dependent on the type of epithelia. However, it remains unknown how the complex manner of claudin assembly, including heteromeric or heterotypic assembly, affects this property.

VIII. *IN VIVO* FUNCTIONS OF CLAUDINS: CLAUDIN-DEFICIENT MICE AND HEREDITARY DISEASES WITH CLAUDIN MUTATIONS

The identification of claudins has enabled analyses of TJ functions *in vivo* using molecular biological approaches by generating knockout or knockdown mice for claudin genes. In addition, positional cloning analyses have identified mutations in the genes of several claudins in hereditary diseases, thus demonstrating critical roles for TJs in various organs. Since the claudin family consists of more than 20 subtypes, each of which shows a unique expression pattern *in vivo*, the pathologies of various tissues have been investigated depending on the disruption of individual claudins. Importantly, and similar to the classification of claudins into barrier and channel-forming subtypes, the pathologies of claudin deficiency also appear to be derived from a barrier defect or a loss of channel activity for the paracellular pathway, although the detailed mechanisms of these pathologies are not necessarily well understood. In claudin knockout, knockdown, or mutant mice, the barrier defects include defects in the epidermal barrier (claudin-1) (Furuse et al., 2002), blood–brain barrier (claudin-5) (Nitta et al., 2003), cochlear barrier (claudin-9, claudin-11, and claudin-14) (Ben-Yosef et al., 2003; Gow et al., 2004; Kitajiri, Miyamoto, et al., 2004; Nakano et al., 2009), blood–testis barrier (claudin-11) (Gow et al., 1999), and myelin sheath barrier (claudin-11 and claudin-19) (Gow et al., 1999; Miyamoto et al., 2005), while the channel abnormalities include reduced ion conductance in intestinal epithelial cells (claudin-15) (Tamura et al., 2008) and defects in cation selectivity of the TAL of Henle (claudin-16 and claudin-19) (Hou et al., 2007, 2009) and the proximal tubules (claudin-2) (Muto et al., 2010) in nephron segments. Among these, deficiencies in claudin-14, claudin-16, and claudin-19 in mice exhibit phenocopies of human hereditary diseases with mutations in the corresponding claudins (Konrad et al., 2006; Simon et al., 1999; Wilcox et al., 2001). Human familial syndrome caused by claudin-1 mutations includes abnormalities in the skin and liver (Baala et al., 2002; Hadj-Rabia et al., 2004). A liver defect was not observed in claudin-1-deficient mice, probably because they died on the day of birth, accompanied by excessive water loss from the skin (Furuse et al., 2002), although a large amount of claudin-1 is expressed in the mouse liver. Since claudin-11 is the major claudin expressed in TJs in oligodendrocytes, Sertoli cells, and basal cells in the stria vascularis in the organ of Corti, at least in mice (Gow et al., 1999, 2004; Kitajiri, Furuse, et al., 2004; Morita, Sasaki, et al., 1999), claudin-11-deficient mice show a wide variety of phenotypes, including hind limb weakness, male sterility, and deafness (Gow et al., 1999, 2004; Kitajiri, Miyamoto, et al., 2004). However, a human syndrome that exhibits these pathologies has not yet been reported (Tables II and III).

TABLE II

Claudin Gene Knockouts, Transgenics, and Mutations in Mice

Gene	Phenotype	Reference
Cldn-1 KO	Skin barrier defect	Furuse et al. (2002)
Cldn-5 KO	Blood–brain barrier defect	Nitta et al. (2003)
Cldn-6 TG	Skin barrier defect	Turksen and Troy (2002)
Cldn-7 KO	Renal salt wasting and chronic dehydration	Tatum et al. (2009)
Cldn-9 MT	Deafness	Nakano et al. (2009)
Cldn-11 KO	Male sterility and hind limb weakness	Gow et al. (1999)
	Deafness (reduced endocochlear potential)	Gow et al. (2004), Kitajiri, Miyamoto, et al. (2004)
Cldn-14 KO	Deafness	Ben-Yosef et al. (2003)
Cldn-15 KO	Megaintestine	Tamura et al. (2008)
Cldn-16 KD	Renal wasting of Mg^{2+} and Ca^{2+}, nephrocalcinosis	Hou et al. (2007)
Cldn-19 KO	Reduced nerve conduction of peripheral myelinated fibers	Miyamoto et al. (2005)
Cldn-19 KD	Phenocopy of cldn-16 KD	Hou et al. (2009)

TABLE III
Hereditary Diseases Caused by Claudin Gene Mutations

Gene	Disease	Reference
Cldn-1	Neonatal ichthyosis-sclerosing colangitis (NISCH)	Hadj-Rabia et al. (2004)
Cldn-14	Nonsyndromic deafness, *DFNB29*	Wilcox et al. (2001)
Cldn-16	Familial hypomagnesemia and hypercalciuria and nephrocalcinosis (FHHNC) (Human)	Simon et al. (1999)
	Cholonic interstitial nephritis (Bovine)	Hirano et al. (2000)
Cldn-19	FHHNC with visual impairment	Konrad et al. (2006)

IX. DYNAMIC BEHAVIOR OF CLAUDIN-BASED TJS

Recent advances in GFP technology have enabled analyses of the dynamic behavior of claudin-based TJs in living cells. In L fibroblasts expressing GFP-tagged claudin-1, individual-reconstituted TJ strands were visualized by time-lapse fluorescence microscopy. Interestingly, GFP-tagged claudin-1-based TJ strands were occasionally broken and then annealed dynamically (Sasaki et al., 2003). If such reorganization of TJ strands occurs in epithelial cells, solutes including those with higher molecular weights may pass across TJs gradually while retaining the structural integrity of the TJs as a whole. This manner may provide a pathway for solutes that differs from claudin-based channels or pores for small molecules such as inorganic ions, and may explain the unsolved discrepancy that TER measurements are not necessarily correlated with paracellular flux measurements (Balda et al., 1996). Unfortunately, the aspect of whether TJ strands in epithelial cells show similar behavior has not been clarified because it is technically difficult to visualize individual TJ strands in these cells using the same technique, partly because the density of the TJ strands in these cells is too high for them to be clearly distinguished by light microscopy and partly because the plane of TJ strands in epithelial cells is parallel to the observation axis.

Fluorescence recovery after photobleaching (FRAP) analyses have also revealed the dynamic behavior of proteins in TJ strands. FRAP analyses of GFP-tagged claudin-1 and occludin proteins in MDCK cells revealed that TJs undergo constant remodeling and that claudin-1 is less mobile than occludin in TJs (Shen, Weber, & Turner, 2008), indicating that the behaviors of these two TJ-associated membrane proteins differ within TJ strands. These observations may provide clues for how occludin is localized within claudin-based TJ strands for future studies.

The internalization of claudins from TJs by endocytosis is also a dynamic process. The inflammatory cytokine interferon-gamma induces endocytic internalization of TJ-associated integral membrane proteins, including occludin, JAM, and claudin-1, into actin-coated large vesicles in cultured intestinal epithelial cells in a myosin II-dependent manner (Bruewer et al., 2005; Utech et al., 2005). On the other hand, time-lapse imaging of GFP-tagged claudins in cultured epithelial cells clarified a peculiar internalization of claudins during the remodeling of TJs caused by cell movement within the cellular sheet (Matsuda, Kubo, Furuse, & Tsukita, 2004). In this process, two apposed TJ membranes were coendocytosed without detaching. Moreover, claudins appeared to be dissociated from other TJ-associated proteins, including occludin, JAM, and ZO-1, and selectively internalized. Ubiquitination may be involved in this selective internalization of claudins from TJs. Recently, an E3 ubiquitin ligase, LNX1, was found to bind to the C-terminus of claudin-1 and to ubiquitinate claudins when expressed in HEK293 cells (Takahashi et al., 2009). Interestingly, the overexpression of LNX1 in MDCK cells caused internalization and lysosomal degradation of claudins, accompanied by loss of claudins in TJs and a remarkable decrease in the TJ structure, while occludin remained in the TJs (Takahashi et al., 2009). The physiological function of LNX1 *in vivo* has not yet been demonstrated.

X. SUMMARY AND FUTURE PERSPECTIVES

Intensive investigations of claudins over the past decade have rapidly clarified their functions in TJs, but various questions still remain unsolved. One of the simple questions is why so many subtypes of claudins exist with complex tissue-specific combinations and expression patterns. Further generation and analyses of knockout mice for each claudin gene will determine which claudins are essential and which are redundant, and may provide hints toward understanding this issue. In claudin-1-deficient mice, the TJs in epidermal keratinocytes become leaky, although claudin-4 still generates TJs (Furuse et al., 2002). These observations imply that the overlapping expressions of different claudins have some important meanings for the establishment of the barrier function of TJs.

One of the most important contributions of studies on claudins to biology is to show the concrete nature of the paracellular pathway. Analyses of the coupling of paracellular transport with transcellular transport will allow us to figure out epithelial transport in a more comprehensive way than before, especially in leaky epithelia. For a complete understanding of the mechanisms of claudin-based barriers and pores, the fine structures of the claudins in TJ strands must be determined by structural biological approaches, although no crystal structures of claudins have yet been reported. These analyses would also clarify the unit structure of TJ strands and provide information about the functional contributions of the heteromeric and heterotypic assembly of different claudins to the barrier property of TJs.

The mechanism of the turnover of TJ strands is also of interest. During cell division, accompanied by dynamic cell shape changes or morphogenetic processes in which cells move with one another, cell–cell junctions including TJs must undergo dynamic rearrangements. In these processes, however, the integrity of TJs should be maintained to keep the barrier function of the cellular sheet. The mechanisms for how the internalization of claudins and their incorporation into TJ strands are coordinated and balanced are interesting issues for future studies.

Acknowledgments

M. F. is supported by a Grant-in-Aid for Cancer Research from the Ministry of Education, Culture, Sports, Science and Technology of Japan.

References

Alexandre, M. D., Jeansonne, B. G., Renegar, R. H., Tatum, R., & Chen, Y. H. (2007). The first extracellular domain of claudin-7 affects paracellular Cl^- permeability. *Biochemical and Biophysical Research Communications, 357*, 87–91.

Alexandre, M. D., Lu, Q., & Chen, Y. H. (2005). Overexpression of claudin-7 decreases the paracellular Cl^- conductance and increases the paracellular Na^+ conductance in LLC-PK1 cells. *Journal of Cell Science, 118*, 2683–2693.

Amasheh, S., Meiri, N., Gitter, A. H., Schoneberg, T., Mankertz, J., Schulzke, J. D., et al. (2002). Claudin-2 expression induces cation-selective channels in tight junctions of epithelial cells. *Journal of Cell Science, 115*, 4969–4976.

Angelow, S., Ahlstrom, R., & Yu, A. S. (2008). Biology of claudins. *American Journal of Physiology. Renal Physiology, 295*, F867–F876.

Angelow, S., El-Husseini, R., Kanzawa, S. A., & Yu, A. S. (2007). Renal localization and function of the tight junction protein, claudin-19. *American Journal of Physiology. Renal Physiology, 293*, F166–F177.

Angelow, S., & Yu, A. S. (2009). Cysteine mutagenesis to study the structure of claudin-2 paracellular pores. *Annals of the New York Academy of Sciences, 1165*, 143–147.

Aono, S., & Hirai, Y. (2008). Phosphorylation of claudin-4 is required for tight junction formation in a human keratinocyte cell line. *Experimental Cell Research, 314*, 3326–3339.

Baala, L., Hadj-Rabia, S., Hamel-Teillac, D., Hadchouel, M., Prost, C., Leal, S. M., et al. (2002). Homozygosity mapping of a locus for a novel syndromic ichthyosis to chromosome 3q27-q28. *The Journal of Investigative Dermatology, 119*, 70–76.

Balda, M. S., Whitney, J. A., Flores, C., Gonzalez, S., Cereijido, M., & Matter, K. (1996). Functional dissociation of paracellular permeability and transepithelial electrical resistance and disruption of the apical-basolateral intramembrane diffusion barrier by expression of a mutant tight junction membrane protein. *The Journal of Cell Biology, 134*, 1031–1049.

Bamforth, S. D., Kniesel, U., Wolburg, H., Engelhardt, B., & Risau, W. (1999). A dominant mutant of occludin disrupts tight junction structure and function. *Journal of Cell Science, 112*(Pt 12), 1879–1888.

Ben-Yosef, T., Belyantseva, I. A., Saunders, T. L., Hughes, E. D., Kawamoto, K., Van Itallie, C. M., et al. (2003). Claudin 14 knockout mice, a model for autosomal recessive deafness DFNB29, are deaf due to cochlear hair cell degeneration. *Human Molecular Genetics, 12*, 2049–2061.

Bronstein, J. M., Micevych, P. E., & Chen, K. (1997). Oligodendrocyte-specific protein (OSP) is a major component of CNS myelin. *Journal of Neuroscience Research, 50*, 713–720.

Bruewer, M., Utech, M., Ivanov, A. I., Hopkins, A. M., Parkos, C. A., & Nusrat, A. (2005). Interferon-gamma induces internalization of epithelial tight junction proteins via a macropinocytosis-like process. *FASEB Journal, 19*, 923–933.

Citi, S., Sabanay, H., Jakes, R., Geiger, B., & Kendrick-Jones, J. (1988). Cingulin, a new peripheral component of tight junctions. *Nature, 333*, 272–276.

Colegio, O. R., Van Itallie, C., Rahner, C., & Anderson, J. M. (2003). Claudin extracellular domains determine paracellular charge selectivity and resistance but not tight junction fibril architecture. *American Journal of Physiology. Cell Physiology, 284*, C1346–C1354.

Colegio, O. R., Van Itallie, C. M., McCrea, H. J., Rahner, C., & Anderson, J. M. (2002). Claudins create charge-selective channels in the paracellular pathway between epithelial cells. *American Journal of Physiology. Cell Physiology, 283*, C142–C147.

Daugherty, B. L., Ward, C., Smith, T., Ritzenthaler, J. D., & Koval, M. (2007). Regulation of heterotypic claudin compatibility. *The Journal of Biological Chemistry, 282*, 30005–30013.

D'Souza, T., Agarwal, R., & Morin, P. J. (2005). Phosphorylation of claudin-3 at threonine 192 by cAMP-dependent protein kinase regulates tight junction barrier function in ovarian cancer cells. *The Journal of Biological Chemistry, 280*, 26233–26240.

Farquhar, M. G., & Palade, G. E. (1963). Junctional complexes in various epithelia. *Journal of Cell Biology, 17*, 375–412.

Fujita, K., Katahira, J., Horiguchi, Y., Sonoda, N., Furuse, M., & Tsukita, S. (2000). *Clostridium perfringens* enterotoxin binds to the second extracellular loop of claudin-3, a tight junction integral membrane protein. *FEBS Letters, 476*, 258–261.

Furuse, M., Fujita, K., Hiiragi, T., Fujimoto, K., & Tsukita, S. (1998). Claudin-1 and -2: Novel integral membrane proteins localizing at tight junctions with no sequence similarity to occludin. *Journal of Cell Biology, 141*, 1539–1550.

Furuse, M., Furuse, K., Sasaki, H., & Tsukita, S. (2001). Conversion of zonulae occludentes from tight to leaky strand type by introducing claudin-2 into Madin-Darby canine kidney I cells. *Journal of Cell Biology, 153*, 263–272.

Furuse, M., Hata, M., Furuse, K., Yoshida, Y., Haratake, A., Sugitani, Y., et al. (2002). Claudin-based tight junctions are crucial for the mammalian epidermal barrier: A lesson from claudin-1-deficient mice. *Journal of Cell Biology, 156*, 1099–1111.

Furuse, M., Hirase, T., Itoh, M., Nagafuchi, A., Yonemura, S., & Tsukita, S. (1993). Occludin: A novel integral membrane protein localizing at tight junctions. *Journal of Cell Biology, 123*, 1777–1788.

Furuse, M., Sasaki, H., Fujimoto, K., & Tsukita, S. (1998). A single gene product, claudin-1 or -2, reconstitutes tight junction strands and recruits occludin in fibroblasts. *Journal of Cell Biology, 143*, 391–401.

Furuse, M., Sasaki, H., & Tsukita, S. (1999). Manner of interaction of heterogeneous claudin species within and between tight junction strands. *Journal of Cell Biology, 147*, 891–903.

Gow, A., Davies, C., Southwood, C. M., Frolenkov, G., Chrustowski, M., Ng, L., et al. (2004). Deafness in Claudin 11-null mice reveals the critical contribution of basal cell tight junctions to stria vascularis function. *The Journal of Neuroscience, 24*, 7051–7062.

Gow, A., Southwood, C. M., Li, J. S., Pariali, M., Riordan, G. P., Brodie, S. E., et al. (1999). CNS myelin and sertoli cell tight junction strands are absent in Osp/claudin-11 null mice. *Cell, 99*, 649–659.

Gumbiner, B., Lowenkopf, T., & Apatira, D. (1991). Identification of a 160-kDa polypeptide that binds to the tight junction protein ZO-1. *Proceedings of the National Academy of Sciences of the United States of America, 88*, 3460–3464.

Gumbiner, B., & Simons, K. (1986). A functional assay for proteins involved in establishing an epithelial occluding barrier: Identification of a uvomorulin-like polypeptide. *Journal of Cell Biology, 102*, 457–468.

Hadj-Rabia, S., Baala, L., Vabres, P., Hamel-Teillac, D., Jacquemin, E., Fabre, M., et al. (2004). Claudin-1 gene mutations in neonatal sclerosing cholangitis associated with ichthyosis: A tight junction disease. *Gastroenterology, 127*, 1386–1390.

Hamazaki, Y., Itoh, M., Sasaki, H., Furuse, M., & Tsukita, S. (2002). Multi-PDZ domain protein 1 (MUPP1) is concentrated at tight junctions through its possible interaction with claudin-1 and junctional adhesion molecule. *The Journal of Biological Chemistry, 277*, 455–461.

Haskins, J., Gu, L., Wittchen, E. S., Hibbard, J., & Stevenson, B. R. (1998). ZO-3, a novel member of the MAGUK protein family found at the tight junction, interacts with ZO-1 and occludin. *Journal of Cell Biology, 141*, 199–208.

Hirano, T., Kobayashi, N., Itoh, T., Takasuga, A., Nakamaru, T., Hirotsune, S., et al. (2000). Null mutation of PCLN-1/claudin-16 results in bovine chronic interstitial nephritis. *Genome Research, 10*, 659–663.

Hou, J., Gomes, A. S., Paul, D. L., & Goodenough, D. A. (2006). Study of claudin function by RNA interference. *The Journal of Biological Chemistry, 281*, 36117–36123.

Hou, J., Paul, D. L., & Goodenough, D. A. (2005). Paracellin-1 and the modulation of ion selectivity of tight junctions. *Journal of Cell Science, 118*, 5109–5118.

Hou, J., Renigunta, A., Gomes, A. S., Hou, M., Paul, D. L., Waldegger, S., et al. (2009). Claudin-16 and claudin-19 interaction is required for their assembly into tight junctions and for renal reabsorption of magnesium. *Proceedings of the National Academy of Sciences of the United States of America, 106*, 15350–15355.

Hou, J., Renigunta, A., Konrad, M., Gomes, A. S., Schneeberger, E. E., Paul, D. L., et al. (2008). Claudin-16 and claudin-19 interact and form a cation-selective tight junction complex. *Journal of Clinical Investigation, 118*, 619–628.

Hou, J., Shan, Q., Wang, T., Gomes, A. S., Yan, Q., Paul, D. L., et al. (2007). Transgenic RNAi depletion of claudin-16 and the renal handling of magnesium. *The Journal of Biological Chemistry, 282*, 17114–17122.

Ikari, A., Matsumoto, S., Harada, H., Takagi, K., Hayashi, H., Suzuki, Y., et al. (2006). Phosphorylation of paracellin-1 at Ser217 by protein kinase A is essential for localization in tight junctions. *Journal of Cell Science, 119*, 1781–1789.

Inai, T., Kobayashi, J., & Shibata, Y. (1999). Claudin-1 contributes to the epithelial barrier function in MDCK cells. *European Journal of Cell Biology, 78*, 849–855.

Ishizaki, T., Chiba, H., Kojima, T., Fujibe, M., Soma, T., Miyajima, H., et al. (2003). Cyclic AMP induces phosphorylation of claudin-5 immunoprecipitates and expression of claudin-5 gene in blood-brain-barrier endothelial cells via protein kinase A-dependent and -independent pathways. *Experimental Cell Research, 290*, 275–288.

Itoh, M., Furuse, M., Morita, K., Kubota, K., Saitou, M., & Tsukita, S. (1999). Direct binding of three tight junction-associated MAGUKs, ZO-1, ZO-2, and ZO-3, with the COOH termini of claudins. *Journal of Cell Biology, 147*, 1351–1363.

Katahira, J., Inoue, N., Horiguchi, Y., Matsuda, M., & Sugimoto, N. (1997). Molecular cloning and functional characterization of the receptor for *Clostridium perfringens* enterotoxin. *Journal of Cell Biology, 136*, 1239–1247.

Keon, B. H., Schafer, S., Kuhn, C., Grund, C., & Franke, W. W. (1996). Symplekin, a novel type of tight junction plaque protein. *Journal of Cell Biology, 134*, 1003–1018.

Kitajiri, S., Miyamoto, T., Mineharu, A., Sonoda, N., Furuse, K., Hata, M., et al. (2004). Compartmentalization established by claudin-11-based tight junctions in stria vascularis is required for hearing through generation of endocochlear potential. *Journal of Cell Science, 117*, 5087–5096.

Kitajiri, S. I., Furuse, M., Morita, K., Saishin-Kiuchi, Y., Kido, H., Ito, J., et al. (2004). Expression patterns of claudins, tight junction adhesion molecules, in the inner ear. *Hearing Research, 187*, 25–34.

Konrad, M., Schaller, A., Seelow, D., Pandey, A. V., Waldegger, S., Lesslauer, A., et al. (2006). Mutations in the tight-junction gene claudin 19 (CLDN19) are associated with renal magnesium wasting, renal failure, and severe ocular involvement. *American Journal of Human Genetics, 79*, 949–957.

Kubota, K., Furuse, M., Sasaki, H., Sonoda, N., Fujita, K., Nagafuchi, A., et al. (1999). Ca(2+)-independent cell-adhesion activity of claudins, a family of integral membrane proteins localized at tight junctions. *Current Biology, 9*, 1035–1038.

Matsuda, M., Kubo, A., Furuse, M., & Tsukita, S. (2004). A peculiar internalization of claudins, tight junction-specific adhesion molecules, during the intercellular movement of epithelial cells. *Journal of Cell Science, 117*, 1247–1257.

McCarthy, K. M., Skare, I. B., Stankewich, M. C., Furuse, M., Tsukita, S., Rogers, R. A., et al. (1996). Occludin is a functional component of the tight junction. *Journal of Cell Science, 109*(Pt 9), 2287–2298.

Mitic, L. L., Unger, V. M., & Anderson, J. M. (2003). Expression, solubilization, and biochemical characterization of the tight junction transmembrane protein claudin-4. *Protein Science, 12*, 218–227.

Miyamoto, T., Morita, K., Takemoto, D., Takeuchi, K., Kitano, Y., Miyakawa, T., et al. (2005). Tight junctions in Schwann cells of peripheral myelinated axons: A lesson from claudin-19-deficient mice. *Journal of Cell Biology, 169*, 527–538.

Morita, K., Furuse, M., Fujimoto, K., & Tsukita, S. (1999). Claudin multigene family encoding four-transmembrane domain protein components of tight junction strands. *Proceedings of the National Academy of Sciences of the United States of America, 96*, 511–516.

Morita, K., Sasaki, H., Fujimoto, K., Furuse, M., & Tsukita, S. (1999). Claudin-11/OSP-based tight junctions of myelin sheaths in brain and Sertoli cells in testis. *Journal of Cell Biology, 145*, 579–588.

Moroi, S., Saitou, M., Fujimoto, K., Sakakibara, A., Furuse, M., Yoshida, O., et al. (1998). Occludin is concentrated at tight junctions of mouse/rat but not human/guinea pig Sertoli cells in testes. *American Journal of Physiology, 274*, C1708–C1717.

Muller, D., Kausalya, P. J., Claverie-Martin, F., Meij, I. C., Eggert, P., Garcia-Nieto, V., et al. (2003). A novel claudin 16 mutation associated with childhood hypercalciuria abolishes binding to ZO-1 and results in lysosomal mistargeting. *American Journal of Human Genetics, 73*, 1293–1301.

Muto, S., Hata, M., Taniguchi, J., Tsuruoka, S., Moriwaki, K., Saitou, M., et al. (2010). Claudin-2-deficient mice are defective in the leaky and cation-selective paracellular permeability properties of renal proximal tubules. *Proceedings of the National Academy of Sciences of the United States of America, 107*, 8011–8016. Epub 2010 Apr 12.

Nakano, Y., Kim, S. H., Kim, H. M., Sanneman, J. D., Zhang, Y., Smith, R. J., et al. (2009). A claudin-9-based ion permeability barrier is essential for hearing. *PLoS Genetics, 5*, e1000610.

Nitta, T., Hata, M., Gotoh, S., Seo, Y., Sasaki, H., Hashimoto, N., et al. (2003). Size-selective loosening of the blood-brain barrier in claudin-5-deficient mice. *Journal of Cell Biology, 161*, 653–660.

Nunes, F. D., Lopez, L. N., Lin, H. W., Davies, C., Azevedo, R. B., Gow, A., et al. (2006). Distinct subdomain organization and molecular composition of a tight junction with adherens junction features. *Journal of Cell Science, 119*, 4819–4827.

Pinto da Silva, P., & Kachar, B. (1982). On tight-junction structure. *Cell, 28*, 441–450.

Piontek, J., Winkler, L., Wolburg, H., Muller, S. L., Zuleger, N., Piehl, C., et al. (2008). Formation of tight junction: Determinants of homophilic interaction between classic claudins. *FASEB Journal, 22*, 146–158.

Powell, D. W. (1981). Barrier function of epithelia. *American Journal of Physiology, 241*, G275–G288.

Roh, M. H., Liu, C. J., Laurinec, S., & Margolis, B. (2002). The carboxyl terminus of zona occludens-3 binds and recruits a mammalian homologue of discs lost to tight junctions. *The Journal of Biological Chemistry, 277*, 27501–27509.

Ruffer, C., & Gerke, V. (2004). The C-terminal cytoplasmic tail of claudins 1 and 5 but not its PDZ-binding motif is required for apical localization at epithelial and endothelial tight junctions. *European Journal of Cell Biology, 83*, 135–144.

Saitou, M., Fujimoto, K., Doi, Y., Itoh, M., Fujimoto, T., Furuse, M., et al. (1998). Occludin-deficient embryonic stem cells can differentiate into polarized epithelial cells bearing tight junctions. *Journal of Cell Biology, 141*, 397–408.

Saitou, M., Furuse, M., Sasaki, H., Schulzke, J. D., Fromm, M., Takano, H., et al. (2000). Complex phenotype of mice lacking occludin, a component of tight junction strands. *Molecular Biology of the Cell, 11*, 4131–4142.

Sas, D., Hu, M., Moe, O. W., & Baum, M. (2008). Effect of claudins 6 and 9 on paracellular permeability in MDCK II cells. *American Journal of Physiology: Regulatory, Integrative and Comparative Physiology, 295*, R1713–R1719.

Sasaki, H., Matsui, C., Furuse, K., Mimori-Kiyosue, Y., Furuse, M., & Tsukita, S. (2003). Dynamic behavior of paired claudin strands within apposing plasma membranes. *Proceedings of the National Academy of Sciences of the United States of America, 100*, 3971–3976.

Shen, L., Weber, C. R., & Turner, J. R. (2008). The tight junction protein complex undergoes rapid and continuous molecular remodeling at steady state. *Journal of Cell Biology, 181*, 683–695.

Simon, D. B., Lu, Y., Choate, K. A., Velazquez, H., Al-Sabban, E., Praga, M., et al. (1999). Paracellin-1, a renal tight junction protein required for paracellular Mg^{2+} resorption. *Science, 285*, 103–106.

Sonoda, N., Furuse, M., Sasaki, H., Yonemura, S., Katahira, J., Horiguchi, Y., et al. (1999). *Clostridium perfringens* enterotoxin fragment removes specific claudins from tight junction strands: Evidence for direct involvement of claudins in tight junction barrier. *Journal of Cell Biology, 147*, 195–204.

Staehelin, L. A. (1973). Further observations on the fine structure of freeze-cleaved tight junctions. *Journal of Cell Science, 13*, 763–786.

Stevenson, B. R., & Goodenough, D. A. (1984). Zonulae occludentes in junctional complex-enriched fractions from mouse liver: Preliminary morphological and biochemical characterization. *Journal of Cell Biology, 98*, 1209–1221.

Stevenson, B. R., Siliciano, J. D., Mooseker, M. S., & Goodenough, D. A. (1986). Identification of ZO-1: A high molecular weight polypeptide associated with the tight junction (zonula occludens) in a variety of epithelia. *Journal of Cell Biology, 103*, 755–766.

Takahashi, S., Iwamoto, N., Sasaki, H., Ohashi, M., Oda, Y., Tsukita, S., et al. (2009). The E3 ubiquitin ligase LNX1p80 promotes the removal of claudins from tight junctions in MDCK cells. *Journal of Cell Science, 122*, 985–994.

Tamura, A., Kitano, Y., Hata, M., Katsuno, T., Moriwaki, K., Sasaki, H., et al. (2008). Megaintestine in claudin-15-deficient mice. *Gastroenterology, 134*, 523–534.

Tatum, R., Zhang, Y., Salleng, K., Lu, Z., Lin, J. J., Lu, Q., Jeansonne, B. G., Ding, L., & Chen, Y. H. (2009). Renal salt wasting and chronic dehydration in claudin-7-deficient mice. *American Journal of Physiology. Renal Physiology, 298*, F24–34.

Turksen, K., & Troy, T. C. (2002). Permeability barrier dysfunction in transgenic mice over-expressing claudin 6. *Development, 129*, 1775–1784.

Umeda, K., Ikenouchi, J., Katahira-Tayama, S., Furuse, K., Sasaki, H., Nakayama, M., et al. (2006). ZO-1 and ZO-2 independently determine where claudins are polymerized in tight-junction strand formation. *Cell, 126*, 741–754.

Utech, M., Ivanov, A. I., Samarin, S. N., Bruewer, M., Turner, J. R., Mrsny, R. J., et al. (2005). Mechanism of IFN-gamma-induced endocytosis of tight junction proteins: Myosin II-dependent vacuolarization of the apical plasma membrane. *Molecular Biology of the Cell, 16*, 5040–5052.

Van Itallie, C. M., & Anderson, J. M. (2006). Claudins and epithelial paracellular transport. *Annual Review of Physiology, 68*, 403–429.

Van Itallie, C. M., Colegio, O. R., & Anderson, J. M. (2004). The cytoplasmic tails of claudins can influence tight junction barrier properties through effects on protein stability. *The Journal of Membrane Biology, 199*, 29–38.

Van Itallie, C. M., Fanning, A. S., & Anderson, J. M. (2003). Reversal of charge selectivity in cation or anion-selective epithelial lines by expression of different claudins. *American Journal of Physiology. Renal Physiology, 285*, F1078–F1084.

Van Itallie, C. M., Gambling, T. M., Carson, J. L., & Anderson, J. M. (2005). Palmitoylation of claudins is required for efficient tight-junction localization. *Journal of Cell Science, 118*, 1427–1436.

Van Itallie, C. M., Holmes, J., Bridges, A., Gookin, J. L., Coccaro, M. R., Proctor, W., et al. (2008). The density of small tight junction pores varies among cell types and is increased by expression of claudin-2. *Journal of Cell Science, 121*, 298–305.

Van Itallie, C., Rahner, C., & Anderson, J. M. (2001). Regulated expression of claudin-4 decreases paracellular conductance through a selective decrease in sodium permeability. *Journal of Clinical Investigation, 107*, 1319–1327.

Van Itallie, C. M., Rogan, S., Yu, A., Vidal, L. S., Holmes, J., & Anderson, J. M. (2006). Two splice variants of claudin-10 in the kidney create paracellular pores with different ion selectivities. *American Journal of Physiology. Renal Physiology, 291*, F1288–F1299.

Wen, H., Watry, D. D., Marcondes, M. C., & Fox, H. S. (2004). Selective decrease in paracellular conductance of tight junctions: Role of the first extracellular domain of claudin-5. *Molecular and Cellular Biology, 24*, 8408–8417.

Wilcox, E. R., Burton, Q. L., Naz, S., Riazuddin, S., Smith, T. N., Ploplis, B., et al. (2001). Mutations in the gene encoding tight junction claudin-14 cause autosomal recessive deafness DFNB29. *Cell, 104*, 165–172.

Wong, V., & Gumbiner, B. M. (1997). A synthetic peptide corresponding to the extracellular domain of occludin perturbs the tight junction permeability barrier. *Journal of Cell Biology, 136*, 399–409.

Yu, A. S., Cheng, M. H., Angelow, S., Gunzel, D., Kanzawa, S. A., Schneeberger, E. E., et al. (2009). Molecular basis for cation selectivity in claudin-2-based paracellular pores: Identification of an electrostatic interaction site. *The Journal of General Physiology, 133*, 111–127.

Yu, A. S., Enck, A. H., Lencer, W. I., & Schneeberger, E. E. (2003). Claudin-8 expression in Madin-Darby canine kidney cells augments the paracellular barrier to cation permeation. *The Journal of Biological Chemistry, 278*, 17350–17359.

Zhong, Y., Saitoh, T., Minase, T., Sawada, N., Enomoto, K., & Mori, M. (1993). Monoclonal antibody 7H6 reacts with a novel tight junction-associated protein distinct from ZO-1, cingulin and ZO-2. *Journal of Cell Biology, 120*, 477–483.

CHAPTER 2

Morphological Studies of Claudins in the Tight Junction

Eveline E. Schneeberger
Molecular Pathology Unit, Department of Pathology, Massachusetts General Hospital, Charlestown, Massachusetts, USA

I. OVERVIEW

Claudins are small (20–27 kDa), integral tight junction proteins that span the plasma membrane four times, forming two external loops that are joined by a short cytoplasmic segment and with both N- and C-termini located in the cytoplasm. To date, 24 distinct claudins have been identified. Site-directed mutagenesis experiments indicate that specific amino acid sequences in the first external loop of the claudins determines the ion selectivity of tight junctions, while a cluster of aromatic residues on the second loop appears to provide a strong binding site between claudins expressed on adjacent cells. Combinations of different claudins are expressed in the cell-specific tight

Current Topics in Membranes, Volume 65 1063-5823/10 $35.00

DOI: 10.1016/S1063-5823(10)65002-8

junction strands of polarized cells, where they appear to form heterologous ion pores in the paracellular space. Moreover, fluorescence recovery after photobleaching studies in epithelia with established tight junctions suggest that, in contrast to occludin, claudin-1 appears largely immobile in the plane of the membrane. Although a great deal has been learned regarding the molecular organization of the tight junction, there still remain gaps in our knowledge with regard to the regulation, under physiological and pathological conditions, of this important structure.

II. BACKGROUND

Over a century ago, physiologists suspected that a structural permeability barrier exists in the paracellular pathway, a concept that was confirmed in 1963 when Farquhar and Palade provided the first ultrastructural description of the tight junction (Farquhar & Palade, 1963). They reported that, at the level of the tight junction, there are multiple foci in which the outer lipid leaflets of the adjacent cell membranes appear to merge. The subsequent freeze-fracture studies of Staehelin and colleagues revealed that in the apical region of the intercellular space, the tight junction in fact forms a belt-like structure consisting of multiple parallel, interconnected rows of particles (Staehelin, Mukherjee, & Williams, 1969). These measure 10 nm in diameter with an ~18 nm center-to-center spacing (Anderson, 2001), and are located in the plane of the plasma membrane; their composition, however, was not known.

In vitro experiments, combined with freeze-fracture, had shown that linear arrays of "lipidic particles" form when divalent cations are added to a variety of lipid mixtures (Verkleij, 1984). This gave rise to speculations that the tight junction particles might represent inverted lipid micelles. However, with the discovery of the integral tight junction proteins, occludin (Furuse et al., 1993), claudins (Furuse, Fujita, Hiragi, Fujimoto, & Tsukita, 1998), and tricellulin (Ikenouchi et al., 2005), the pure lipidic model of the tight junction became untenable. Nevertheless, these three classes of integral tight junction proteins have in common that they traverse the membrane bilayer four times, indicating that segments of these molecules are in intimate contact with membrane lipids. Moreover, palmitoylation of claudins (Van Itallie, Gambling, Carson, & Anderson, 2005), but not occludin (Lynch et al., 2007), was shown to be required for their efficient insertion into the tight junction. These observations, together with additional data, led Lee and his colleagues to propose a lipid-protein, hybrid model of the tight junction (Lee, Jamgotchian, Allen, Abeles, & Ward, 2008), in which the lipidic characteristics of the tight junction and its association with lipid rafts are incorporated. These investigators also provided data to suggest that annexin A2 heterotetramers are involved in tight junction assembly (Lee, Jamgotchian, Allen, Kan, & Hale, 2004). In addition, the experimental modulation of membrane cholesterol content in cultured cells, either by treatment with Lovastatin (Stankewich, Francis, Vu, Schneeberger, & Lynch, 1996) or methyl-β-cyclodextrin (Francis, McCarthy, McCormack, Lynch, & Schneeberger, 2001; Francis et al., 1999; Lynch, Tkachuk, Ji, Rabito, & Schneeberger, 1993), has been shown to reversibly alter tight junction barrier function. Using density gradient centrifugation and expanding on an earlier study showing the association of occludin with detergent insoluble glycolipid rafts (Nusrat et al., 2000), Lynch and colleagues observed that both occludin and claudins reside in cholesterol/sphingolipid-rich domains of the plasma membrane (Lynch et al., 2007). Taken together, these observations indicate that the lipid environment of the tight junction proteins is important for optimal function of the tight junction.

To define and characterize the integral tight junction proteins, Stevenson and Goodenough (1984) isolated junctional complexes from mouse liver and used these to generate tight junction-specific antibodies. This led to the discovery of the first cytoplasmic, tight junction-associated protein, zonula occludens-1 (ZO-1) (Stevenson, Siliciano, Mooseker, & Goodenough, 1986). Identification of the integral tight junction proteins, however, remained elusive. It was not until Tsukita and his colleagues reasoned that, because integral tight junction proteins are evolutionarily highly conserved (Ando-Akatsuka et al., 1996), it might require membrane preparations from a non-mammalian species in order to generate the necessary immunological reagents. Using cell membranes isolated from avian livers, these investigators successfully raised antibodies that recognized occludin, thus leading to the discovery of the first integral tight junction protein (Furuse et al., 1993). However, when occludin was expressed in insect cells, intracellular multilamellar structures were observed. Freeze-fracture images revealed short segments of tight junction-like particles in the walls of these multilamellar structures, but no tight junction networks were detected (Furuse et al., 1996). When occludin-deficient embryonic stem cells were found to differentiate into polarized epithelial cells with well-developed tight junctions (Saitou et al., 1998) and occludin knockout mice survived to adulthood, albeit with a subtle, and as yet unexplained, phenotype (Saitou et al., 2000), it prompted the Tsukita group to continue the search for additional integral tight junction proteins.

Using a combination of guanidine–HCl extraction and sucrose density centrifugation of avian liver cell membrane preparations, Furuse and colleagues identified two novel tight junction membrane proteins that were structurally related but with no sequence similarity to occludin. These were designated as claudins-1 and -2 (Furuse, Fujita, et al., 1998) from the Latin "to close." In further studies by the Tsukita group, the transfection of Eph4 epithelial cells with Snail, a transcription repressor, followed by the screening of a high-density oligonucleotide microarray of 24,000 mouse genes, revealed that the transcript of a third tight junction protein appeared to be suppressed by Snail. After raising the requisite antibodies to its NH_2-terminus, this new tight junction protein was found to be located in the area of the tight junction where three cells abut and was designated tricellulin (Ikenouchi et al., 2005). These three distinct groups of integral tight junction proteins share several structural features. All have a tetraspan conformation with their C- and N-termini located in the cytoplasm where, in the case of occludin, they interact with ZO-1, ZO-2, and ZO-3 (Furuse et al., 1994; Haskins, Lijie, Wittchen, Hibbard, & Stevenson, 1998; Itoh, Kazumasa, & Tsukita, 1999), and in the case of selected claudins, an interaction with ZO-1, ZO-2, ZO-3, MUPP-1, and PATJ has been reported (Hamazaki, Itoh, Sasaki, Furuse, & Tsukita, 2002; Itoh et al., 1999; Roh & Margolis, 2003; Schneeberger & Lynch, 2004).

III. CLAUDINS BUT NOT OCCLUDIN FORM TIGHT JUNCTION STRANDS

When mouse L-fibroblasts, lacking tight junctions, were transfected with the cDNA of claudin-1 or -2, tight junction strands formed in the plane of the membrane, a phenomenon that did not occur when occludin cDNA was introduced. Furthermore, when fibroblasts were cotransfected with occludin and claudin-1 cDNA, occludin was readily incorporated into the tight junction strands formed by claudin-1 (Furuse, Fujita, et al., 1998). Support for these observations was provided by a subsequent study with MDCK cells in which either occludin-VSV-G or claudin-1-myc was expressed. Twofold overexpression of claudin-1-myc resulted in the proliferation of tight junction strands in the basolateral membrane below the tight junction, while a 5.6-fold overexpression of occludin-VSV-G induced only subtle changes in tight junction morphology (McCarthy et al., 2000). Taken together, these studies indicate that claudins are not only an essential structural component of tight junction strands but are also capable of self-assembly.

To date, 24 distinct claudins have been identified, none of which share any sequence similarity with either occludin or tricellulin (Ikenouchi et al., 2005; Morita, Furuse, Fujimoto, & Tsukita, 1999; Tsukita, Furuse, & Itoh, 2001). Hydropathy plots indicate that, like tricellulin and occludin, claudins have four transmembrane domains. However, in contrast to occludin, claudins have a short, intracellular amino terminus (7 aa), a large first extracellular loop (52 aa), that is connected via a short, cytoplasmic segment to a smaller second extracellular loop (16–33 aa) and a carboxy terminus (21–63 aa) that

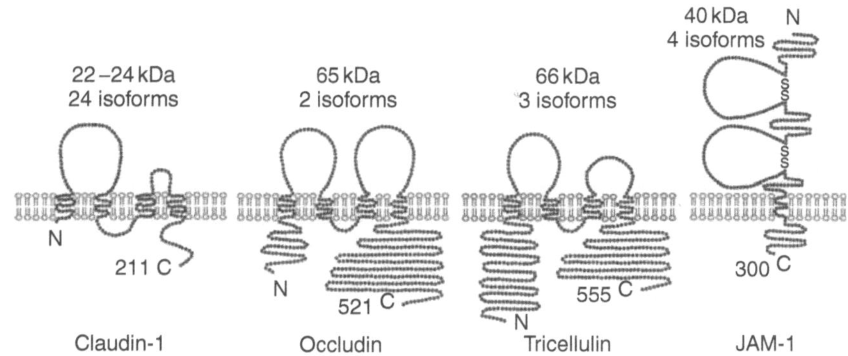

FIGURE 1 Diagram of integral tight junction membrane proteins. The N- and C-termini of claudins, here represented by claudin-1, are significantly shorter than those of occludin or tricellulin. In contrast to occludin, claudin-1 and tricellulin, JAM-1 traverses the cell membrane only once. The indicated figures represent the number of amino acid residues in murine occludin, claudin-1, tricellulin, and JAM-1. Modified from Schneeberger and Lynch (2004), with permission from the *American Journal of Physiology*.

varies in length among the 24 different claudins (Angelow, Ahlstrom, & Yu, 2008; Van Itallie & Anderson, 2006) (Fig. 1). Interestingly, the amino acid composition of each of the extracellular loops of the claudins is predicted to generate a unique pK_i (Mitic & Van Itallie, 2001). When the net charge of the tight junction was changed by the exogenous expression of claudins with different isoelectric points, the permeability for Na^+ was altered (Van Itallie, Fanning, & Anderson, 2003), indicating that the claudins in the tight junction form a charge selective barrier in the intercellular space. With the identification of the large array of unique claudins, the question arises not only how these molecules associate to form the network of strands in the plane of the membrane but also how the two extracellular loops of the claudins interact with those on the apposing cell membranes. To address these questions, a number of morphological approaches have been utilized.

IV. MORPHOLOGICAL TOOLS FOR THE STUDY OF TIGHT JUNCTIONS

A variety of morphological techniques have been applied to the study of tight junction structure and function. These include *immunofluorescence microscopy* for the localization of integral tight junction proteins in the plasma membrane, as well as cytoplasmic proteins that bind specifically to the N- or C-termini of claudins, occludin, and tricellulin. At the ultrastructural level, *immunogold-labeling* methods are used to identify tight junction proteins in thin sections of plastic-embedded tissues or cell monolayers. Since the integral tight junction proteins span the plasma membrane four times, they are in intimate contact with the lipids of the plasma membrane (Lee et al., 2008), a property that lends itself to their ultrastructural analysis by *freeze-fracture* and *fracture-labeling* methods. All of these morphological techniques are dependent on the availability and specificity of the antibodies that recognize individual claudins, as well as occludin and tricellulin.

A. Localization of Tight Junction Proteins by Immunofluorescence Microscopy

The development of immunological reagents that uniquely recognize individual tight junction proteins, together with secondary antibodies tagged with fluorophores that are excited at different, specific wavelengths, facilitates the localization of integral tight junction proteins and their cytoplasmic binding partners, in cultured cells (Fig. 2) and tissues, by immunofluorescence microscopy. Furthermore, the transfection of cultured cells with native or mutated cDNA provides a versatile tool with which to examine specific molecular associations among integral tight junction proteins, as well as with their cytoplasmic binding partners (Furuse, Sasaki, Fujimoto, & Tsukita, 1998; Furuse, Sasaki, & Tsukita, 1999; Furuse et al., 1996; McCarthy et al., 2000;

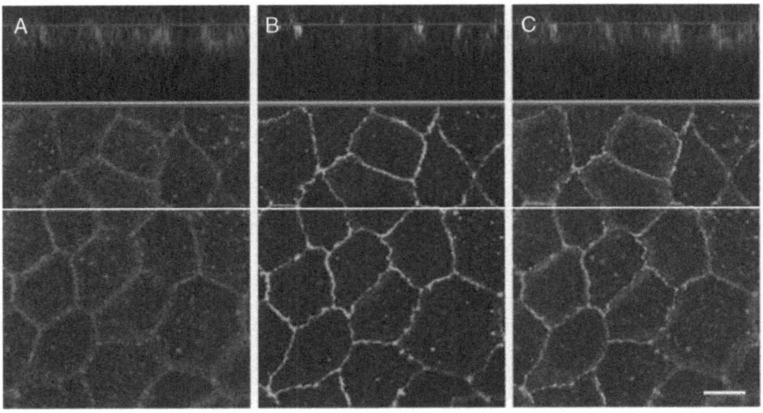

FIGURE 2 Confocal microscopy of MDCK II cell monolayers. Monolayers are labeled by immunofluorescence for (A) claudin-7 (CY3, red) and (B) occludin (Alexa 488, green). (C) Both of these tight junction proteins colocalize at the tight junction (yellow). In the *XZ* view of the tight junction, claudin-1 appears to extend slightly more basally than occludin. Scale bar equals 10 m. (See Color Insert.)

Van Itallie, Rahner, & Anderson, 2001; Yu, Enck, Lencer, & Schneeberger, 2003). The introduction of point mutations in the cDNA of specific claudins used to transfect cells, combined with electrophysiological measurements, provides a means to identify the specific amino acid residues that form the ion selective pores in the tight junction (Angelow & Yu, 2009; Piontek et al., 2008; Van Itallie et al., 2003). An important step in these transfection studies is to verify that the mutated protein is in fact inserted into the plasma membrane and localized at the tight junction. For this purpose, immunofluorescence microscopy is an indispensable tool for monitoring the success of cDNA transfections.

B. Localization of Tight Junction Proteins at the Ultrastructural Level

To localize tight junction proteins at the ultrastructural level, two approaches are commonly utilized. In the first approach, specific tight junction proteins are localized at the ultrastructural level by immunogold labeling of thin sections obtained from aldehyde-fixed, plastic-embedded specimens (Fig. 3A and B) (Hyatt, 1995). In these preparations it is possible to assess the distribution of tight junction proteins in both intracellular compartments

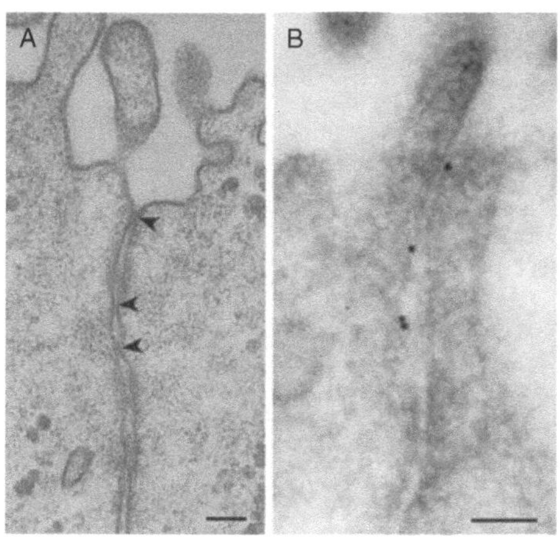

FIGURE 3 (A) Electron micrograph of the tight junction between two MDCK cells, fixed in glutaraldehyde. Arrow heads indicate the site of the tight junction strands where the adjacent cell membranes are in close apposition. (B) Electron micrograph of an epithelial tight junction, immunogold-labeled for claudin-3. Four gold particles label sites in the tight junction that were accessible to the anti-claudin-3 antibody. For immunogold labeling, cells were fixed in 1% paraformaldehyde in PBS. Scale bars equal 100 nm.

and in the plasma membrane at the site of the tight junction. However, this approach provides a somewhat limited view of the tight junction, since sections are usually cut perpendicular to the plane of the tight junction. In such images, immunogold particles label those tight junction proteins that are accessible at the discrete sites where neighboring membranes are in close apposition (Fig. 3B).

Freeze-fracture replicas are used to assess the organization of the tight junction strand network in the plane of the membrane (Fig. 4A) and, when combined with immunogold labeling, they provide a means to localize specific proteins to individual tight junction strands (Fujimoto, 1995) (Fig. 4B). Briefly, to produce these replicas, rapidly frozen, fixed or unfixed, cryoprotected biological specimens are fractured under vacuum at −180 °C, a condition under which the fracture plane preferentially traverses along the hydrophobic core of cell membranes (Fig. 5A). As a result, the membrane bilayer is cleaved in such a manner that the inner half remains associated with the cell and is designated the protoplasmic (P) face, while the outer half is designated the exoplasmic (E) face (Fig. 5B). The freshly fractured surface is coated at a 45° angle with a layer of platinum and stabilized with a layer of

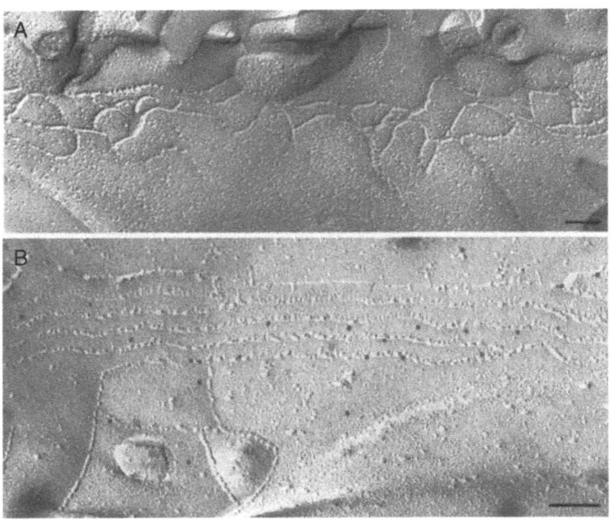

FIGURE 4 (A) Freeze-fracture replica of the tight junction between two MDCK cells, fixed in glutaraldehyde. The tight junction consists of a network of interconnected strands in the plane of the membrane. (B) Fracture labeling of S180 fibroblasts transfected with claudin-16. Cell pellets, fixed in paraformaldehyde, were freeze-fractured and the replicas incubated with anti-claudin-16 antibody, followed by protein-A coated gold. Immunogold particles are aligned with the tight junction strands. Scale bars equal 100 nm.

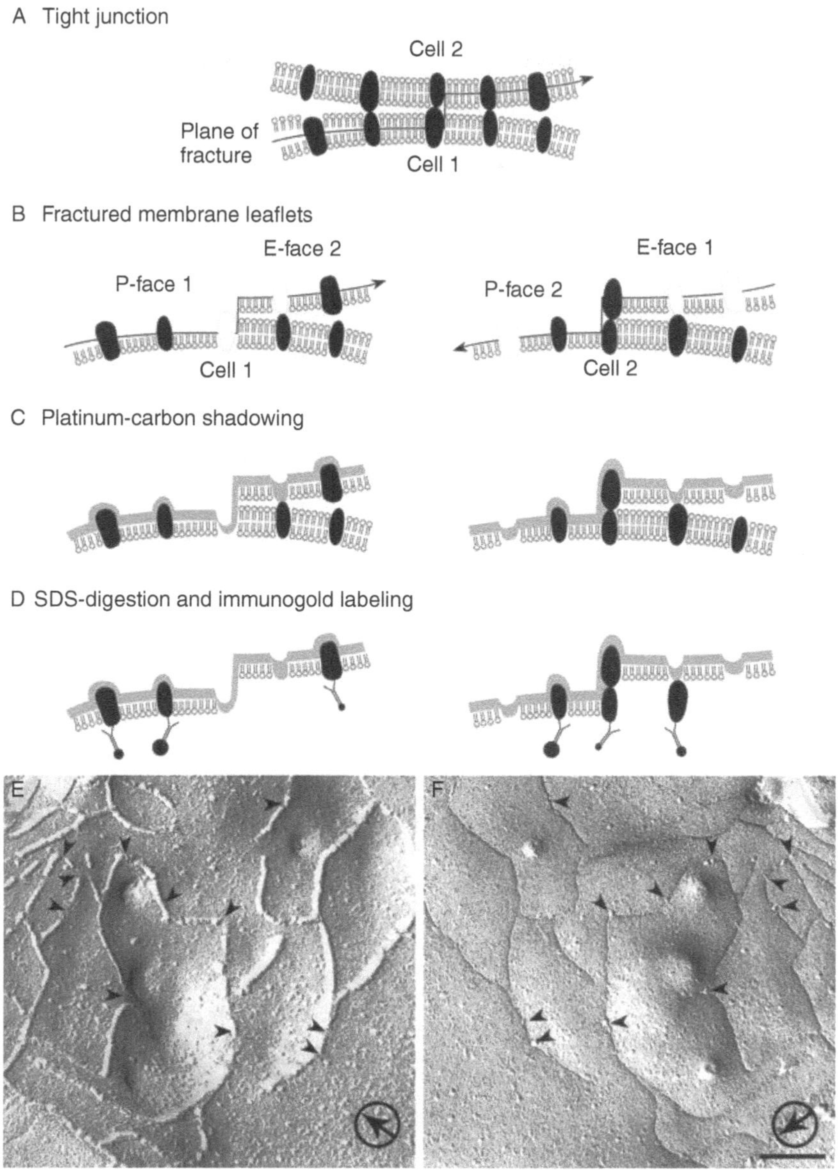

FIGURE 5 Diagrammatic representation of the fracture-labeling process, as it pertains to the tight junction, using a double replica device. (A) At low temperatures, the plasma membrane is preferentially cleaved along the hydrophobic plane of the lipid bilayer, indicated by the arrow. Note that the fracture plane skips from the plasma membrane of cell 1 to that of cell 2. (B) The resulting mirror images of the cleaved lipid bilayer, including a short segment of noncleaved membrane, are shown. The black ovals represent integral tight junction proteins and the empty

carbon (Fig. 5C). The underlying cellular material adherent to the replica is either completely removed using sodium hypochlorite for conventional replicas (Schneeberger & Lynch, 2001) or partially removed using a buffered SDS solution, and tight junction proteins that remain adherent to the replica are subsequently labeled with immunogold reagents (Fujimoto, 1995) (Fig. 5D). The replicas are then examined by electron microscopy.

In the double replica images shown in Fig. 5E and F, the tight junction strands on the P face appear to be discontinuous (Fig. 5E) (arrow heads), whereas on the E face (arrow heads), these sites contain small segments of tight junction proteins that have been pulled from the P face. The manner in which membrane particles partition onto either the E or P face is speculated to depend, in part, on the strength of the bond between the integral tight junction protein and its cytoplasmic binding partner. It is also dependent on the type and concentration of the fixative, as well as the length of time used to fix the cells. In general, buffered glutaraldehyde fixatives strengthen the protein–protein cross-linking, while paraformaldehyde-based fixatives result in weaker protein cross-linking, better preservation of antigenic sites, and are the preferred fixative for immunogold labeling.

To obtain semiquantitative data, electron micrographs of freeze-fractured tight junction images have been analyzed morphometrically by counting the number of parallel strands at 1 cm intervals along the length of the tight junction electron micrograph. This was based on the assumption that each strand acts as a "resistor" in the paracellular pathway and that there is a relationship between the number of parallel strands and transepithelial electrical resistance (TER) (Claude & Goodenough, 1973). However, the fact that TER does not correlate with the number of tight junction strands became apparent when two strains of MDCK cells were compared. While the measured TER of MDCK I and MDCK II cell monolayers was $>10{,}000$ and $\sim 200\ \Omega\ cm^2$, respectively, the number of parallel tight junction strands was in fact similar in both strains of cells (Stevenson, Anderson, Goodenough, & Mooseker, 1988). This apparent paradox was subsequently found to be due to the expression of claudin-2 in the low-resistance MDCK II

---

spaces represent the site from which the protein was pulled out during the fracturing procedure. (C) The fractured surface is coated with a layer of platinum, followed by a layer of carbon. Fragments of attached cytoplasmic material are solubilized with SDS. (D) Integral membrane proteins of interest are labeled by immunogold techniques. As indicated, immunogold labeling may occur on both E and P fracture surfaces. (E, F) Double replica of an epithelial tight junction in which arrow heads in (E) indicate gaps in the P face of the tight junction strands, while in (F), the arrow heads point to aggregates of tight junction proteins that have partitioned into the E face half of the membrane. Scale bars equal 100 nm. Reprinted by permission from Schneeberger et al. (1978).

cells and its absence in MDCK I cells (Furuse, Furuse, Sasaki, & Tsukita, 2001). Moreover, when MDCK I cells were transfected with claudin-2 cDNA, TER fell to that observed in MDCK II cells (Furuse et al., 2001) and paracellular sodium flux increased (Amasheh et al., 2002). These observations clearly demonstrate that TER and ion selectivity of the tight junction are determined by the type of expressed claudins and not by the number of parallel tight junction strands.

V. CLAUDINS ARE RELATIVELY STABLE COMPONENTS OF THE TIGHT JUNCTION

While freeze-fracture images provide information regarding the organization of the tight junction strands in a given cell monolayer or tissue (Schneeberger & Lynch, 1992), they do not convey the dynamic behavior of certain integral tight junction proteins within individual strands. That tight junction strands themselves might be mobile was first observed when L-fibroblasts were transfected with cDNA that encoded green fluorescent protein (GFP)-labeled claudin-1 (Sasaki et al., 2003). The GFP-labeled claudin-1 polymerized into paired strands that were found to be surprisingly mobile, associating with each other in both an end-to-side and a side-to-side manner. Fluorescence recovery after photobleaching (FRAP) data, however, indicated that within individual strands, claudin-1 appeared immobile. It is of note that since these studies were conducted in fibroblasts that lack endogenous tight junctions, it was unclear whether in epithelial cells with well-established tight junctions, a similar dynamic behavior of the tight junction strands is detectable.

To address this question, Turner and colleagues expressed enhanced green fluorescent protein (EGFP)-labeled occludin or claudin-1 in live MDCK epithelial cells with established tight junctions. Using FRAP it was observed that, unlike their behavior in transfected fibroblasts, there did not appear to be any detectable movement of the tight junction strands themselves (Shen, Weber, & Turner, 2008). Furthermore, within the strands, approximately 76% of claudin-1 was stably localized at the tight junction, confirming the earlier observations by the Tsukita group (Sasaki et al., 2003). Surprisingly, however, the majority (71%) of occludin appeared to diffuse within the tight junction strands, while ZO-1 was found to exchange between the membrane and intracellular pools of ZO-1 in an energy-dependent manner (Shen et al., 2008). These observations suggest that the tight junction is a dynamic structure that may be controlled and/or altered by both physiological and pathological stimuli.

VI. CLAUDIN–CLAUDIN INTERACTIONS

It is now well established that the expression of selected claudins and their alternatively spliced isoforms determines the ion selectivity of the tight junctions in a particular epithelium (Guenzel et al., 2009; Hou et al., 2008; Simon et al., 1999; Van Itallie et al., 2003; Yu et al., 2003, 2008). Since multiple claudins may be expressed in the tight junctions of a given cell monolayer or tissue, the question is, are there any molecular constraints on the ability of individual claudins to interact side-to-side within the tight junction strands and head-to-head with those in the tight junctions of neighboring cells? In early studies, Furuse and colleagues analyzed side-to-side interactions in L-fibroblasts that were cotransfected with two of the following three claudins, claudin-1, -2, or -3 (Furuse et al., 1999). In all combinations examined by fracture labeling, the three claudins were readily coexpressed in the newly formed tight junction strands of transfected fibroblasts, indicating that there are no restrictions in the side-to-side interaction among the claudins tested. However, when L-fibroblasts were transfected with one of these three claudins and then cocultured with L-fibroblasts expressing one of the other two claudins, it was observed that claudin-3 interacts head-to-head with claudins-1 and -2, but claudin-1 does not interact with claudin-2. These observations indicate that while heteropolymers of different claudins are formed within the tight junction strands, there appear to be constraints and/or preferences in the head-to-head interactions of paired heterologous claudins. The precise amino acid composition of the extracellular loops of the claudins that either promotes or inhibits claudin–claudin interactions between adjacent cells is only beginning to be defined.

To examine homophilic interactions among claudins, Piontek and colleagues transfected human embryonal kidney cells that lack endogenous tight junction proteins with claudin-5 cDNA (Piontek et al., 2008). In this model it appeared that extracellular loop 2 is involved in strand formation via its trans-interaction with loop 2 of claudin-5 on the adjacent cells. A combination of site-directed mutagenesis, live cell imaging, electron microscopy, and molecular modeling, led to a proposed antiparallel homodimer model in which a cluster of conserved aromatic (F147, Y148, and Y158) and hydrophilic (Q156 and E159) residues on loop 2 could potentially provide a strong binding site for claudins on apposing cells. This conclusion was further strengthened by the fact that these residues are conserved among most claudins and suggest a possible mechanism whereby mechanical stability is gained by this trans-junctional interaction among claudins.

Specific heterophilic interactions among claudins have been observed in the recently identified human renal disorder that is characterized by hypomagnesemia, hypercalciuria, and nephrocalcinosis with progressive renal

Mg^{2+} and Ca^{2+} wasting. The disease has been linked to mutations in the gene-encoding claudin-16, also known as paracellin-1 (Simon et al., 1999). However, when claudin-16 was expressed in LLC-PK1 cells, the permeability of the tight junctions to Na^{+} increased, but there was relatively little effect on Mg^{2+} flux, suggesting that claudin-16 by itself does not form a Mg^{2+} selective channel (Hou, Paul, & Goodenough, 2005). Interestingly, in a subsequent study in which both claudin-16 and -19 were introduced into MDCK or LLC-PK1 cells, neither of which expressed these claudins endogenously, it was observed that for claudin-16 to be incorporated into the tight junction strands, the presence of claudin-19 was required. Furthermore, it was shown that claudin-16 and -19 interact synergistically to form a cation-selective pore in the tight junction (Hou et al., 2008). This was further supported by the observation that siRNA-induced knockdown of claudin-19 in mice caused claudin-16 to be lost from the thick ascending limb of Henle (Hou et al., 2009). These observations indicate that in order to form cation-selective pores, a heteromeric interaction between claudin-16 and -19 is required.

VII. CLAUDINS, REGULATORS OF PARACELLULAR ION SELECTIVITY

In view of the central role that claudins play in determining the ion selectivity of tight junctions, studies have begun to focus on the molecular basis underlying this ion selectivity. In a recent study, Angelow and Yu (2009) examined four potential pore-lining amino acids (Tyr^{35}, His^{57}, Asp^{65}, and Ile^{66}) in loop 1 of claudin-2, using a novel cysteine-scanning mutagenesis approach. Briefly, each of these four amino acid residues were mutated to cysteines and then screened for their ability to interact with thiol-reactive reagents of differing size and charge. Cysteine mutagenesis did not affect the localization of claudin-2 at the tight junction, as evaluated by confocal microscopy. Furthermore, all claudin-2 mutants maintained functional pores, except D65C, which formed intermolecular disulfide bonds, leading to a loss of both charge and size selectivity, indicating that Asp^{65} is close to a protein–protein interface. Addition of methanethiosulfonate reagents of different size and charge decreased paracellular ion permeation only in I66C mutants, suggesting that Ile^{66} is situated in a narrow segment of the pore. In addition, I66C was weakly reactive, while Y35C was strongly reactive with *N*-biotinylaminoethylmethanethiosulfonate, indicating that Tyr^{35} is located external to the ion pores of the tight junctions. Detailed biochemical studies such as these, provide vital molecular information regarding the structure and organization of claudins in the tight junction.

VIII. SUMMARY

By combining the tools of molecular biology, electrophysiology, immunocytochemistry, and fracture labeling, new information has been generated regarding the structure/function relationships of selected claudins in the tight junction. The ion selectivity of the tight junction pores appears to be governed by the amino acid composition of the specific claudins that form these pores. With the application of novel biochemical reagents, the structure of these tight junction pores within the tight junction is beginning to be characterized. To date, relatively few of the 24 known claudins have been analyzed for their pore-forming function and detailed tissue distribution. However, recently discovered mutations in a number of claudin genes that are associated with human disease, have yielded valuable new insight regarding the function of several specific claudins.

References

Amasheh, S., Meiri, N., Gitter, A. H., Schoeneberg, T., Mankertz, J., Schulzke, J. D., et al. (2002). Claudin-2 expression induces cation-selective channels in tight junctions of epithelial cells. *Journal of Cell Science, 115*, 4969–4976.

Anderson, J. M. (2001). Molecular structure of tight junctions and their role in epithelial transport. *News Physiological Science, 16*, 126–130.

Ando-Akatsuka, Y., Saitou, M., Hirase, T., Kishi, M., Sakakibara, A., Itoh, M., et al. (1996). Interspecies diversity of the occludin sequence: cDNA cloning of human, mouse, dog and rat-kangaroo homologues. *Journal of Cell Biology, 133*, 43–47.

Angelow, S., Ahlstrom, R., & Yu, A. S. L. (2008). Biology of claudins. *American Journal of Physiology Renal Physiology, 295*, F867–F876.

Angelow, S., & Yu, A. L. S. (2009). Structure-function studies of claudin extracellular domains by cysteine-scanning mutagenesis. *Journal of Biological Chemistry, 284*, 29205–29217.

Claude, P., & Goodenough, D. A. (1973). Fracture faces of zonulae occludentes from "tight" and "leaky" epithelia. *Journal of Cell Biology, 58*, 390–400.

Farquhar, M. G., & Palade, G. E. (1963). Junctional complexes in various epithelia. *Journal of Cell Biology, 17*, 375–412.

Francis, S. A., Kelly, J. M., McCormack, J. M., Rogers, R. A., Lai, J., Schneeberger, E. E., et al. (1999). Rapid reduction of MDCK cell cholesterol by methyl-β-cyclodextrin alters steady state transepithelial electrical resistance. *European Journal of Cell Biology, 78*, 473–484.

Francis, S. A., McCarthy, K. M., McCormack, J. M., Lynch, R. D., & Schneeberger, E. E. (2001). Depletion of membrane cholesterol alters detergent solubility, phosphorylation and susceptibility to proteolysis of tight junction proteins. *Molecular Biology of the Cell, 12*, 217a.

Fujimoto, K. (1995). Freeze-fracture replica electron microscopy combined with SDS digestion for cytochemical labeling of integral membrane proteins. Application to the immunogold labeling of intercellular junctional complexes. *Journal of Cell Science, 108*, 3443–3449.

Furuse, M., Fujimoto, K., Sato, N., Hirase, T., Tsukita, S., & Tsukita, S. (1996). Overexpression of occludin, a tight junction-associated integral membrane protein, induces the formation of intracellular multilamellar bodies bearing tight junction-like structures. *Journal of Cell Science, 109*, 429–435.

Furuse, M., Fujita, K., Hiragi, T., Fujimoto, K., & Tsukita, S. (1998). Claudin 1 and 2: Novel integral membrane proteins localizing at tight junctions with no sequence similarity to occludin. *Journal of Cell Biology, 141*, 1539–1550.

Furuse, M., Furuse, K., Sasaki, H., & Tsukita, S. (2001). Conversion of *zonulae occludentes* from tight to leaky strand type by introducing claudin-2 into Madin-Darby canine kidney I cells. *Journal of Cell Biology, 153*, 263–272.

Furuse, M., Hirase, T., Itoh, M., Nagafuchi, A., Yonemura, S., Tsukita, S., et al. (1993). Occludin: A novel integral membrane protein localizing at tight junctions. *Journal of Cell Biology, 123*, 1777–1788.

Furuse, M., Itoh, M., Hirase, T., Nagafuchi, A., Yonemura, S., Tsukita, S., et al. (1994). Direct association of occludin with ZO-1 and its possible involvement in the localization of occludin at tight junctions. *Journal of Cell Biology, 127*, 1617–1626.

Furuse, M., Sasaki, H., Fujimoto, K., & Tsukita, S. (1998). A single gene product, claudin-1 or -2, reconstitutes tight junction strands and recruits occludin in fibroblasts. *Journal of Cell Biology, 143*, 391–401.

Furuse, M., Sasaki, H., & Tsukita, S. (1999). Manner of interaction of heterogeneous claudin species within and between tight junction strands. *Journal of Cell Biology, 147*, 891–903.

Guenzel, D., Stulver, M., Kausalya, P. J., Halsch, L., Krug, S. M., Rosenthal, R., et al. (2009). Claudin-10 exists in six alternatively spliced isoforms that exhibit distinct localization and function. *Journal of Cell Science, 122*, 1507–1517.

Hamazaki, Y., Itoh, H., Sasaki, M., Furuse, M., & Tsukita, S. (2002). Multi-PDZ domain protein 1 (MUPP1) is concentrated at tight junctions through its possible interaction with claudin-1 and junctional adhesion molecule. *Journal of Biological Chemistry, 277*, 455–461.

Haskins, J., Lijie, G., Wittchen, E. S., Hibbard, J., & Stevenson, B. R. (1998). ZO-3, a novel member of the MAGUK protein family found at the tight junction, interacts with ZO-1 and occludin. *Journal of Cell Biology, 141*, 199–208.

Hou, J., Paul, D. L., & Goodenough, D. A. (2005). Paracellin-1 and the modulation of ion selectivity of tight junctions. *Journal of Cell Science, 118*, 5109–5118.

Hou, J., Renigunta, A., Gomes, A. S., Hou, M., Paul, D. L., Waldegger, S., et al. (2009). Claudin-16 and claudin-19 interaction is required for their assembly into tight junctions and for renal reabsorption of magnesium. *Proceedings of the National Academy of Sciences of the United States of America, 106*(36), 15350–15355.

Hou, J., Renigunta, A., Konrad, M., Gomes, A. S., Schneeberger, E. E., Paul, D. L., et al. (2008). Claudin-16 and claudin-19 interact and form a cation-selective tight junction complex. *Journal of Clinical Investigation, 118*, 619–628.

Koehler, J. K. (1972). The Freeze-Etching Technique. In M. A. Hayat (Ed.), *Principles and techniques of electron microscopy, Vol. 2* (pp. 53–82). New York: Van Nostrand Reinhold Co.

Ikenouchi, J., Furuse, M., Furuse, K., Sasaki, H., Tsukita, S., & Tsukita, S. (2005). Tricellulin constitutes a novel barrier at tricellular contacts of epithelial cells. *Journal of Cell Biology, 171*, 939–945.

Itoh, M., Furuse, M., Morita, K., Kubota, K., Saitou, M., & Tsukita, S. (1999). Direct binding of three tight junction-associated MAGUKs, ZO-1, ZO-2 and ZO-3, with the COOH termini of claudins. *Journal of Cell Biology, 147*, 1351–1363.

Itoh, M., Kazumasa, M., & Tsukita, S. (1999). Characterization of ZO-2 as a MAGUK family member associated with tight as well as adherens junctions with a binding affinity to occludin and α catenin. *Journal of Biological Chemistry, 274*, 5981–5986.

Lee, D. B. N., Jamgotchian, N., Allen, S. G., Abeles, M. B., & Ward, H. J. (2008). A lipid-protein hybric model for tight junction. *American Journal of Physiology Renal Physiology, 295*, F1601–F1612.

Lee, D. B. N., Jamgotchian, N., Allen, S. G., Kan, F. W. K., & Hale, I. L. (2004). Annexin A2 heterotetramer: Role in tight junction assembly. *AJP: Renal Physiology, 287*, F481–F491.

Lynch, R. D., Francis, S. A., McCarthy, K. M., Casas, E., Thiele, C., & Schneeberger, E. E. (2007). Cholesterol depletion alters detergent-specific solubility profiles of selected tight junction proteins and the phosphorylation of occludin. *Experimental Cell Research, 313*, 2597–2610.

Lynch, R. D., Tkachuk, L. J., Ji, X., Rabito, C. A., & Schneeberger, E. E. (1993). Depleting cell cholesterol alters calcium-induced assembly of tight junctions by monolayers of MDCK cells. *European Journal of Cell Biology, 60*, 21–30.

McCarthy, K. M., Francis, S. A., McCormack, J. M., Lai, J., Rogers, R. A., Skare, I. B., et al. (2000). Inducible expression of claudin-1-myc but not occludin-VSVG results in aberrant tight junction strand formation in MDCK cells. *Journal of Cell Science, 113*, 3387–3398.

Mitic, L. L., & Van Itallie, C. M. (2001). Occludin and claudins: Transmembrane proteins of the tight junction. In M. Cereijido & J. M. Anderson (Eds.), *Tight junctions* (pp. 213–230). (2nd ed.). Boca Raton, FL: CRC Press.

Morita, K., Furuse, M., Fujimoto, K., & Tsukita, S. (1999). Claudin multigene family encoding four-transmembrane domain protein components of tight junction strands. *Proceedings of the National Academy of Sciences of the United States of America, 96*, 511–516.

Nusrat, A., Parkos, C. A., Verkade, P., Foley, C. S., Liang, T. W., Innis-Whitehouse, W., et al. (2000). Tight junctions are membrane microdomains. *Journal of Cell Science, 113*, 1771–1781.

Piontek, J., Winkler, L., Wolburg, H., Mueller, S. L., Zuleger, N., Piehl, C., et al. (2008). Formation of tight junction: Determinants of homophilic interaction between classic claudins. *The FASEB Journal, 22*, 146–158.

Roh, M. H., & Margolis, B. (2003). Composition and function of PDZ protein complexes during cell polarization. *American Journal of Physiology, 285*, F377–F387.

Saitou, M., Fujimoto, K., Doi, Y., Itoh, M., Fujimoto, T., Furuse, M., et al. (1998). Occludin-deficient embryonic stem cells can differentiate into polarized epithelial cells bearing tight junctions. *Journal of Cell Biology, 141*, 397–408.

Saitou, M., Furuse, M., Sasaki, H., Schulzke, J. K., Fromm, M., Takano, H., et al. (2000). Complex phenotype of mice lacking occludin, a component of tight junction strands. *Molecular Biology of the Cell, 11*, 4131–4142.

Sasaki, H., Matsui, C., Furuse, K., Mimori-Kiyosue, Y., Furuse, M., & Tsukita, S. (2003). Dynamic behavior of paired claudin strands within apposing plasma membranes. *Proceedings of the National Academy of Sciences of the United States of America, 100*, 3971–3976.

Schneeberger, E. E., & Lynch, R. D. (1992). Tight junctions in the lung. In M. Cereijido (Ed.), *Tight junctions* (pp. 337–351) (1st ed.). Boca Raton, FL: CRC Press.

Schneeberger, E. E., & Lynch, R. D. (2001). *Ultrastructure and immuno-labeling of the tight junction* (2nd ed.). Boca Raton, FL: CRC Press.

Schneeberger, E. E., & Lynch, R. D. (2004). The tight junction: A multifunctional complex. *AJP: Cell Physiology, 286*, C1213–C1228.

Shen, L., Weber, C. R., & Turner, J. R. (2008). The tight junction protein complex undergoes rapid and continuous molecular remodeling at steady state. *Journal of Cell Biology, 181*, 683–695.

Simon, D. B., Lu, Y., Choate, K. A., Velazquez, H., Al-Sabban, E., Praga, M., et al. (1999). Paracellin-1, a renal tight junction protein required for paracellular Mg^{2+} resorption. *Science, 285*, 103–106.

Staehelin, L. A., Mukherjee, T. M., & Williams, A. W. (1969). Freeze-etch appearance of the tight junctions in the epithelium of small and large intestine of mice. *Protoplasma, 67*, 165–184.

Stankewich, M. C., Francis, S. A., Vu, Q. U., Schneeberger, E. E., & Lynch, R. D. (1996). Alterations in cell cholesterol content modulate of calcium induced tight junction assembly by MDCK cells. *Lipids, 31*, 817–828.

Stevenson, B. R., Anderson, J. M., Goodenough, D. A., & Mooseker, M. S. (1988). Tight junction structure and ZO-1 content are identical in two strains of Madin-Darby canine kidney cells which differ in transepithelial electrical resistance. *Journal of Cell Biology, 107*, 2401–2408.

Stevenson, B. R., & Goodenough, D. A. (1984). Zonulae occludentes in junctional complex-enriched fractions from mouse liver: Preliminary morphological and biochemical characterization. *Journal of Cell Biology, 98*, 1209–1221.

Stevenson, B. R., Siliciano, J. D., Mooseker, J. D., & Goodenough, D. A. (1986). Identification of ZO-1: A high molecular weight polypeptide associated with the tight junction (zonula occludens) in a variety of epithelia. *Journal of Cell Biology, 103*, 755–766.

Tsukita, S., Furuse, M., & Itoh, M. (2001). Multifunctional strands in tight junctions. *Nature Reviews Molecular Cell Biology, 2*, 285–293.

Van Itallie, C. M., & Anderson, J. M. (2006). Claudins and epithelial paracellular transport. *Annual Review of Physiology, 68*, 403–429.

Van Itallie, C. M., Fanning, A. S., & Anderson, J. M. (2003). Reversal of charge selectivity in cation or anion-selective epithelial lines by expression of different claudins. *American Journal of Physiology, 285*, F1078–F1084.

Van Itallie, C. M., Gambling, T. M., Carson, J. L., & Anderson, J. M. (2005). Palmitoylation of claudins is required for efficient tight junction localization. *Journal of Cell Science, 118*, 1427–1436.

Van Itallie, C., Rahner, C., & Anderson, J. M. (2001). Regulated expression of claudin-4 decreases paracellular conductance through a selective decrease in sodium permeability. *Journal of Clinical Investigation, 107*, 1319–1327.

Verkleij, A. J. (1984). Lipidic intramembranous particles. *Biochimica Biophysica Acta, 779*, 43–63.

Yu, A. L. S., Cheng, M. H., Angelow, S., Guenzel, D., Kanzawa, S. A., Schneeberger, E. E., et al. (2008). Molecular basis for cation selectivity in claudin-2-based paracellular pores: Identification of an electrostatic interaction site. *Journal of General Physiology, 133*(1), 111–127.

Yu, A. S. L., Enck, A. H., Lencer, W. I., & Schneeberger, E. E. (2003). Claudin-8 expression in Madin-Darby canine kidney cells augments the paracellular barrier to cation permeation. *Journal of Biological Chemistry, 278*, 17350–17359.

CHAPTER 3

Biophysical Methods to Study Tight Junction Permeability

Dorothee Günzel, Susanne M. Krug, Rita Rosenthal, and Michael Fromm
Institute of Clinical Physiology, Campus Benjamin Franklin,
Charité, Berlin, Germany

Current Topics in Membranes, Volume 65 1063-5823/10 $35.00

DOI: 10.1016/S1063-5823(10)65003-X

I. OVERVIEW

Transepithelial resistance (TER) can be determined, as a repetitive screening method, by chopstick electrodes directly on cell culture filters or, as a more reliable technique, in Ussing chambers. Although TER often serves as a rough indicator of tight junction permeability, only more advanced biophysical methods like impedance spectroscopy and conductance scanning allow to quantify changes in paracellular resistance and to separate these changes from those occurring in transcellular or subcellular resistance. These techniques are thus superior to conventional TER measurements which provide combined effects only. In addition, tight junction permeabilities can be determined by flux or diffusion potential measurements. It has to be kept in mind, however, that there is not one single permeability of a specimen, but that permeabilities are different for every ion or uncharged molecule of distinct size, as exemplarily shown for tricellulin which differentially modulates bicellular and tricellular tight junction permeabilities for ions and macromolecules. Paracellular permeabilities to water can be determined by transepithelial measurements on claudin-perturbed cells, as has been done for claudin-2 and claudin-10b. Finally, biophysical measurements provide information about tight junction channel properties such as pore size ion charge preferences, and may help to identify independent and parallel pathways.

II. INTRODUCTION

More than half a century ago, the terms “zonula occludens” and “tight junction” (TJ) were introduced after it had been shown in electron microscopic studies that these intercellular structures form a border against the passage of solutes across epithelia (Farquhar & Palade, 1963). Morphological differences were found between TJs of “leaky” and “tight” epithelia (Claude & Goodenough, 1973). The observation that sealing is dependent on the amount of horizontally oriented strands (Claude, 1978) is in line with the idea that this structure generally seals the paracellular space but in a more or less perfect way.

During the 1990s, only shortly after the discovery of the claudins, it turned out that many members of this protein family, such as claudin-1 and claudin-5, indeed seal the TJ, but there are also exceptions that behave differently, for example, claudin-2. Claudin-2 causes TJ strands to become leaky (Furuse, Furuse, Sasaki, & Tsukita, 2001) by forming paracellular cation-selective ion channels (Amasheh et al., 2002). This and similar findings of other channel-forming claudins restimulated research on permeability properties of the TJ with special focus on the understanding of specific barrier function and of mechanisms concerning disturbed barrier in epithelial and endothelial diseases. Disturbed tight junctional barrier may have two functional consequences, (i) unwanted absorption of potentially harmful large solutes or (ii) excessive secretion of ions, followed by an osmotically equivalent amount of water which, in intestines, causes leak flux diarrhea.

Four points may be stressed regarding permeability studies of TJs:

First, conventional TER (transepithelial resistance) measurements do not adequately describe the resistance provided by the TJ. Perturbations of the TJ regularly cause a much larger change in paracellular resistance than is reflected in TER changes. Perturbations or pathologies may not only alter the paracellular but also the transcellular barrier, that is, cell membrane ion channels.

Second, in simple resistance measurements on native tissues like intestinal epithelia, adhering subepithelial layers add to total resistance. For inflammatory diseases, it is typical that subepithelial layers proliferate and thus increase in resistance while epithelial resistance decreases due to a TJ barrier defect. In the worst case, both changes compensate each other, resulting in an unchanged total resistance as detected by TER measurement.

Third, simply to state "the permeability" of a barrier is not a meaningful statement. In contrast to resistance, permeability always refers to a substance. This can be anything between small ions and macromolecules. However, frequently only one tracer is measured to determine "the permeability." Two opposing examples may illustrate that this is insufficient: Claudin-2-dominated TJs are permeable to small cations and water but not to anions or larger solutes (Amasheh et al., 2002; Rosenthal et al., 2010). In contrast, tricellulin-a abolishes the permeability of tricellular TJs to macromolecules but not to ions (Krug et al., 2009). Thus, at least the permeability to several solutes of different sizes should be determined in order to obtain a permeability profile of a barrier.

Fourth, permeabilities should be given as such (common unit: cm/s), not as marker flux rates or even concentration changes within the bathing fluid, because flux rates or concentration changes do not allow comparisons of results from different studies or for different markers.

In the following paragraphs, we describe strategies and methods allowing a detailed description of TJ resistance and permeabilities to various solutes and to water.

III. RESISTANCE MEASUREMENTS

A. Transepithelial Resistance (TER, R^t)

The most common measure of TJ permeability to ions is the "transepithelial electrical resistance" (TER or, in the symbolic language of electrophysiology, R^t). To be exact, R^t represents the reciprocal of the sum of the permeabilities of all ions of the adjacent bath solution times their respective concentrations. In a simplified form, including only Na^+ and Cl^-, this relation is detailed in Eqs. (3) and (4).

In effect, in a nonselective epithelium, R^t mainly represents the permeabilities to Na^+, Cl^-, K^+ and, if present, HCO_3^-. The reciprocal of R^t is G^t, the transepithelial conductance, given in mS/cm^2 of gross epithelial area. It is clear that conductance is proportional to area, and consequently that resistance decreases with increasing area. This is the reason why R^t is expressed as $\Omega\ cm^2$, not Ω/cm^2. In general, R^t is measured by applying a current pulse, ΔI, across the cell layer and recording the resulting voltage change ΔV. According to Ohm's law, the resistance is calculated as the voltage difference between both electrodes divided by the amplitude of the current pulse ($\Delta V/\Delta I$). Since any resistor located between the voltage-sensing electrodes is included in this raw resistance measurement, the contribution of the bathing fluid has to be determined without epithelium and then subtracted in order to obtain R^t.

B. Chopstick Electrodes

In confluent cell cultures grown on permeable filter supports, R^t can be conveniently and repetitively measured by commercially available "chopstick electrode" systems. However, chopstick measurements often yield values different from those obtained in standard Ussing-type chambers. There are several reasons for this discrepancy.

First, due to a nonuniform current field produced by the chopstick electrodes across the filter insert, R^t results may differ depending on the position of the electrode above the cell layer (Jovov, Wills, & Lewis, 1991). Especially in low-resistance cells, the access resistance from the lower side of the filter causes current to pass the filter mainly near its edge and, to a lesser degree, at the center of the filter. This effect is dramatically worsened in setups where the cells grow directly on metal electrode plates. It is clear from this that the electrodes should be positioned over the cell monolayers in a standardized way, for example, by a motorized adjusting apparatus.

A second disadvantage of the chopstick systems arises if solute net fluxes occur. Because the bathing fluids are static, unstirred layers develop and cause different concentrations of the transported solute on the apical and the basolateral side, which in turn affect flux rates and apparent permeabilities.

Furthermore, chopstick systems are usually driven by alternating clamp currents (AC) which, in part, short-circuit the ohmic-resistor elements of the cell layer. This causes a systematic underestimation of the true R^t proportional to both frequency and R^t.

Finally, chopstick systems without temperature control do not allow prolonged measurements because metabolic energy-driven transport and hence epithelial resistance is strongly temperature-dependent.

In conclusion, "chopstick" systems have to be used with caution, although they are good tools for screening the state of confluence of cell cultures, because they make repetitive measurements of identical filters possible.

C. Ussing Chamber

A reliable way to measure R^t is by means of four-electrode chambers, which are named after their inventor Hans Ussing (Ussing, 1949). Here, an almost homogenously distributed electric current (or DC current step; ΔI) is generated by two electrodes which are positioned at a distance from the epithelium (Fig. 1). This current is applied on the whole area of the epithelial layer and results in a voltage change (ΔV) that is sensed by two further electrodes close to the epithelium. The fluids on both sides of the epithelium are permanently stirred and temperature-controlled by a circulating bubble lift, which, at the same time, provides equilibration of these fluids with O_2 and CO_2, thus making experiments of several hours duration possible. Ussing chamber techniques, including the short-circuit technique for measuring active transport, are explained in detail, for example, by Clarke (2009) and Brown and O'Grady (2008).

D. Impedance Spectroscopy

In an attempt to investigate the specific function of single TJ proteins, simple R^t measurements on cell layers transfected with such proteins may give first crude indications, whether these proteins tighten the paracellular barrier, such as claudin-1 (Furuse et al., 2002) or claudin-5 (Amasheh et al., 2005; Nitta et al., 2003), or whether they increase paracellular ion permeability, as, for example, claudin-2 (Amasheh et al., 2002) or claudin-10 (Günzel, Stuiver, et al., 2009; Van Itallie et al., 2006).

However, R^t measurements are only of limited use to quantify these effects, as they do not discriminate between different resistance components of an epithelium. As illustrated in Fig. 2B, R^t consists of three major components. Two of them lie in parallel, the transcellular resistance (R^{trans}, dominated by the resistances of the basolateral and apical plasma membrane) and the paracellular resistance (R^{para}) which is formed by the TJ. A third resistor, the subepithelial resistance (R^{sub}), lies in series to R^{trans} and R^{para} and represents all resistances built by nonepithelial tissues or areas underneath the epithelia. In cell culture systems, it is present due to the filter support on which the cells are grown.

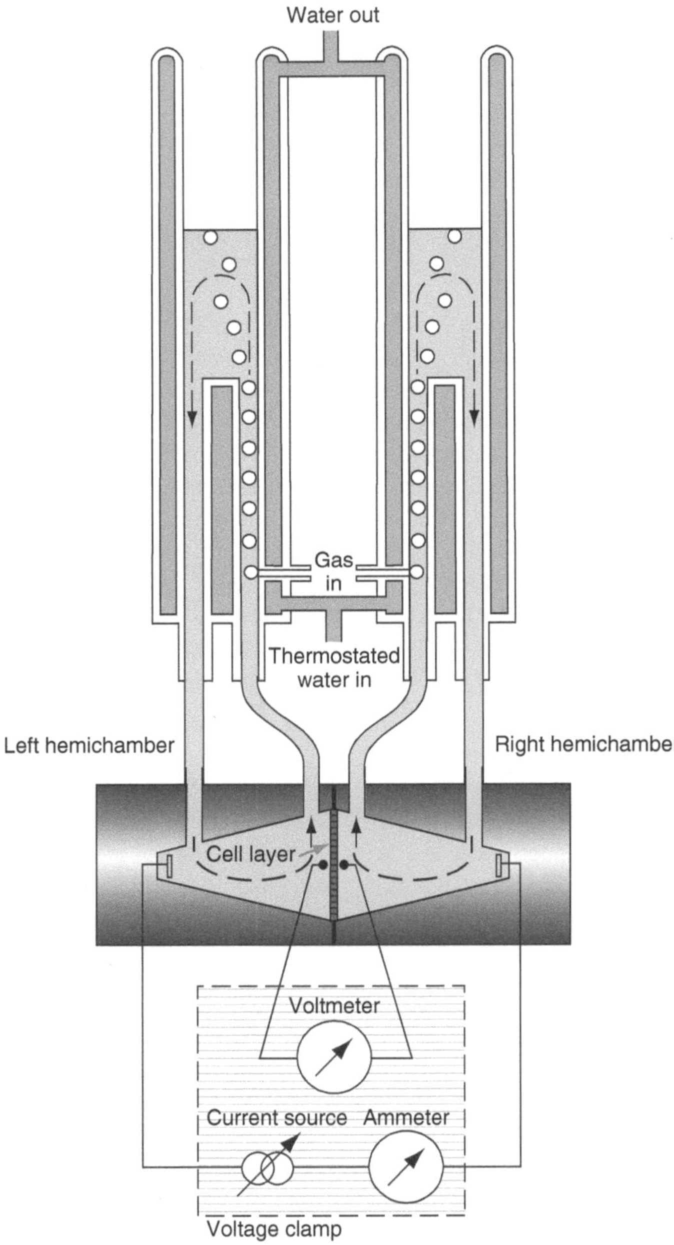

FIGURE 1 Schematic of an Ussing chamber. The chamber is named after the Danish physiologist Hans Ussing (1911–2000) and consists of two fluid-filled hemi-chambers which are separated by a filter-grown cell culture monolayer or a native epithelium. Two fluid reservoirs with gas lifts provide recirculation, gassing, and tempering of the bathing fluids. Driven by a voltage clamp unit, current is applied via two distant electrodes and voltage is sensed by two electrodes positioned close to the cell layer.

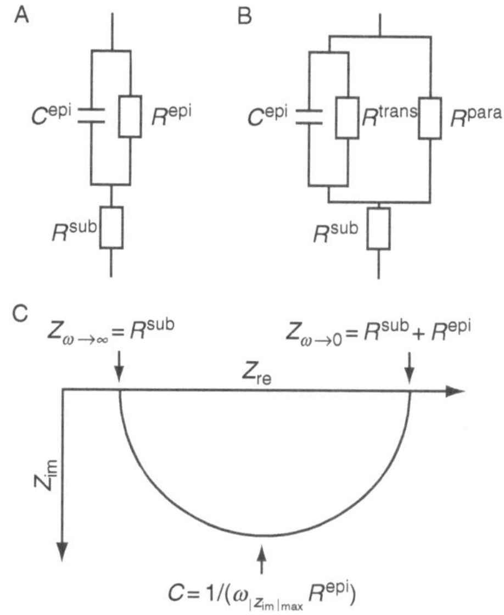

FIGURE 2 Equivalent electrical circuit of epithelia: (A) "One path impedance" measurements are based on an electrical model that only discriminates between subepithelial (R^{sub}) and epithelial (R^{epi}) resistance, and the epithelial capacity C^{epi}. (B) "Two-path impedance" measurements are based on the assumption that the epithelial resistance R^{epi} is composed of two resistances lying in parallel: a paracellular ($R = {}^{para}$) and a transcellular (R^{trans}) resistance. Under DC conditions or AC conditions at very low frequencies R^{epi} equals $R^{para} \times R^{trans}/(R^{para} + R^{trans})$. (C) Nyquist diagram (plot of the real and the imaginary portion of the impedance, Z_{re}, Z_{im}) for models shown in (A) and (B). For impedance values measured at different AC frequencies (angular frequency, ω), both models result in a semicircle. For high frequencies ($\omega \to \infty$), Z_{re} approaches R^{sub}, and for low frequencies ($\omega \to 0$), Z_{re} approaches $R^{t} = R^{sub} + R^{epi}$. The capacitance C^{epi} can be calculated from the frequency at which $|Z_{im}|$ reaches a maximum ($C^{epi} = 1/(\omega_{|Z_{im}|max} \times R^{epi})$. Adapted from Krug, Fromm, & Günzel (2009).

Impedance measurements have been used for almost a century to investigate the membrane properties of living cells (Cole & Curtis, 1938a, 1938b, 1938c; Curtis & Cole, 1938; Fricke, 1925; Höber, 1910; McClendon, 1927; McClendon, 1936) and since the 1940s to investigate the properties of epithelia (Teorell, 1946). Over the past decades, a great variety of equivalent electrical circuits with varying degrees of complexity were used to model epithelial properties. Impedance measurements make use of the fact that cell membranes act like capacitors and that under AC conditions capacitive reactances are strongly dependent on the frequencies employed (typically up to 50 different frequencies in the range of about 0.1 Hz up to several 10,000 Hz).

The inadequacy of conventional R^t measurements to compare epithelia under control and diseased states, or to study the potential barrier effects of drugs or hormones, and the necessity to employ impedance spectroscopy are illustrated by the following two examples.

First, when investigating the epithelial barrier in colonic biopsies from patients with inflammatory bowel diseases, it was found that R^t remained almost unchanged, although these patients clearly suffered from diarrhea caused by barrier loss and although claudin expression patterns were greatly changed. Further investigations employing impedance spectroscopy uncovered that a substantial decrease in R^{para} indeed occurred but was masked by a concomitant increase in R^{sub} (Bürgel et al., 2002; Kroesen, Dullat, Schulzke, Fromm, & Buhr, 2008; Zeissig et al., 2007).

Second, the application of the hormone aldosterone to colon epithelium increases expression of the epithelial Na^+ channel ENaC and thus causes a reduction in R^{trans}. However, a recent study (Amasheh et al., 2009) demonstrated that at the same time claudin-8 expression increased. Impedance measurements proved that this resulted in a several-fold increase in R^{para}. Despite this strong tightening of the paracellular barrier, the overall effect on R^t was a decrease.

These examples show clearly that simple R^t measurements do not necessarily reflect the barrier properties (i.e., paracellular tightness or leakiness) of an epithelium and that impedance measurements are a potent technique to differentiate between transcellular, paracellular, and subcellular effects.

E. One-Path Impedance Spectroscopy

In the simplest "lumped" model (Fig. 2A), epithelia are represented by three parameters: a membrane capacitance, C, an epithelial resistance, R^{epi}, and a subepithelial resistance, R^{sub}. All three parameters can be determined directly from impedance spectra by the equations given in the legend to Fig. 2C. In this model, only one single pathway for the current across the epithelium is supposed to exist. Therefore, impedance measurements based on this model have recently been dubbed "one-path impedance spectroscopy" (1PI) (Fromm et al., 2009).

1PI measurements were successfully applied in investigations on various inflammatory bowel diseases to uncover barrier loss, which, in simple transepithelial resistance measurements, is masked by a simultaneous increase in subepithelial resistance (Bürgel et al., 2002; Kroesen et al., 2008; Zeissig et al., 2007). Under physiological conditions, the thickness of the subepithelium does not contribute toward the transepithelial tightness because blood capillaries make their way through this subepithelial tissue and lie in direct vicinity to the epithelium. Under all inflammatory conditions, the leakiness of the epithelium was increased by an upregulation of pore-forming claudins (e.g., claudin-2) and a concomitant downregulation of barrier-forming claudins (e.g., claudin-3, -4, -5, or -8).

F. Two-Path Impedance Spectroscopy

In the model underlying two-path impedance spectroscopy (2PI, Fig. 2B), R^{epi} is represented by a transcellular (R^{trans}) and a paracellular (R^{para}) fraction, that is, the current is assumed to cross the epithelium along two distinct pathways. To discriminate between these two pathways, additional measurements are necessary, as the equation system derived from impedance measurements is underdetermined (legend of Fig. 2B, for details, see Krug, Fromm, & Günzel, 2009).

In order to determine all four parameters (C, R^{sub}, R^{trans}, and R^{para}), studies from the 1980s employed intracellular impedance measurements using glass microelectrodes to gain information on the resistance and capacitance of the apical and basolateral plasma membrane (Kottra & Frömter, 1984). In an alternative approach, several studies used the high-resolution conductance scanning technique described later to analyze paracellular versus cellular resistances (Frömter & Diamond, 1972; Gitter, Bertog, Schulzke, & Fromm, 1997) and thus obtain sufficient information to solve the 2PI equation. As pointed out in the next section, conductance scanning is technically challenging, especially when used at its highest resolution and therefore not useful as a routine tool. A further approach to derive sufficient information to define all parameters within the epithelial equivalent circuit was a permeabilization of the plasma membrane by the application of ionophores (Clausen, Lewis, & Diamond, 1979; Lewis, Eaton, Clausen, & Diamond, 1977; Wills, Lewis, & Eaton, 1979).

In recent studies from our own lab, we used a combination of impedance spectroscopy and flux measurements of paracellular marker substances (e.g., fluorescein) while modulating the paracellular pathway, for example, by the removal of extracellular Ca^{2+} (Krug, Fromm, & Günzel, 2009; Reiter et al., 2006). The method is based on three assumptions: First, the reduction in the free extracellular Ca^{2+} concentration, induced by an application of EGTA, reduces the paracellular resistance without affecting the transcellular resistance (Ca^{2+} switch); second, fluorescein is a true paracellular marker, that is, it is exclusively transported along the paracellular pathway, and the resulting flux is proportional to the paracellular conductance (G^{para}); and third, the subepithelium does not affect fluorescein flux. If these assumptions are valid, then the following equation is valid:

$$G^{epi} = G^{trans} + G^{para} = G^{trans} + (s \times \text{fluorescein flux}) \qquad (1)$$

with s being a constant; G^{epi}, epithelial conductance; G^{trans}, transcellular conductance; and G^{para}, paracellular conductance.

This is a linear equation. Plotting fluorescein flux values (x-axis) against G^{epi} values (y-axis) obtained from experiments during which only the paracellular pathway was modulated, should therefore result in a straight line with a y-intercept equaling G^{trans}.

As R^{sub}, G^{epi} ($=1/R^{epi}$), and C can be directly obtained from the impedance data and G^{trans} ($=1/R^{trans}$) and G^{para} ($=1/R^{para}$) can be determined from Eq. (1), the combination of EGTA switch experiments with a concomitant flux and impedance measurement allows a full description of all four epithelial parameters C, R^{sub}, R^{trans}, and R^{para} (Krug, Fromm, & Günzel, 2009). Applicability of this method could be confirmed in several cell lines such as the human colonic carcinoma cell line HT-29/B6 or the canine kidney cell line MDCK with its various subtypes (MDCK I, MDCK II, MDCK C7, and MDCK C11). It cannot be used on the human colonic carcinoma cell line Caco-2, as in these cells, fluorescein appears to be at least partially transported along the transcellular route.

Despite its being developed only recently, the 2PI spectroscopy technique has already been successfully employed in several studies to differentiate para- from transcellular effects on R^t (Amasheh et al., 2009; Krug et al., 2009; Krug, Fromm, & Günzel, 2009; Mankertz et al., 2009).

1. Example: Tricellulin and Tricellular Tight Junction

Tricellulin, a TJ protein discovered in 2005 (Ikenouchi et al., 2005) and predominantly localized in the tricellular meeting points of TJs, the tricellular TJ (tTJ), was assumed to be a barrier-forming and -stabilizing TJ protein. In overexpression studies in MDCK II cells, two types of clones were used. One differed solely in tricellular localization of tricellulin, the other differed in bicellular as well as tricellular localization. In this latter type, an increase in R^t was observed, which lacked in clones with overexpression restricted to the tTJ (Krug et al., 2009). 2PI spectroscopy disclosed that the threefold rise in R^{epi} in the bi- and tricellularly overexpressing clone was due to a 14-fold increase in R^{para}, while R^{trans} remained unchanged (Fig. 3). Although 2PI spectroscopy allows high resolution of even small changes in R^{epi}, there was no increase detectable in any of the three resistance parameters of the tricellularly overexpressing clone, indicating that the tricellular central tube has no measurable influence on R^t. Calculations using estimations of upper and lower limits of conductance underlined this finding, as in MDCK II cells the contribution of the tTJ on R^{epi} would only be $\sim$1% of the contribution of the bTJ.

2. Example: Claudin-10b

As recently published, MDCK C7 cells transfected with claudin-10b exhibit a strong paracellular cation permeability that leads to a reduction in R^{epi} (Günzel, Stuiver, et al., 2009). Performance of 2PI spectroscopy seemed to be

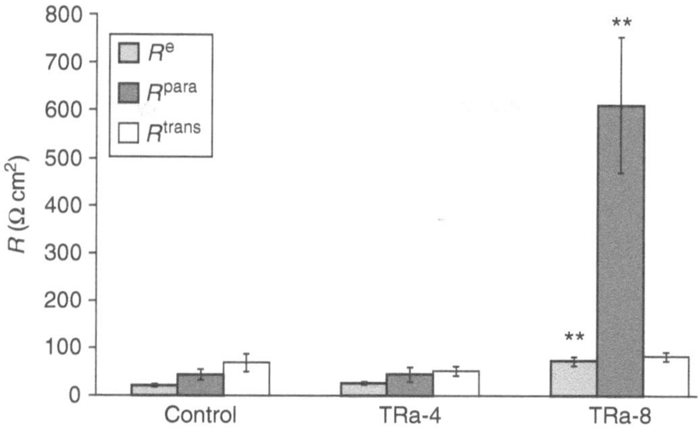

FIGURE 3 Two-path impedance spectroscopy of clones overexpressing TRIC-a. Tricellular overexpression of TRIC-a in TRa-4 cells did not change epithelial resistance (R^{epi}) in comparison to controls, while the bicellular overexpression of TRIC-a in TRa-8 cells induced a threefold increase in R^e ($^{**}p < 0.01$; $n = 6$, 7, and 6, respectively). This increase was caused by a 14-fold rise of paracellular resistance (R^{para}, $^{**}p < 0.01$), which is determined by the ion permeability of the TJ. Transcellular resistance (R^{trans}) was not significantly changed. From Krug et al. (2009).

suitable to analyze these findings employing fluorescein, Cl^-, and Na^+ as paracellular flux markers (Krug, Fromm, & Günzel, 2009). Plotting G^{epi} against the permeabilities of those flux markers yielded linear relationships. Extrapolation of the y-intercept for all three flux markers resulted in the same G^{trans}, while the slopes of the three relationships differed remarkably (Fig. 4). The slope of G^{epi}/Na^+ permeability was reduced to less than a half compared to G^{epi}/Cl^- permeability, indicating an increased P_{Na}/P_{Cl} as it was already observed by dilution potential measurements (Günzel, Stuiver, et al., 2009).

3. Example: Tumor Necrosis Factor-α and Claudin-2

Tumor necrosis factor-α (TNFα) is a proinflammatory cytokine which is elevated, for example, in active Crohn's disease. After the application of TNFα to HT-29/B6 cells, the performance of 2PI spectroscopy indicated a distinct decrease in R^{epi} which was solely due to a decrease in R^{para}, while the transcellular resistance R^{trans} remained unchanged (Mankertz et al., 2009; Fig. 5). Further examinations revealed an increased expression of claudin-2 after TNFα application to be responsible for the decrease of R^{para}. This claudin is also often upregulated in chronic intestinal inflammation such as Crohn's disease (Zeissig et al., 2007), supporting the finding that there is an influence of TNFα on tight junctional barrier function.

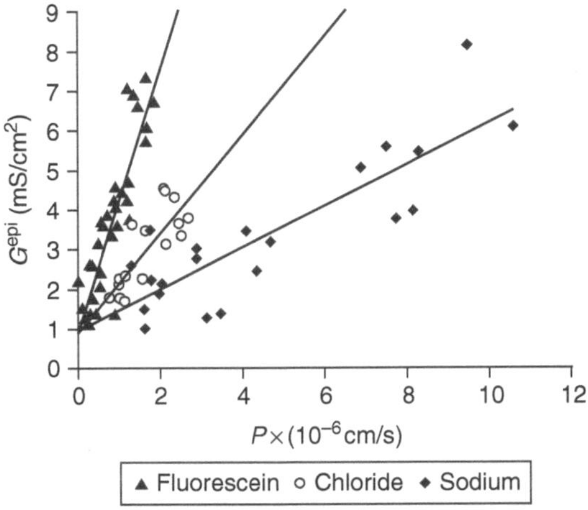

FIGURE 4 Plot of G^{epi} versus permeabilities for fluorescein, Cl^-, and Na^+ in clones expressing claudin-10b. Plotting of G^{epi} versus fluorescein (▲), Cl^- (○), and Na^+ (◆) permeability yields G^{trans} values of 0.94, 0.93, and 0.98 mS/cm^2, respectively, which have to be corrected by 0.6 mS/cm^2 for values in the absence of EGTA. R^{trans} thus amounts to 2860 Ω cm^2. The slope of G^{epi} versus Na^+ (0.50×10^3 S × s/cm^3) is lower than for Cl^- (1.12×10^3 S × s/cm^3) and fluorescein (3.3×10^3 S × s/cm^3), mirroring the claudin-10b-induced increase in cation permeability. From Krug, Fromm, & Günzel (2009).

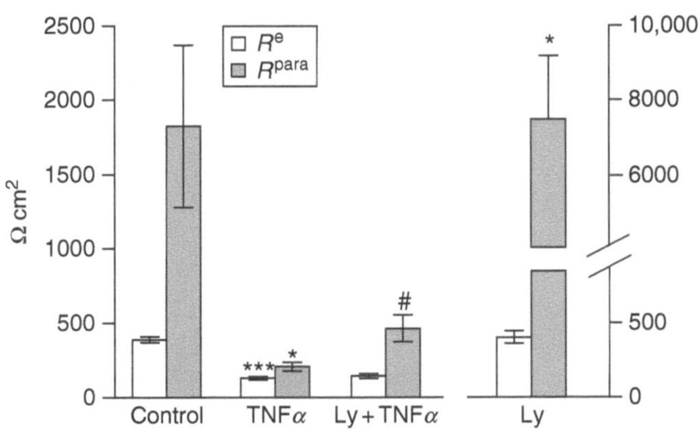

FIGURE 5 Two-path impedance spectroscopy of HT-29/B6 cells treated with TNFα. TNFα reduces the epithelial as well as the paracellular resistance. The effect can be attenuated by LY294002 (Ly) (R^{epi}, epithelial resistance; R^{para}, paracellular resistance). $*p < 0.05$, $***p < 0.001$, $\#p < 0.05$. From Mankertz et al. (2009).

G. Conductance Scanning

Conductance scanning (also named voltage scanning) techniques are methods based on the analysis of local differences in current density, recorded in the supraepithelial bath solution by glass microelectrodes during the application of a transepithelial clamp current. Since the inhomogeneous conductivity of an epithelium is evaluated, these methods allow to determine the size and distribution of local conductivities in the plane of flat epithelia. Thus, differentiation between para- and transcellular conductivities is also possible.

The general principle of the conductance scanning technique was pioneered by Frömter, who applied large direct currents and detected voltage peaks above TJs (Frömter, 1972; Frömter & Diamond, 1972). For our variant of the technique, we combined the high spatial resolution of a scanning glass microelectrode, as used by Frömter (1972), Hudspeth (1975), and Cereijido, Stefani, and Palomo (1980), with the superior signal-to-noise ratio of the lock-in principle (Foskett & Scheffey, 1989; Jaffe & Nuccitelli, 1974).

The setup consists of a four-electrode setup in a horizontal chamber plus a pair of mobile-scanning microelectrodes (Fig. 6). Epithelia or confluent cultured cell monolayers are mounted and viewed through a 20× or 40×

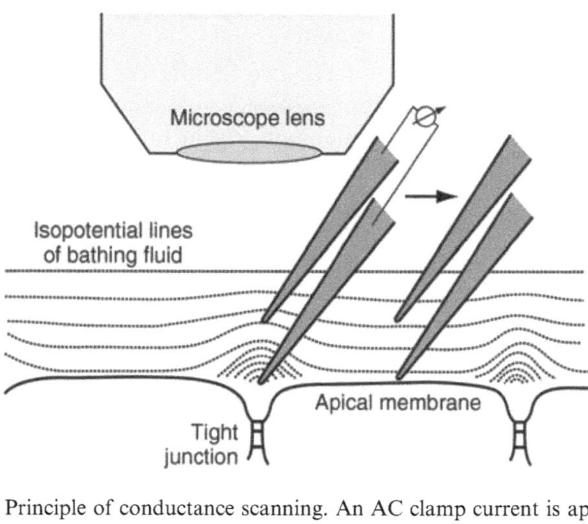

FIGURE 6 Principle of conductance scanning. An AC clamp current is applied across the cell monolayer and induces—depending on the resistances of the respective site—a major voltage drop across the cells and a very small one (μV range) across the bathing fluids, indicated by isopotential lines. The cell layer is scanned horizontally by a pair of jointed microelectrodes. The signal is detected by a phase-locked loop amplifier. Adapted from Fromm et al. (2009).

water-immersion objective lens. Alternating electric current of 0.2 mA/cm^2 and 24 Hz is clamped across the tissue and the electric field generated in the mucosal bath solution is measured with a pair of microcapillaries above the epithelial surface. In the present development, the two microcapillaries are glued together, yielding a tip distance of approximately 30–50 μm. The measured voltage between the tip openings varies in proportion to the local tissue conductance below the electrodes.

The position of the microelectrodes in relation to the tissue is adjusted by moving the experimental chamber with an electrically driven micromanipulator. The distance between the tissue and the lower scanning electrode is adjusted between 200 μm and virtually zero, depending on the intended spatial resolution. With the measured electric field and the known specific resistivity of the bath solution, the local current density is calculated. Together with the transepithelial voltage, this yields the local conductivity.

Depending on the distance between epithelium and scanning electrodes, different spatial resolutions are obtained. A list of applications, ordered with increasing resolution, include

- areas of ulceration (Gitter, Wullstein, Fromm, & Schulzke, 2001);
- colon surface epithelium versus crypt openings (Grotjohann et al., 1998; Köckerling & Fromm, 1993; Köckerling, Sorgenfrei, & Fromm, 1993);
- focal leaks as, for example, generated by bacteria (Troeger et al., 2007),
- single-cell wound repair (Florian, Schöneberg, Schulzke, Fromm, & Gitter, 2002; Günzel et al., 2006);
- leaks caused by epithelial apoptoses (Bojarski et al., 2001; Gitter, Bendfeldt, Schulzke, & Fromm, 2000a);
- paracellular versus transcellular pathway (Gitter, Bendfeldt, Schulzke, & Fromm, 2000b; Gitter et al., 1997; Schulzke et al., 2005).

A characteristic feature of the conductance scanning technique lies in the fact that it provides distinct local data. If conductance distribution of an epithelium is homogenous, this is no advantage, but in case of unequal, for example, focal, distributions, this technique is favorable over techniques which yield averaged data from the entire chamber opening.

At highest resolution, the resistance across the TJ, R^{para}, is determined. This requires the use of scanning electrodes with tip diameters of ~1 μm and hence comparatively high resistances (resistance ~4–5 MΩ when filled with 0.5 M KCl), and scanning very close (vertical distance < 3 μm) to the surface of the cells (Gitter et al., 1997). Both requirements limit the applicability of the conductance scanning technique: Cell layers have to be very even and must not be covered with mucus in order to accurately position the scanning electrodes so close above the cells. Furthermore, the high resistance of the scanning electrodes induces a high level of noise above which very small changes in potential have to be detected. For this reason, conductance scanning is only applicable if tight junctional conductance greatly differs from transcellular conductance.

Gitter et al. (1997) successfully used low-resistance MDCK C11 cell monolayers and found that the paracellular and transcellular pathways contributed approximately 80% and 20%, respectively, toward the total transepithelial conductance of $\sim$13 mS/cm^2 ($R^{epi} = 76\ \Omega\ cm^2$, $R^{para} = 95\ \Omega\ cm^2$, $R^{trans} = 385\ \Omega\ cm^2$). A later study on the same cell type (Amasheh et al., 2002) yielded similar values ($R^{epi} = 52\ \Omega\ cm^2$, $R^{para} = 68\ \Omega\ cm^2$, $R^{trans} = 221\ \Omega\ cm^2$). These values are somewhat higher than those recently obtained by 2PI spectroscopy ($R^{epi} = 41\ \Omega\ cm^2$, $R^{para} = 105\ \Omega\ cm^2$, $R^{trans} = 98\ \Omega\ cm^2$) and it yet needs to be established whether these differences are due to the different methods or whether they reflect natural variability in this cell type. Differences in R^{epi}, however, point toward the latter assumption, as in both cases, R^{epi} was measured using the same conventional Ussing chamber method described earlier.

In order to study the effects of claudin-2 transfected into high-resistance MDCK C7 cells, Amasheh et al. (2002) used a combination of impedance and conductance scanning techniques. In this study, they were able to demonstrate that the claudin-2-induced 3.6-fold reduction in R^{epi} was almost entirely due to a reduction in R^{para} from $\sim$2700 to 485 Ω cm^2.

IV. ION PERMEABILITY MEASUREMENTS

Although conventional Ussing chamber experiments do not allow a direct discrimination of R^{trans} and R^{para}, they allow the analysis of ion selectivities and permeabilities of the paracellular pathway, and estimates of the paracellular pore size. These data have recently provided the basis for a modeling of the paracellular pore architecture (Yu et al., 2009).

Various methods have been established to determine epithelial ion permeabilities and three of these methods are briefly outlined in the following paragraphs. A major difficulty of all methods is finding a way to distinguish between the paracellular and transcellular pathways. Generally, it is assumed that paracellular ion movement is nonrectifying, that is, independent of the direction of measurement. If flux in the apical to basolateral direction differs from that in the opposite direction, it is usually assumed that the difference is due to transcellular ion transport. Ways to distinguish between paracellular and transcellular ion movements include pharmacological inhibition of the transcellular route or carrying out the experiments at low temperature, as active ions transport mechanisms show a stronger temperature dependence than passive (diffusive) transport mechanism.

A. Ion Flux Measurements

Ion permeabilities can be measured directly if suitable radioactive isotopes are available. This is the case for Na^+ (^{22}Na, $t_{1/2} = 2.6$ years), Cl^- (^{36}Cl, $t_{1/2} = 3.0$ years), and Ca^{2+} (^{45}Ca, $t_{1/2} = 162.7$ days). ^3H- or ^{14}C-labeled organic ions may also be available. Difficulties arise with K^+ (^{40}K is not available, but sometimes ^{86}Rb is used as a surrogate; however, see below for differences in permeability) and with Mg^{2+}. ^{28}Mg ($t_{1/2} = 21$ days) is practically unavailable, ^{27}Mg ($t_{1/2} = 9.46$ min) is so short-lived that it is only of very limited use (however, see Bijvelds, Kolar, Wendelaar-Bonga, & Flik, 1996; Bijvelds, Kolar, Bonga, & Flik, 1997).

As a rule, a small (micromolar) amount of a radioactive tracer is added to the saline on the apical or basolateral side (=donor side) and aliquots are collected from the other side (=acceptor side) at regular intervals to determine flux as the transported amount per time and epithelial area. Flux data can easily be converted into permeabilities by multiplication with the concentration of the investigated ion in the saline.

Flux measurements are usually carried out under equilibrium exchange conditions, that is, the stable isotope or unlabeled organic ion is present in the saline at physiological (millimolar) concentrations on both sides of the cell layer. These unidirectional flux measurements have the advantage that they are carried out under equilibrium conditions and thus reflect the physiological situation.

If no radioactive isotope or labeled compound is available, flux measurements can be carried out under "zero-trans" conditions, that is, the acceptor side initially does not contain the ion under investigation and its accumulation is detected by an appropriate analytical technique (ion-selective electrode, colorimetric reaction, atomic absorption spectrometry, HPLC, etc.). Results obtained under zero-trans conditions may differ from those obtained under equilibrium exchange conditions, as transport kinetics under both conditions are different. Furthermore, TJ properties may be directly affected by the ion gradient used under zero-trans conditions. For example, withdrawal of Ca^{2+} from one side of the epithelium in order to measure Ca^{2+} flux may cause a loss of TJ barrier function (Martinez-Palomo, Meza, Beaty, & Cereijido, 1980). Application of a pH-gradient to determine paracellular H^+ transport has been applied successfully (Angelow, Kim, & Yu, 2006);

however, it has to be kept in mind that changes in extracellular pH may titrate charged side-chains within the tight junctional pore and thus cause a loss of pore selectivity.

B. Dilution and Biionic Potentials

Ion selectivities of epithelia can be estimated from diffusion potentials that build up across these epithelia if ion gradients are applied. To determine, for example, if Na^+ and Cl^- permeabilities (P_{Na}, P_{Cl}) of an epithelial barrier differ, part of the NaCl in the bath solution on one side of the epithelium is replaced iso-osmotically by an uncharged substance, for example, mannitol. Thus, NaCl is "diluted" on one side of the epithelium. If $P_{Na} \neq P_{Cl}$, a transepithelial potential difference (dilution potential, ΔE) will develop which, for $P_{Na} > P_{Cl}$, is positive on the diluted side relative to the unchanged side, while it is negative for $P_{Na} < P_{Cl}$. Ideally, P_{Na}/P_{Cl} can be calculated using the Eq. (2).

$$\frac{P_{Na}}{P_{Cl}} = \frac{10^{(\Delta E/s)} \cdot a_{Cl}^{ap} - a_{Cl}^{bl}}{a_{Na}^{ap} - 10^{(\Delta E/s)} \cdot a_{Na}^{bl}} \quad \text{with}\,\Delta E = E_{bl} - E_{ap}; \quad s = 2.303(RT/F) \quad (2)$$

a_{ion}^{ap}, a_{ion}^{bl} apical and basolateral ion activities; R, universal gas constant; T, absolute temperature; F, Faraday constant.

Absolute permeabilities can be calculated from P_{Na}/P_{Cl} and the epithelial conductance using the following two equations (Eqs. (3) and (4), see Hou, Paul, & Goodenough, 2005):

$$P_{Cl} = \frac{G}{[\mathrm{NaCl}]} \frac{RT}{F^2} \frac{1}{(1 + P_{Na}/P_{Cl})} \quad (3)$$

$$P_{Na} = P_{Cl} \frac{P_{Na}}{P_{Cl}} = \frac{G}{[\mathrm{NaCl}]} \frac{RT}{F^2} \frac{1}{1 + (P_{Na}/P_{Cl})} \frac{P_{Na}}{P_{Cl}} \quad (4)$$

where G, transepithelial conductance (1/(transepithelial resistance)); [NaCl], NaCl concentration.

Similarly, relative permeabilities of other monovalent cations can be determined by replacing part of the Na^+ on one side of the epithelium by the cation of interest (X^+) and measuring the resulting transepithelial potential ("biionic potential") and using Eq. (5).

$$\frac{P_X}{P_{Na}} = \frac{10^{(\Delta E/s)} a_{Na}^{bl} - a_{Na}^{ap}}{a_X^{ap} - 10^{(\Delta E/s)} a_X^{bl}} \quad (5)$$

However, Eq. (5) is only valid if $P_X \gg P_{Cl}$. If this is not the case, Cl^- will diffuse across the epithelium along the electrical gradient and thus dissipate at least part of the biionic potential. Therefore, in most cases, it is advisable to use Eq. (6) instead of Eq. (5), which takes into consideration the effects of Na^+, Cl^-, and the cation X^+:

$$\frac{P_X}{P_{Na}} = \frac{a_{Na}^{ap} + (P_{Cl}/P_{Na})a_{Cl}^{bl} - 10^{(\Delta E/s)}(a_{Na}^{bl} + (P_{Cl}/P_{Na})a_{Cl}^{ap})}{10^{(\Delta E/s)}a_X^{bl} - a_X^{ap}a_{X-ap}} \quad (6)$$

The ratio P_{Cl}/P_{Na} within Eq. (6) has to be determined from separate dilution potential measurements, preferably from the same cell layer. Subsequently, absolute permeabilities can be calculated from P_{Na} values obtained from Eq. (4).

Although the general principle of these measurements is relatively simple, there are several pitfalls that have to be avoided to obtain correct permeability values.

First, active (transcellular), electrogenic transport may be activated by the solutions used to determine dilution/biionic potentials. Therefore, measurements should always be carried out in the apical to basolateral as well as in the basolateral to apical direction to determine whether the resulting potentials are symmetrical. If not, the pharmacological inhibition of active transport components may be considered. Alternatively, experiments may be carried out at lower temperatures in order to inhibit active transport.

Second, insufficient movement of the bath solution during the experiment may cause so-called "unstirred layer effects," that is, an enrichment/depletion of the transported ion in close vicinity to the cell membrane that will affect local ion gradients and thus the resulting potentials.

Third, Eqs. (2), (5), and (6) require the use of ion activities rather than concentrations. At physiological ion strengths, activity coefficients are clearly different from unity and therefore have to be taken into consideration. However, activity coefficients of complex solutions are difficult to estimate. Several approaches, such as the Debye–Hückel formalism, may be used for an approximation (for a detailed discussion, see e.g., Ammann, 1986; Barry, 2006; Sugiharto, Lewis, Moorhouse, Schofield, & Barry, 2008).

Fourth, liquid junction potentials (LJP) occur at the interface between the apical and basolateral solutions in the Ussing chamber and the salt bridges used to connect the bath to the voltage electrodes. As long as the apical and basolateral side of the chamber contain the same solution, LJPs on both sides should be of the same magnitude, but of opposite sign, and thus cancel each other out. As soon as the ion compositions on both sides differ, the LJPs will differ and the resulting potential difference will add to the diffusion potential and distort the results. It is the "first law of electrophysiology" (Thomas, 1978) that 3 M KCl electrodes yield correct results as they minimize LJPs. However, agar bridges used in Ussing chambers usually have large diameters and therefore lose KCl to the surrounding solution, causing "history effects" (Barry & Diamond, 1970), that is, potentials that depend on the duration of exposure to different solutions. Alternatively, salt bridges may be used that contain NaCl concentrations close to those used during the experiments, to minimize leakage and then to measure or calculate the LJP that will occur, when the solution on one side of the chamber is changed. Again, this is not trivial. Often, measurements across blank filter membranes are used to measure the overall effect at the two salt bridges; however, a further LJP will develop at this membrane and render the result useless (see supplement to Yu et al., 2009 for a detailed discussion).

In summary, it is preferable to use 150 mM NaCl bridges and to calculate LJPs using the Henderson formalism instead of relying on salt bridges containing high KCl concentrations or attempting to measure LJP (Barry & Diamond, 1970; Ammann, 1986, supplement to Yu et al., 2009):

$$V_\mathrm{j} = V_\mathrm{sol} - V_\mathrm{ref} = \frac{RT}{F}\frac{\sum_i\left(z_i u_i\left(a_i^\mathrm{sol} - a_i^\mathrm{ref}\right)\right)}{\sum_i\left(z_i^2 u_i\left(a_i^\mathrm{sol} - a_i^\mathrm{ref}\right)\right)}\ln\frac{\sum_i\left(z_i^2 u_i a_i^\mathrm{ref}\right)}{\sum_i\left(z_i^2 u_i a_i^\mathrm{sol}\right)} \tag{7}$$

where V_j is the liquid junction potential between bath solution (sol) and reference or agar bridge (ref), respectively; u_i is the absolute mobility of ion i (tables: see e.g., Ammann, 1986; Barry & Lynch, 1991; Meier, Ammann, Morf, & Simon, 1980; Morf, 1981; Ng & Barry, 1995; Robinson & Stokes, 1959; Yu et al., 2009); a_i is the single ion activity of ion i within bath solution (sol) and reference or agar bridge (ref), respectively; z_i is the charge number of ion i; R is the universal gas constant; T is the absolute temperature; and F is the Faraday constant.

C. Conductance Measurements

Changes in conductance induced by changes in ionic composition of the bath solution have been interpreted in terms of differences in the permeabilities of the ions involved (Tang & Goodenough, 2003; Yu et al., 2009). For the interpretation of such data, it has to be kept in mind that part of the conductance measured is due to the transcellular pathway and that this pathway might also be affected by the imposed changes in ionic composition of the bath solution. Yu et al. (2009), therefore, only investigated differences in conductance in cells transfected with Tet-Off/claudin-2 in the presence and absence of doxycycline (Dox). Both Tang and Goodenough (2003) and

Yu et al. (2009) find that data obtained from conductance measurements do not necessarily match data obtained from flux or dilution potential measurements and conclude that all inorganic cations investigated compete for the same pore.

1. Implications: Channel Properties of Claudins

First indications that members of the claudin family may affect paracellular ion permeability were published about a decade ago (claudin-2, Furuse et al., 2001; claudin-4, Van Itallie, Rahner, & Anderson, 2001; claudin-16, Simon et al., 1999). Amasheh et al. (2002) further investigated claudin-2-induced changes in paracellular ion permeability using dilution potential and flux measurements, and found that claudin-2-overexpression in high-resistance MDCK C7 cells not only greatly decreased R^t but that this effect was specifically due to an increase in P_{Na}, while P_{Cl} remained constant. Amasheh et al. (2002) further demonstrated that the permeabilities of cations of various sizes (Na^+, K^+, $NMDG^+$, choline$^+$) were increased, while permeabilities to anions (Cl^-, Br^-) and uncharged molecules (mannitol, lactulose, and 4-kDa dextran) were unaffected. Claudin-2 thus behaved like a nonselective cation channel. Yu et al. (2009) showed in further investigations that upon the expression of claudin-2, the permeability sequence for alkali metal ions was $K^+ > Rb^+ > Na^+ > Li^+ \gg Cs^+$ and thus resembled Eisenman sequence V to VIII. The ratio between the permeabilities of the most permeable cation (K^+) and the least permeable (Cs^+) was 1.6, which is relatively narrow and suggests that the interaction sites within the pore are widely spaced (Eisenman, 1962) and/or partially hydrated (Yu et al., 2009). Permeability measurements for various organic ions of different sizes further indicated a pore diameter of approximately 6.5 Å at the narrowest point of the claudin-2 pore (Yu et al., 2009). Based on these data, Yu et al. (2009) performed Brownian dynamics simulation and obtained a pore model as best fit that assumed a channel length of 32 Å and funnel-like entrances of 16 Å in diameter, narrowing to 6.5 Å in diameter at the center of the pore. From this model, they were able to estimate single-channel conductance to be in the order of 100 pS.

Similar to claudin-2, the overexpression of mouse claudin-10b in MDCK C7 cells greatly increases cation permeability. However, in contrast to claudin-2, claudin-10b changed the permeability sequence to $Na^+ > Li^+ > K^+ > Rb^+ > Cs^+$, corresponding to Eisenman sequence X (Günzel, Stuiver, et al., 2009). In this study, P_{Na}/P_{Cs} was in the order of 3, indicating that the spacing of electric charges within the pore is narrower than in claudin-2 pores. This correlates well with the higher number of negative charges within the first extracellular loop of claudin-10b (five negatively charged amino acids) compared to claudin-2 (three negatively charged amino acids, of which only one, D65, appears to reside within the pore, Yu et al., 2009).

In contrast to claudin-10b, the first extracellular loop of claudin-10a contains only one (human) or two (mouse) negatively charged, but seven positively charged amino acids. Overexpression of claudin-10a conveys an anion preference to the TJ, mainly through the arginine residues R33 and R62 (Van Itallie et al., 2006).

Claudin-16 had been assumed to form paracellular pores for divalent cations for two reasons: (i) claudin-16 mutations cause renal loss of Ca^{2+} and Mg^{2+} and (ii) claudin-16 is expressed in the thick ascending limb of Henle's loop which has previously been identified as the major location for (paracellular) Mg^{2+} absorption in the kidney. As recently reviewed by Günzel and Yu (2009), however, data from claudin-16 overexpression studies are conflicting. Ca^{2+} and Mg^{2+} permeabilities were measured using the various methods described earlier (equilibrium and zero-trans flux measurements, biionic potential measurements), but, if present at all, were only minor (Günzel, Amasheh, et al., 2009; Ikari et al., 2004; Kausalya et al., 2006; Hou et al., 2005). While Hou et al. (2005) found an increased Na^+ permeability in transfected LLC-PK1 cells, Kausalya et al. (2006) and Günzel, Amasheh, et al. (2009) failed to do so in MDCK C7 cells. Thus, the role of claudin-16 as a pore-forming claudin yet remains unsolved.

V. FLUXES OF UNCHARGED PARACELLULAR TRACERS

Fluxes of different-sized hydrophilic tracers are often measured to determine paracellular permeability properties. To this end, radioactively labeled substances as urea, mannitol, inulin, or polyethylene glycols (PEGs; Ghandehari, Smith, Ellens, Yeh, & Kopecek, 1997) and/or fluorescence-labeled dextrans (Sanders, Madara, McGuirk, Gelman, & Colgan, 1995) have been used.

Passage through paracellular pores is a passive process and, as described earlier, flux measurements are usually carried out under equilibrium exchange conditions. It is therefore attempted to abolish gradients between donor and acceptor sides by adding equal amounts of the unlabeled species of the analyzed probe to the apical and basolateral bathing solution. Beforehand, however, probes such as dextrans or PEGs should be dialyzed. Although labeled with an average molecular weight, these molecules do not have a precisely defined size but contain a range of different sizes. In addition, these rather fragile probes may decay over time. Preferential permeation of the resulting smaller fragments would consequently distort the results and overestimates of paracellular pore sizes.

The fluxes (J) of the measured probes are then calculated by

$$J = \frac{\Delta c V_{ch}}{\Delta t A} \tag{8}$$

where Δc is the concentration difference of the probe in the acceptor chamber at time t_1 and t_2; Δt is the time difference $t_2 - t_1$; V_{ch} is the volume of hemi-chamber; and A is the area of the specimen.

Fluxes are expressed as nmol/h/cm^2 and give a first insight into the characteristics of paracellular pores formed by TJ proteins. However, due to the concentration dependence, fluxes from different studies cannot simply be compared. It is therefore preferable to convert fluxes into apparent permeabilities, P_{app}. The unit is formally that of a speed, but originates from the division of the flux by the concentration of the measured solute:

$$P_{app} = \frac{J}{c_{fin}} (10^{-6}\,\text{cm/s}) = (\mu\text{mol/h/3.6/cm}^2)\;/\;(\text{mmol/l}) \tag{9}$$

where J is the flux of the measured probe and c_{fin} is the concentration of the probe in the donor chamber.

Monitoring individual paracellular probes give rough estimations of the character of paracellular pores formed by TJ proteins.

First examinations using membrane-impermeant tracers suggested tight junctional pore diameters of ~6 Å (Spring, 1998), which is similar to some conventional ion channels (Balasubramanian, Lynch, & Barry, 1997; Goulding, Tibbs, Liu, & Siegelbaum, 1993).

Measurement of permeability for ethanolamine with a diameter of ~4.9 Å (Dwyer, Adams, & Hille, 1980) and for mannitol with ~7.2 Å (Madara & Dharmsathaphorn, 1985) refined the paracellular pore diameter to at least 4.9 Å, but not much bigger than 7 Å (Tang & Goodenough, 2003). Mannitol appeared not to move freely through the paracellular pore as it had been observed that there was a disproportionately larger permeability for compounds smaller than mannitol (Artursson, Ungell, & Lofroth, 1993; Knipp, Ho, Barsuhn, & Borchardt, 1997; Tang & Goodenough, 2003; Tavelin et al., 2003).

In 2001, an elegant technique measuring the permeabilities for 24 PEG sizes was developed simultaneously (Watson, Rowland, & Warhurst, 2001). Separation of the oligomers by liquid chromatography–mass spectrometry (LC–MS) allowed detailed functional profiling and mathematical modeling of the paracellular route. Profiling of fluxes across T84 and Caco-2 cells showed permeability to be biphasic, consisting of one size-restrictive, high-capacity pathway due to abundant, but small (diameter of ~8 Å) pores, and a low-capacity pathway which is size-independent and can be explained by sparse but much larger pores.

Profiling of the paracellular pathway employing a separation of modified PEGs by HPLC, where retention times are proportional to the molecular mass, showed that the pore aperture of ~8 Å in diameter was similar in all cell types investigated (MDCK II cells, MDCK C7 cells, Caco-2 cells, and porcine ileum) although those cell types differ significantly in electrical resistance (Van Itallie et al., 2008; Fig. 7).

Comparison of P_{app} from PEG profiles allowed the estimation of relative pore numbers in various cell types. Expression of claudin-2 resulted in a selective increase in the number of small pores as the permeability for PEGs up to ~4 Å increases, while a knockdown of claudin-2 did not alter the number and size of these small pores (Van Itallie et al., 2008). From these findings, the permeability to small solutes was suggested to be proportional to the pore number and the profile of TJ proteins expressed, which would explain the dissociation between P_{app} for noncharged solutes and electrical resistance.

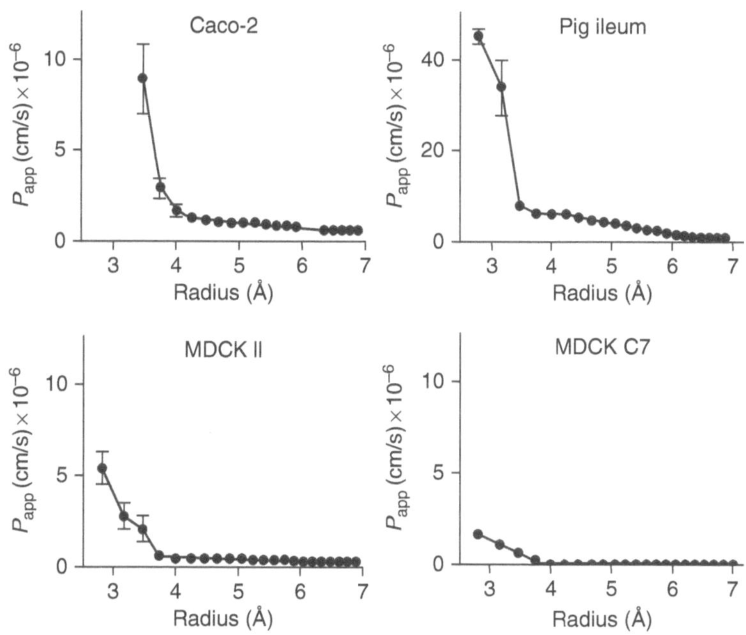

FIGURE 7 P_{app} as a function of PEG radius in Caco-2, MDCK II, and MDCK C7 monolayers and *ex vivo* pig ileum. All cell types show a size-restrictive pore calculated to be of radius ~4 Å. Ileum appears to have an additional pore cutoff at ~6.5 Å. In contrast to their similar pore size, the number of pores (reflected in the P_{app} < 4 Å) is highly variable. Caco-2 cells have the largest numbers of pores as well as the greatest permeation through the size-independent pathway; MDCK II cells have an intermediate number and MDCK C7 cells have few pores and little permeation through the second pathway. The relative pore number in pig ileum cannot be compared with that of cell lines because of the difference in the amplified surface area in intact tissue compared with the flat cultured cell monolayers. From Van Itallie et al. (2008).

Widely consistent with this model of two different pathways of passage, it was found that after the overexpression of tricellulin in tTJs of MDCK II cells, the permeability for macromolecules was remarkably decreased, while additional overexpression in bTJs led to increasing electrical resistance and decreasing permeabilities for ions (Krug et al., 2009). From these findings, different localizations of the two pathways were deduced. The high-capacity "small pores" were suggested to be an integral part of the bicellular junctions and thus frequent enough to carry ~99% of the paracellular permeability for ions, while the size-independent "large pores" were located at the tricellular central tube, which is wide enough to allow the passage of large solutes but rare enough to contribute only ~1% toward paracellular ion permeability.

VI. PARACELLULAR WATER TRANSPORT

Not only ions or macromolecules but also large volumes of water may be transported across epithelia not only through transcellular but also through paracellular routes (Rosenthal et al., 2010).

The transcellular route mediated by aquaporin water channels (Agre et al., 1993) is well described, but there is an ongoing dispute concerning the existence of significant water flow across the TJ (Spring, 1998). Although TJs are often named aqueous pores, there was yet no direct experimental evidence for a paracellular water flux because it is difficult to separate the TJ-controlled paracellular from transcellular water flux.

Technically, the obvious method for the determination of paracellular water flux would be to measure the rate of transepithelial water flux before and after blocking the TJ pathway. However, this approach has proven unsuccessful because the transcellular side effects of paracellular permeability inhibitors could not be excluded (Poler & Reuss, 1987). The converse procedure, blocking the transcellular water channels and measuring the resulting decrease in transepithelial water flux, is inapplicable since mercury, the only potent blocker of aquaporins, is not effective on all aquaporin isoforms (Knepper, 1994) and additionally, mercury applied in effective concentrations is cytotoxic in many cells and tissues.

Since suitable inhibitors of water movement across either pathway are not available, the permeability of the paracellular pathway has been determined indirectly by several methods. One method is to measure the water permeability of the apical and basolateral cell membrane and the transepithelial water permeability, and to calculate the paracellular water permeability from these data (Carpi-Medina & Whittembury, 1988; Flamion, Spring, & Abramow, 1995). The apical and basolateral cell membrane water permeability was calculated from the rate of cell swelling or shrinkage after changing to an anisosmotic solution (apical or basolateral) in combination with the determination of the cell surface area, while an analysis of epithelial cell volume was performed with light microscopy. From a comparison of transepithelial and cell membrane water permeability, it was concluded that in rabbit renal proximal tubule, about 50% of the water passes through the paracellular pathway.

Another indirect approach involves the flux measurements of nonelectrolyte radiolabeled tracers across the epithelial layer for the estimation of paracellular water movement (Hernandez, Gonzalez, & Whittembury, 1995; Shachar-Hill & Hill, 1993; Steward, 1982; Whittembury, Malnic, Mello-Aires, & Amorena, 1988). As already discussed earlier, the fluxes of labeled tracers of definite size and molecular weight are also used for the evaluation of the TJ pore size. A precondition of this method is that the tracers cannot use the transcellular pathway and the assumption is that paracellular water movement induces a solvent drag of these paracellular probes (Spring, 1998). The net flux of these solutes is affected when the rate of the transepithelial volume flow is altered by changing the osmotic gradient across the epithelial layer. Thus, the flux data are used to calculate the fraction of water flow across the TJ. The fraction of paracellular water flow estimated by this method varies between 50% and 100% of the transepithelial flow.

Apart from these indirect measurements, optical microscopic approaches have been reported: One method used confocal microscopy in combination with fluid-phase fluorescent tracer technique for visualizing water secretion and differentiating the routes of water transport across epithelial layers (Segawa, Yamashina, & Murakami, 2002). In this approach, the acinar lumen of parotid and submandibular glands were perfused with fluorescent tracers and the intensity of the luminal fluorescence was observed before and after the stimulation of fluid secretion. Stimulation of fluid secretion caused a rapid decline of luminal fluorescence intensity, indicating that the secreted water washed out the fluorescent tracer in the luminal space. From the pattern of fluorescence decline under different experimental conditions, the authors conclude that water secretion occurs via the trans- and paracellular route.

Another optical microscopic technique was developed for a direct visualization of the fluid movement within the lateral intercellular spaces (LIS) of low-resistance MDCK cells, which represent a fluid-absorptive renal cell line (Kovbasnjuk, Leader, Weinstein, & Spring, 1998). Fluid movement within the LIS can be visualized by introducing a fluorescent dye which is trapped in the LIS and observing the concentration profile of this dye along the LIS. Since the flow velocity was near zero adjacent to the TJ and could not be augmented by the imposition of an osmotic gradient for the induction of transepithelial fluid movement, the authors conclude that a significant transjunctional flow does not occur.

As described earlier, approaches to distinguish between para- and transcellular water permeability are technically very difficult and these investigations yielded, in part, contrary results. None of these studies related the results to the molecular composition of the TJ.

1. Example: Claudin-2 and Paracellular Water Transport

In an alternative approach, the overall transepithelial water flux was measured before and after selective molecular perturbation of the TJ (Rosenthal et al., 2010). Chosen perturbators were claudin-2 and claudin-10b, both of which form cation channels when overexpressed in the TJ of MDCK C7 cells which lacks endogenous expression of these TJ proteins (Amasheh et al., 2002; Günzel, Stuiver, et al., 2009). Transepithelial water permeability was measured in a modified Ussing chamber with two separated silanized glass tubes instead of the gas lifts. Fluid movement was induced by an osmotic or ionic (Na^+) gradient and the fluid level in both tubes was monitored by a video-optic system at different times over a period of 2 h. The water flux was calculated (Fig. 8A) from the difference between the menisci at the registration times. The study revealed that water flux in claudin-2-transfected cells was elevated under all experimental conditions compared to control cells, whereas claudin-10b transfection did not alter water flux, although both claudins are permeable for small cations (Fig. 8B). In claudin-2-expressing cells, water flux could not only be induced by an osmotic gradient but also by a sodium flux through the TJ without any osmotic gradient. From these data, the authors conclude that claudin-2, but not claudin-10b, forms a paracellular water channel and by this mediates the paracellular part of water transport in leaky epithelia.

VII. EXPERIMENTAL STRATEGIES FOR TJ PERTURBATION

For perturbing the TJ experimentally, different approaches are used, such as overexpression or knockdown of TJ proteins that can be achieved in *in vitro* models using cell cultures, while functional studies *in vivo* are usually performed in knockout (KO) animal models.

A. Cell Culture Models: Overexpression and Knockdown

Characterization of a TJ protein can be readily achieved by its overexpression in cell lines that show no or only weak endogenous expression of this protein. Conversely, knockdowns may be performed in cell lines showing

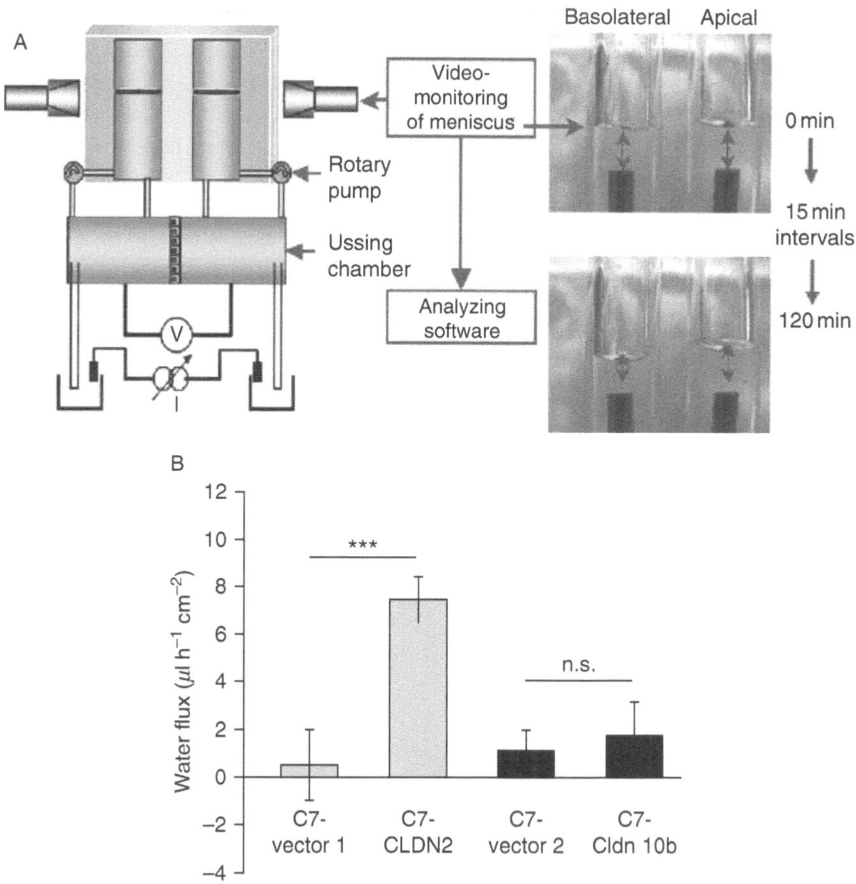

FIGURE 8 Water flux measurements in MDCK C7 cells: (A) The experimental setup consists of a modified Ussing chamber in combination with a video-optic system for monitoring the menisci in the glass tubes connected with the Ussing chamber for a period of 2 h. (B) Water flux was measured in MDCK C7 cells transfected with the cation-permeable claudins claudin-2 C7-CLDN2) or claudin-10b (C7-cldn10b) and the corresponding vector controls (C7-vector 1, 2). Water flux was stimulated by a NaCl gradient (80 mM) with the high NaCl concentration in the basolateral compartment and osmotic compensation at the apical side ($n = 7$–9). Water flux was increased in C7-CLDN2 cells in comparison to the vector control, whereas no effect could be observed in C7-cldn10b cells. Thus, only claudin-2 forms a water-permeable channel in the TJ of epithelial cells. Part (B) adapted from Rosenthal et al. (2010).

strong endogenous expression of the protein of interest. For overexpression, cDNA of the protein is cloned into an expression vector, while for knock-down, specific short-hairpin RNA (shRNA), which activates the RNA interference system (for reviews, see Almeida & Allshire, 2005; Ouellet, Perron, Plante, & Provost, 2006), is introduced into the expression vector. As transient transfection is only affecting a fraction of cells and therefore is not suitable for a detailed and complete analysis of the protein's function, stable transfections are performed, keeping the transfected cells under antibiotic selective pressure.

Two different expression models are generally used for characterization. In constitutive expression models, the created clones are permanently affected by the transgenic expression or knockdown. Although this is a convenient method to create clones of interest, the expression pattern of other proteins may strongly differ due to the artificial environment and its influences created by the overexpression or absence of the target protein. Therefore, clones have to be screened not only for being positive for the protein of interest but also for being unchanged in all other relevant parameters. As it is impossible to analyze the whole proteome, characterization should be performed on several clones in parallel. A further complication may be the induction of lethality due to absence or high overexpression of the protein of interest which, obviously, would prevent the generation of useful clone.

Alternatively, conditional or inducible system can be used. The most widely used externally regulatable transgenic system is based on tetracycline-controlled transcriptional regulation (Gossen & Bujard, 1992; Gossen et al., 1995) and is differentiated in two basic variants, the tetracycline-controlled transcriptional activator system (tTA system; "Tet-Off") and the reverse tetracycline-controlled transcriptional activator system (rtTA system; "Tet-On").

In the "Tet-Off" system, doxycycline (Dox) is used to switch off transcription. It allows high levels of induction when in the "On" state and is comparable to the constitutive system because permanent transcription occurs in the absence of Dox. A major drawback is that this system may also lead to strong changes in endogenous protein expression and regulation.

The "Tet-On" system permits rapid activation in some systems within hours and usually no change in the endogenous expression patterns of other proteins, unless due to the transfection itself, because here Dox is used to start transcription. A disadvantage of this system is its "leakiness," that is, a permanent basal level of transgenic expression, possibly due to weak binding affinities.

B. *In Vivo* Models: Knockout Mice

Although cell cultures as *in vitro* models provide functional information on TJ proteins, their usefulness is limited as they cannot mimic the conditions of a complete tissue or whole organ. As endogenous TJ protein expression may affect the TJ protein of interest, the resulting interactions and TJ changes may differ from one cell culture model to another. Similarly, the barrier properties of TJs and their components vary among different types of epi- and endothelia depending on their physiological function. This makes *in vivo* models for TJ protein and barrier characterization indispensable.

A well-established model is the KO mouse (Gordon, Scangos, Plotkin, Barbosa, & Ruddle, 1980). Here, similar to the described cell culture models, constitutive and conditional mice models are used because constitutive KOs alone may lead to lethality if expression of the target gene is important during embryonic development or if the KO results in disorders that cause early death of the offspring.

KO mice are generated using different methods and constructs, which in conditional KO, may allow time- and tissue-specific analysis of the protein of interest. Here, differently activated ligands for the regulation of gene expression are used (Cre *loxP* system; Feil, Wagner, Metzger, & Chambon, 1997; Gu, Marth, Orban, Mossmann, & Rajewsky, 1994; Metzger & Chambon, 2001).

C. *Established Mouse Models*

Several knockout/knockdown mouse models already exist that directly or indirectly affect the TJ. The lack of some TJ-related proteins, such as ZO-3 or claudin-6, causes no apparent phenotype in mice (Anderson et al., 2008; Hunziker, Kiener, & Xu, 2009), whereas the lack of others, such as ZO-1 and ZO-2, causes embryonic lethality (Katsuno et al., 2008; Xu et al., 2008). In contrast, knockout/knockdown animals for other TJ proteins show a distinct phenotype but are viable and may therefore be accessible to investigations with biophysical techniques. Proteins targeted in these mouse models comprise occludin, claudin-1, -5, -7, -11, -14, -16, -19, and E-cadherin.

Thorough biophysical investigations have not yet been carried out on claudin-5-, claudin-7-, claudin-11-, and claudin-14-deficient mice. Claudin-5 knockout is neonatally lethal due to blood–brain barrier loss, impressively shown by tracer movement into the brain of Cldn5$^{-/-}$ mice (Nitta et al., 2003). Claudin-7$^{-/-}$ mice show a clear growth retardation, renal salt wasting, and chronic dehydration (Tatum et al., 2010). Deficiency in claudin-11 and/or claudin-14 causes deafness, as both proteins are involved in the maintenance of inner ear epithelial barrier function (Ben-Yosef et al., 2003; Elkouby-Naor, Abassi, Lagziel, Gow, & Ben-Yosef, 2008; Gow et al., 2004). In addition, male Cldn11$^{-/-}$ mice are sterile due to an arrest of spermatogenesis caused by the lack of claudin-11 in the Sertoli cells of the testis (Gow et al., 1999; Mazaud-Guittot et al., 2010).

1. Example: Occludin

Although it was the first TJ protein discovered (Furuse et al., 1993), for a long time little was known about the function of occludin. Introduction of occludin into cells normally lacking TJs did not generate a typical anastomosing network (Furuse et al., 1996), and the disruption of both occludin alleles in embryonic stem cells resulted in polarized epithelial cells, which also formed an effective barrier to the diffusion of a low-molecular-weight tracer. Moreover, freeze-fracture replicas of these cells displayed well-developed TJ networks (Saitou et al., 1998), indicating that occludin is not required for the formation of TJ strands.

Occludin-deficient mice ($Occl^{-/-}$) displayed an extensive phenotype indicating more complex functions of occludin (Saitou et al., 2000). $Occl^{-/-}$ mice showed postnatal growth retardation, males produced no litter, while females did not suckle their litter. Calcification of the brain, testicular atrophy, thinning of compact bone, and loss of cytoplasmatic granules in striated duct cells of the salivary gland were observed together with chronic inflammation and hyperplasia in the gastric epithelium. Interestingly, no epithelial barrier defects were shown.

Further detailed analysis of the small and large intestine in comparison with stomach epithelia (Schulzke et al., 2005) unveiled a decrease in electrogenic chloride secretion in the small intestine. 1PI spectroscopy measurements showed no change in epithelial or subepithelial resistances, and the performance of conductance scanning disclosed no differences in crypt or surface epithelium of the intestine. Additional examination of the urinary bladder as a very tight epithelium showed no change after occludin knockout. Performance of mannitol flux measurements also showed no difference between wild-type and knockout mice.

In the stomach, acid secretion was found to be almost abolished. This was accompanied by a dramatic change in gastric morphology with mucus cell hyperplasia and a loss of parietal cells (Saitou et al., 2000). Those findings, once again, indicated no essential role for occludin within the TJ itself, but involvement in regulatory pathways during the differentiation of the gastric epithelium.

Another study, using a knockdown of occludin in MDCK II cells by siRNA, later revealed a role of occludin in the transduction of signals of apoptotic cells through the TJ to the actin cytoskeleton via the Rho signaling pathway (Yu et al., 2005).

2. Example: Claudin-1 and E-Cadherin

E-cadherin is an adherens junction rather than a TJ protein; however, its loss causes a redistribution of claudins within the epidermis and therefore similar barrier defects as those observed in claudin-1-deficient mice (Tunggal et al.,

2005). Both claudin-1- and E-cadherin-deficient mice die from water loss across their skin within a few hours after birth. Tracer flux experiments in claudin-1- and E-cadherin-deficient mice emphasized the impairment of epidermal barrier (Furuse et al., 2002; Tunggal et al., 2005). Furthermore, 1PI spectroscopical measurements showed that this barrier loss resulted in a dramatic decrease in R^{epi} in the skin of these mice from more than 4000 Ω cm^2 to less than 1500 Ω cm^2, whereas subepithelial resistance was unaffected (Tunggal et al., 2005).

3. Example: Claudin-15

Effects of claudin-15 loss were investigated by Tamura et al. (2008) in Cldn15$^{-/-}$ mice. A major phenotype in these mice was the development of a mega-intestine, the small intestine in Cldn15$^{-/-}$ mice being twice as long and twice in diameter compared to that of Cldn15$^{+/+}$ mice. Permeability studies disclosed no differences in the permeability of several electrically uncharged tracers between about 0.4 and 20 kDa in molecular weight. In contrast, paracellular ion permeability in the distal part of jejunum was decreased, indicating that claudin-15 specifically acts as a paracellular ion channel.

4. Example: Claudin-16

Patients with defective claudin-16 suffer from familial hypomagnesemia, hypercalciuria, and nephrocalcinosis (FHHNC). Similar symptoms were observed in claudin-16 knockdown mice generated by Hou et al. (2007) through an siRNA approach. Isolated perfused TAL tubules from these mice were used to measure R^t and to carry out ion permeability measurements. To this end, tubules are held and perfused by a concentric glass pipette system developed by Greger (1981) (see Fig. 9). The double-barreled perfusion pipette is used for voltage measurement and constant current injection so that, similar to Ussing chamber experiments, equivalent I_{SC} values can be determined. However, due to the geometry of the tubules, cable equations have to be used to calculate R^t (Greger, 1981), to allow for voltage attenuation along the tubule.

Using this technique, Hou et al. (2007) were able to demonstrate that R^t, length constant, and the transepithelial potential component due to active transport were unchanged. However, in the presence of a NaCl gradient across the epithelium, the resulting dilution potential differed greatly, indicating a reduction in P_{Na}/P_{Cl} from 3.1 to 1.5. In the intact kidney, this would cause a reduction in driving force for the reuptake of divalent cations and thus explain the observed hypomagnesemia and hypercalciuria.

5. Example: Claudin-19

Claudin-19-deficient mice were first generated by Miyamoto et al. (2005). These mice were described to present behavioral abnormalities due to Schwann cell barrier defects that affected the nerve conduction of peripheral

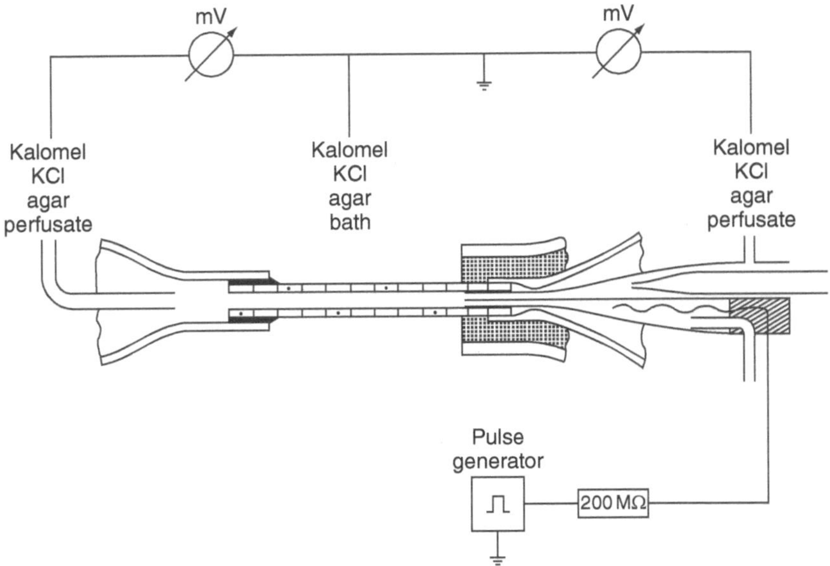

FIGURE 9 Electrophysiology on isolated kidney tubules. The tubule is constantly perfused from right to left through the upper channel of the dual-channel perfusion pipette (right). The lower channel is connected to a pulse generator via Ag-wire for the injection of current pulses. Extratubular bath solution and perfusate (right and left) are in electric contact with agar bridges which connect to calomel electrodes via 3 M KCl solution. The potential difference between these electrodes is measured with two millivoltmeters. From Greger (1981).

myelinated fibers. Recently, Hou et al. (2009) generated a cldn19-siRNA knockdown mouse. In contrast to Miyamoto et al. (2005), they found that claudin-19 knockdown affected claudin-16 distribution in the TJ of the thick ascending limb and, through this mechanism, decreased the cation permeability of these TJs. Conversely, in the absence of claudin-16, claudin-19 failed to assemble in TAL TJ, indicating that both claudins are needed for an intact barrier function of this nephron segment.

VIII. CONCLUSION

During the past 15 years, the discovery of the claudin protein family has revolutionized our understanding of epithelial function. The rapid gain in functional understanding is largely due to the fact that from the very beginning, investigations combined molecular biologic and biophysical techniques. Perturbations of TJs in various cell culture systems, tissues, and whole organisms were achieved by the specific overexpression or downregulation of protein expression, or by treatment with different TJ-influencing agents. The resulting changes in TJ function were then characterized by a multitude of techniques, of which the present review can only give a brief comprehension and summary.

For instance, methods such as impedance spectroscopy allow to quantify changes in paracellular resistance and to separate these changes from those occurring in trans- or subcellular resistance and are thus highly superior to conventional TER measurements which at best give a rough estimate of the combined effects. In an alternative approach, paracellular permeabilities can be determined by flux or diffusion potential measurements. Here, however, it should be kept in mind that there is not THE permeability of a specimen but that there are different permeabilities for every ion or uncharged molecule of different sizes (including water), and that these permeabilities may change independently. Consequently, a closer look into the permeabilities of a broad range of solutes gives information about paracellular pore properties such as pore size, charge preferences in ion passage, and even allows the identification of independent passage pathways.

Taken together, numerous biophysical methods give the opportunity to analyze the TJ and its components under manifold aspects, and complement each other to a detailed view of paracellular pathways regulated by the TJ.

References

Agre, P., Preston, G. M., Smith, B. L., Jung, J. S., Raina, S., Moon, C., et al. (1993). Aquaporin CHIP: The archetypal molecular water channel. *American Journal of Physiology, 265*, 463–476.

Almeida, R., & Allshire, R. C. (2005). RNA silencing and genome regulation. *Trends in Cell Biology, 15*, 251–258.

Amasheh, S., Meiri, N., Gitter, A. H., Schöneberg, T., Mankertz, J., Schulzke, J. D., et al. (2002). Claudin-2 expression induces cation-selective channels in tight junctions of epithelial cells. *Journal of Cell Science, 115*, 4969–4976.

Amasheh, S., Milatz, S., Krug, S. M., Bergs, M., Amasheh, M., Schulzke, J. D., et al. (2009). Na^+ absorption defends from paracellular back-leakage by claudin-8 upregulation. *Biochemical and Biophysical Research Communications, 378*, 45–50.

Amasheh, S., Schmidt, T., Mahn, M., Florian, P., Mankertz, J., Tavalali, S., et al. (2005). Contribution of claudin-5 to barrier properties in tight junctions of epithelial cells. *Cell and Tissue Research, 321*, 89–96.

Ammann, D. (1986). *Ion-selective microelectrode and application.* Berlin: Springer Verlag.

Anderson, W. J., Zhou, Q., Alcalde, V., Kaneko, O. F., Blank, L. J., Sherwood, R. I., et al. (2008). Genetic targeting of the endoderm with claudin-6CreER. *Developmental Dynamics, 237*, 504–512.

Angelow, S., Kim, K. J., & Yu, A. S. (2006). Claudin-8 modulates paracellular permeability to acidic and basic ions in MDCK II cells. *The Journal of Physiology, 571*, 15–26.

Artursson, P., Ungell, A. L., & Lofroth, J. E. (1993). Selective paracellular permeability in two models of intestinal absorption: Cultured monolayers of human intestinal epithelial cells and rat intestinal segments. *Pharmaceutical Research, 10*, 1123–1129.

Balasubramanian, S., Lynch, J. W., & Barry, P. H. (1997). Concentration dependence of sodium permeation and sodium ion interactions in the cyclic AMP-gated channels of mammalian olfactory receptor neurons. *The Journal of Membrane Biology, 159*, 41–52.

Barry, P. H. (2006). The reliability of relative anion–cation permeabilities deduced from reversal (dilution) potential measurements in ion channel studies. *Cell Biochemistry and Biophysics, 46*, 143–154.

Barry, P. H., & Diamond, J. M. (1970). Junction potentials, electrode standard potentials, and other problems in interpreting electrical properties of membranes. *The Journal of Membrane Biology, 3*, 93–122.

Barry, P. H., & Lynch, J. W. (1991). Liquid junction potentials and small cell effects in patch-clamp analysis. *The Journal of Membrane Biology, 121*, 101–117.

Ben-Yosef, T., Belyantseva, I. A., Saunders, T. L., Hughes, E. D., Kawamoto, K., Van Itallie, C. M., et al. (2003). Claudin 14 knockout mice, a model for autosomal recessive deafness DFNB29, are deaf due to cochlear hair cell degeneration. *Human Molecular Genetics, 12*, 2049–2061.

Bijvelds, M., Kolar, Z., Bonga, S., & Flik, G. (1997). Mg^{2+} transport in plasma membrane vesicles of renal epithelium of the Mozambique tilapia (*Oreochromis mossambicus*). *The Journal of Experimental Biology, 200*, 1931–1939.

Bijvelds, M. J. C., Kolar, Z. I., Wendelaar-Bonga, S. E., & Flik, G. (1996). Magnesium transport across the basolateral plasma membrane of the fish enterocyte. *The Journal of Membrane Biology, 154*, 217–222.

Bojarski, C., Gitter, A. H., Bendfeldt, K., Mankertz, J., Schmitz, H., Wagner, S., et al. (2001). Permeability of HT-29/B6 colonic epithelium as a function of apoptosis. *The Journal of Physiology, 535*, 541–552.

Brown, D. R., & O'Grady, S. M. (2008). The Ussing chamber and measurement of drug actions on mucosal ion transport. *Current Protocols in Pharmacology, 41*, 7121–71217.

Bürgel, N., Bojarski, C., Mankertz, J., Zeitz, M., Fromm, M., & Schulzke, J. D. (2002). Mechanisms of diarrhea in collagenous colitis. *Gastroenterology, 123*, 433–443.

Carpi-Medina, P., & Whittembury, G. (1988). Comparison of transcellular and transepithelial water osmotic permeabilities (Pos) in the isolated proximal straight tubule (PST) of the rabbit kidney. *Pflügers Archiv., 412*, 66–74.

Cereijido, M., Stefani, E., & Palomo, A. M. (1980). Occluding junctions in a cultured transporting epithelium: Structural and functional heterogeneity. *The Journal of Membrane Biology, 53*, 19–32.

Clarke, L. L. (2009). A guide to Ussing chamber studies of mouse intestine. *American Journal of Physiology. Gastrointestinal and Liver Physiology, 296*, G1151–G1166.

Claude, P. (1978). Morphological factors influencing transepithelial permeability: A model for the resistance of the zonula occludens. *The Journal of Membrane Biology, 39*, 219–232.

Claude, P., & Goodenough, D. A. (1973). Fracture faces of zonulae occludentes from "tight" and "leaky" epithelia. *The Journal of Cell Biology, 58*, 390–400.

Clausen, C., Lewis, S. A., & Diamond, J. M. (1979). Impedance analysis of a tight epithelium using a distributed resistance model. *Biophysical Journal, 26*, 291–318.

Cole, K. S., & Curtis, H. J. (1938a). Electrical impedance of nerve during activity. *Nature, 142*, 209–210.

Cole, K. S., & Curtis, H. J. (1938b). Electrical impedance of Nitella during activity. *The Journal of General Physiology, 21*, 37–64.

Cole, K. S., & Curtis, H. J. (1938c). Electrical impedance of single marine eggs. *The Journal of General Physiology, 21*, 591–599.

Curtis, H. J., & Cole, K. S. (1938). Transverse electric impedance of the squid giant axon. *The Journal of General Physiology, 21*, 757–765.

Dwyer, T. M., Adams, D. J., & Hille, B. (1980). The permeability of the endplate channel to organic cations in frog muscle. *The Journal of General Physiology, 75*, 469–492.

Eisenman, G. (1962). Cation selective glass electrodes and their mode of operation. *Biophysical Journal, 2*, 259–323.

Elkouby-Naor, L., Abassi, Z., Lagziel, A., Gow, A., & Ben-Yosef, T. (2008). Double gene deletion reveals lack of cooperation between claudin 11 and claudin 14 tight junction proteins. *Cell and Tissue Research, 333*, 427–438.

Farquhar, M. G., & Palade, G. E. (1963). Junctional complexes in various epithelia. *The Journal of Cell Biology, 17*, 375–412.

Feil, R., Wagner, J., Metzger, D., & Chambon, P. (1997). Regulation of Cre recombinase activity by mutated estrogen receptor ligand-binding domains. *Biochemical and Biophysical Research Communications, 237*, 752–757.

Flamion, B., Spring, K. R., & Abramow, M. (1995). Adaptation of inner medullary collecting duct to dehydration involves a paracellular pathway. *American Journal of Physiology, 268*, 53–63.

Florian, P., Schöneberg, T., Schulzke, J. D., Fromm, M., & Gitter, A. H. (2002). Single-cell epithelial defects close rapidly by an actinomyosin purse string mechanism with functional tight junctions. *The Journal of Physiology, 545*, 485–499.

Foskett, J. K., & Scheffey, C. (1989). Scanning electrode localization of transport pathways in epithelial tissues. *Methods in Enzymology, 171*, 792–813.

Fricke, H. (1925). The electric capacity of suspensions of red corpuscles of a dog. *Physical Review, 26*, 682–687.

Fromm, M., Krug, S. M., Zeissig, S., Richter, J. F., Rosenthal, R., Schulzke, J. D., et al. (2009). High resolution analysis of barrier function. *Annals of the New York Academy of Sciences, 1165*, 74–81.

Frömter, E. (1972). The route of passive ion movement through the epithelium of Necturus gallbladder. *The Journal of Membrane Biology, 8*, 259–301.

Frömter, E., & Diamond, J. (1972). Route of passive ion permeation in epithelia. *Nature: New Biology, 235*, 9–13.

Furuse, M., Fujimoto, K., Sato, N., Hirase, T., Tsukita, S., & Tsukita, S. (1996). Overexpression of occludin, a tight junction-associated integral membrane protein, induces the formation of intracellular multilamellar bodies bearing tight junction-like structures. *Journal of Cell Science, 109*, 429–435.

Furuse, M., Furuse, K., Sasaki, H., & Tsukita, S. (2001). Conversion of zonulae occludentes from tight to leaky strand type by introducing claudin-2 into Madin-Darby canine kidney I cells. *The Journal of Cell Biology, 153*, 263–272.

Furuse, M., Hata, M., Furuse, K., Yoshida, Y., Haratake, A., Sugitani, Y., et al. (2002). Claudin-based tight junctions are crucial for the mammalian epidermal barrier: A lesson from claudin-1-deficient mice. *The Journal of Cell Biology, 156*, 1099–1111.

Furuse, M., Hirase, T., Itoh, M., Nagafuchi, A., Yonemura, S., Tsukita, S., et al. (1993). Occludin: A novel integral membrane protein localizing at tight junctions. *The Journal of Cell Biology, 123*, 1777–1788.

Ghandehari, H., Smith, P. L., Ellens, H., Yeh, P. Y., & Kopecek, J. (1997). Size-dependent permeability of hydrophilic probes across rabbit colonic epithelium. *The Journal of Pharmacology and Experimental Therapeutics, 280*, 747–753.

Gitter, A. H., Bendfeldt, K., Schulzke, J. D., & Fromm, M. (2000a). Leaks in the epithelial barrier caused by spontaneous and TNFa-induced single-cell apoptosis. *FASEB Journal, 14*, 1749–1753.

Gitter, A. H., Bendfeldt, K., Schulzke, J. D., & Fromm, M. (2000b). Trans-/paracellular, surface/crypt, and epithelial/subepithelial resistances of mammalian colonic epithelia. *Pflügers Archiv, 439*, 477–482.

Gitter, A. H., Bertog, M., Schulzke, J. D., & Fromm, M. (1997). Measurement of paracellular epithelial conductivity by conductance scanning. *Pflügers Archiv, 434*, 830–840.

Gitter, A. H., Wullstein, F., Fromm, M., & Schulzke, J. D. (2001). Epithelial barrier defects in ulcerative colitis: Characterization and quantification by electrophysiological imaging. *Gastroenterology, 121*, 1320–1328.

Gordon, J. W., Scangos, G. A., Plotkin, D. J., Barbosa, J. A., & Ruddle, F. H. (1980). Genetic transformation of mouse embryos by microinjection of purified DNA. *Proceedings of the National Academy of Sciences of the United States of America, 77*, 7380–7384.

Gossen, M., & Bujard, H. (1992). Tight control of gene expression in mammalian cells by tetracycline-responsive promoters. *Proceedings of the National Academy of Sciences of the United States of America, 89*, 5547–5551.

Gossen, M., Freundlieb, S., Bender, G., Muller, G., Hillen, W., & Bujard, H. (1995). Transcriptional activation by tetracyclines in mammalian cells. *Science, 268*, 1766–1769.

Goulding, E. H., Tibbs, G. R., Liu, D., & Siegelbaum, S. A. (1993). Role of H5 domain in determining pore diameter and ion permeation through cyclic nucleotide-gated channels. *Nature, 364*, 61–64.

Gow, A., Davies, C., Southwood, C. M., Frolenkov, G., Chrustowski, M., Ng, L., et al. (2004). Deafness in Claudin 11-null mice reveals the critical contribution of basal cell tight junctions to stria vascularis function. *The Journal of Neuroscience, 24*, 7051–7062.

Gow, A., Southwood, C. M., Li, J. S., Pariali, M., Riordan, G. P., Brodie, S. E., et al. (1999). CNS myelin and sertoli cell tight junction strands are absent in Osp/claudin-11 null mice. *Cell, 99*, 649–659.

Greger, R. (1981). Cation selectivity of the isolated perfused cortical thick ascending limb of Henle's loop of rabbit kidney. *Pflügers Archiv, 390*, 30–37.

Grotjohann, I., Gitter, A. H., Köckerling, A., Bertog, M., Schulzke, J. D., & Fromm, M. (1998). Localization of cAMP- and aldosterone-induced K^+ secretion in rat distal colon by conductance scanning. *The Journal of Physiology, 507*, 561–570.

Gu, H., Marth, J. D., Orban, P. C., Mossmann, H., & Rajewsky, K. (1994). Deletion of a DNA polymerase beta gene segment in T cells using cell type-specific gene targeting. *Science, 265*, 103–106.

Günzel, D., Amasheh, S., Pfaffenbach, S., Richter, J. F., Kausalya, P. J., Hunziker, W., et al. (2009). Claudin-16 affects transcellular Cl^- secretion in MDCK cells. *The Journal of Physiology (London), 587*, 3777–3793.

Günzel, D., Florian, P., Richter, J. F., Troeger, H., Schulzke, J. D., Fromm, M., et al. (2006). Restitution of single-cell defects in the mouse colon epithelium differs from that of cultured cells. *American Journal of Physiology. Regulatory, Integrative and Comparative Physiology, 290*, R1496–R1507.

Günzel, D., Stuiver, M., Kausalya, P. J., Haisch, L., Krug, S. M., Rosenthal, R., et al. (2009). Claudin-10 exists in six alternatively spliced isoforms which exhibit distinct localization and function. *Journal of Cell Science, 122*, 1507–1517.

Günzel, D., & Yu, A. S. L. (2009). Function and regulation of claudins in the thick ascending limb of Henle. *Pflügers Archiv, 458*, 77–88.

Hernandez, C. S., Gonzalez, E., & Whittembury, G. (1995). The paracellular channel for water secretion in the upper segment of the Malpighian tubule of *Rhodnius prolixus*. *The Journal of Membrane Biology, 148*, 233–242.

Höber, R. (1910). Eine Methode, die elektrische Leitfähigkeit im Innern von Zellen zu messen. *Pflügers Archiv, 133*, 237–253.

Hou, J., Paul, D. L., & Goodenough, D. A. (2005). Paracellin-1 and the modulation of ion selectivity of tight junctions. *Journal of Cell Science, 118*, 5109–5118.

Hou, J., Renigunta, A., Gomes, A. S., Hou, M., Paul, D. L., Waldegger, S., et al. (2009). Claudin-16 and claudin-19 interaction is required for their assembly into tight junctions and for renal reabsorption of magnesium. *Proceedings of the National Academy of Sciences of the United States of America, 106*, 15350–15355.

Hou, J., Shan, Q., Wang, T., Gomes, A. S., Yan, Q., Paul, D. L., et al. (2007). Transgenic RNAi depletion of claudin-16 and the renal handling of magnesium. *The Journal of Biological Chemistry, 282*, 17114–17122.

Hudspeth, A. J. (1975). Establishment of tight junctions between epithelial cells. *Proceedings of the National Academy of Sciences of the United States of America, 72*, 2711–2713.

Hunziker, W., Kiener, T. K., & Xu, J. (2009). Vertebrate animal models unravel physiological roles for zonula occludens tight junction adaptor proteins. *Annals of the New York Academy of Sciences, 1165*, 28–33.

Ikari, A., Hirai, N., Shiroma, M., Harada, H., Sakai, H., Hayashi, H., et al. (2004). Association of paracellin-1 with ZO-1 augments the reabsorption of divalent cations in renal epithelial cells. *The Journal of Biological Chemistry, 279*, 54826–54832.

Ikenouchi, J., Furuse, M., Furuse, K., Sasaki, H., Tsukita, S., & Tsukita, S. (2005). Tricellulin constitutes a novel barrier at tricellular contacts of epithelial cells. *The Journal of Cell Biology, 171*, 939–945.

Jaffe, L. F., & Nuccitelli, R. (1974). An ultrasensitive vibrating probe for measuring steady extracellular currents. *The Journal of Cell Biology, 63*, 614–628.

Jovov, B., Wills, N. K., & Lewis, S. A. (1991). A spectroscopic method for assessing confluence of epithelial cell cultures. *American Journal of Physiology, 261*, C1196–C1203.

Katsuno, T., Umeda, K., Matsui, T., Hata, M., Tamura, A., Itoh, M., et al. (2008). Deficiency of ZO-1 causes embryonic lethal phenotype associated with defected yolk sac angiogenesis and apoptosis of embryonic cells. *Molecular Biology of the Cell, 19*, 2465–2475.

Kausalya, P. J., Amasheh, S., Günzel, D., Wurps, H., Müller, D., Fromm, M., et al. (2006). Disease-associated mutations affect intracellular traffic and paracellular Mg^{2+} transport function of claudin-16. *The Journal of Clinical Investigation, 116*, 878–891.

Knepper, M. A. (1994). The aquaporin family of molecular water channels. *Proceedings of the National Academy of Sciences of the United States of America, 91*, 6255–6258.

Knipp, G. T., Ho, N. F., Barsuhn, C. L., & Borchardt, R. T. (1997). Paracellular diffusion in Caco-2 cell monolayers: Effect of perturbation on the transport of hydrophilic compounds that vary in charge and size. *Journal of Pharmaceutical Sciences, 86*, 1105–1110.

Köckerling, A., & Fromm, M. (1993). Origin of cAMP-dependent Cl^- secretion from both crypts and surface epithelia of rat intestine. *American Journal of Physiology, 264*, C1294–C1301.

Köckerling, A., Sorgenfrei, D., & Fromm, M. (1993). Electrogenic Na^+ absorption of rat distal colon is confined to surface epithelium. A voltage scanning study. *The American Journal of Physiology, 264*, C1285–C1293.

Kottra, G., & Frömter, E. (1984). Rapid determination of intraepithelial resistance barriers by alternating current spectroscopy: Experimental procedures. *Pflügers Archiv, 402*, 409–420.

Kovbasnjuk, O., Leader, J. P., Weinstein, A. M., & Spring, K. R. (1998). Water does not flow across the tight junctions of MDCK cell epithelium. *Proceedings of the National Academy of Sciences of the United States of America, 95*, 6526–6530.

Kroesen, A. J., Dullat, S., Schulzke, J. D., Fromm, M., & Buhr, H. J. (2008). Permanently increased mucosal permeability in patients with backwashileitis after ileoanal pouch for ulcerative colitis. *Scandinavian Journal of Gastroenterology, 43*, 704–711.

Krug, S. M., Amasheh, S., Richter, J. F., Milatz, S., Günzel, D., Westphal, J. K., et al. (2009). Tricellulin forms a barrier to macromolecules in tricellular tight junctions without affecting ion permeability. *Molecular Biology of the Cell, 20*, 3713–3724.

Krug, S. M., Fromm, M., & Günzel, D. (2009). Two-path impedance spectroscopy for measurement of paracellular and transcellular epithelial resistance. *Biophysical Journal, 97*, 2202–2211.

Lewis, S. A., Eaton, D. C., Clausen, C., & Diamond, J. M. (1977). Nystatin as a probe for investigating the electrical properties of a tight epithelium. *The Journal of General Physiology, 70*, 427–440.

Madara, J. L., & Dharmsathaphorn, K. (1985). Occluding junction structure-function relationships in a cultured epithelial monolayer. *The Journal of Cell Biology, 101*, 2124–2133.

Mankertz, J., Amasheh, M., Krug, S. M., Fromm, A., Hillenbrand, B., Tavalali, S., et al. (2009). Tumour necrosis factor alpha up-regulates claudin-2 expression in epithelial HT-29/B6 cells via phosphatidylinositol 3-kinase signaling. *Cell and Tissue Research, 336*, 67–77.

Martinez-Palomo, A., Meza, I., Beaty, G., & Cereijido, M. (1980). Experimental modulation of occluding junctions in a cultured transporting epithelium. *The Journal of Cell Biology, 87*, 736–745.

Mazaud-Guittot, S., Meugnier, E., Pesenti, S., Wu, X., Vidal, H., Gow, A., et al. (2010). Claudin 11 deficiency in mice results in loss of the Sertoli cell epithelial phenotype in the testis. *Biology of Reproduction, 82*, 202–213.

McClendon, J. F. (1927). Colloid properties of the surface of the living cell: III. Electrical impedance and reactance of blood and muscle to alternating currents of 0–1, 500, 000 cycles per second. *American Journal of Physiology, 82*, 525–532.

McClendon, J. F. (1936). Electric impedance and permeability of living cells. *Science, 84*, 184–185.

Meier, P. C., Ammann, D., Morf, W. E., & Simon, W. (1980). Liquid-membrane ion-selective electrodes and their biomedical applications. In J. Koryta (Ed.), *Medical and biological applications of electrochemical applications of electrochemical devices* (p. 13). New York: Wiley.

Metzger, D., & Chambon, P. (2001). Site- and time-specific gene targeting in the mouse. *Methods, 24*, 71–80.

Miyamoto, T., Morita, K., Takemoto, D., Takeuchi, K., Kitano, Y., Miyakawa, T., et al. (2005). Tight junctions in Schwann cells of peripheral myelinated axons: A lesson from claudin-19-deficient mice. *The Journal of Cell Biology, 169*, 527–538.

Morf, W. E. (1981). *The principles of ion-selective electrodes and of membrane transport.* Budapest/Elsevier, Amsterdam, New York: Akadémiai Kiadó.

Ng, B., & Barry, P. H. (1995). The measurement of ionic conductivities and mobilities of certain less common organic ions needed for junction potential corrections in electrophysiology. *Journal of Neuroscience Methods, 56*, 37–41.

Nitta, T., Hata, M., Gotoh, S., Seo, Y., Sasaki, H., Hashimoto, N., et al. (2003). Size-selective loosening of the blood-brain barrier in claudin-5-deficient mice. *The Journal of Cell Biology, 161*, 653–660.

Ouellet, D. L., Perron, M. P., Plante, P., & Provost, P. (2006). MicroRNAs in gene regulation: When the smallest governs it all. *Journal of Biomedicine and Biotechnology, 2006*, 69616.

Poler, S. M., & Reuss, L. (1987). Protamine alters apical membrane K^+ and Cl^- permeability in gallbladder epithelium. *American Journal of Physiology, 253*, 662–671.

Reiter, B., Kraft, R., Günzel, D., Zeissig, S., Schulzke, J. D., Fromm, M., et al. (2006). TRPV4-mediated regulation of epithelial permeability. *FASEB Journal, 20*, 1802–1812.

Robinson, R. A., & Stokes, R. H. (1959). *Electrolyte solutions* (2nd ed.). Mineola, NY: Dover.

Rosenthal, R., Milatz, S., Krug, S. M., Oelrich, B., Schulzke, J. D., Amasheh, S., et al. (2010). Claudin-2, a component of the tight junction, forms a paracellular water channel. *Journal of Cell Science, 123*, 1913–1921.

Saitou, M., Fujimoto, K., Doi, Y., Itoh, M., Fujimoto, T., Furuse, M., et al. (1998). Occludin-deficient embryonic stem cells can differentiate into polarized epithelial cells bearing tight junctions. *The Journal of Cell Biology, 141*, 397–408.

Saitou, M., Furuse, M., Sasaki, H., Schulzke, J. D., Fromm, M., Takano, H., et al. (2000). Complex phenotype of mice lacking occludin, a component of tight junction strands. *Molecular Biology of the Cell, 11*, 4131–4142.

Sanders, S. E., Madara, J. L., McGuirk, D. K., Gelman, D. S., & Colgan, S. P. (1995). Assessment of inflammatory events in epithelial permeability: A rapid screening method using fluorescein dextrans. *Epithelial Cell Biology, 4*, 25–34.

Schulzke, J. D., Gitter, A. H., Mankertz, J., Spiegel, S., Seidler, U., Amasheh, S., et al. (2005). Epithelial transport and barrier function in occludin-deficient mice. *Biochimica et Biophysica Acta, 1669*, 34–42.

Segawa, A., Yamashina, S., & Murakami, M. (2002). Visualization of 'water secretion' by confocal microscopy in rat salivary glands: Possible distinction of para- and transcellular pathway. *European Journal of Morphology, 40*, 241–246.

Shachar-Hill, B., & Hill, A. E. (1993). Convective fluid flow through the paracellular system of Necturus gall-bladder epithelium as revealed by dextran probes. *The Journal of Physiology, 468*, 463–486.

Simon, D. B., Lu, Y., Choate, K. A., Velazquez, H., Al-Sabban, E., Praga, M., et al. (1999). Paracellin-1, a renal tight junction protein required for paracellular Mg^{2+} resorption. *Science, 285*, 103–106.

Spring, K. R. (1998). Routes and mechanism of fluid transport by epithelia. *Annual Review of Physiology, 60*, 105–119.

Steward, M. C. (1982). Paracellular non-electrolyte permeation during fluid transport across rabbit gall-bladder epithelium. *The Journal of Physiology, 322*, 419–439.

Sugiharto, S., Lewis, T. M., Moorhouse, A. J., Schofield, P. R., & Barry, P. H. (2008). Anion-cation permeability correlates with hydrated counterion size in glycine receptor channels. *Biophysical Journal, 95*, 4698–4715.

Tamura, A., Kitano, Y., Hata, M., Katsuno, T., Moriwaki, K., Sasaki, H., et al. (2008). Megaintestine in claudin-15-deficient mice. *Gastroenterology, 134*, 523–534.

Tang, V. W., & Goodenough, D. A. (2003). Paracellular Ion Channel at the Tight Junction. *Biophysical Journal, 84*, 1660–1673.

Tatum, R., Zhang, Y., Salleng, K., Lu, Z., Lin, J. J., Lu, Q., et al. (2010). Renal salt wasting and chronic dehydration in claudin-7-deficient mice. *American Journal of Physiology. Cell Physiology, 298*, F24–F34.

Tavelin, S., Taipalensuu, J., Soderberg, L., Morrison, R., Chong, S., & Artursson, P. (2003). Prediction of the oral absorption of low-permeability drugs using small intestine-like 2/4/A1 cell monolayers. *Pharmaceutical Research, 20*, 397–405.

Teorell, T. (1946). Application of "square wave analysis" to bioelectric studies. *Acta Physiologica Scand., 12*, 235–254.

Thomas, R. C. (1978). *Ion-Sensitive Intracellular Microelectrto Make and Use Them.* London: Academic Press.

Troeger, H., Epple, H. J., Schneider, T., Wahnschaffe, U., Ullrich, R., Burchard, G. D., et al. (2007). Effect of chronic *Giardia lamblia* infection on epithelial transport and barrier function in human duodenum. *Gut, 56*, 328–335.

Tunggal, J. A., Helfrich, I., Schmitz, A., Schwarz, H., Günzel, D., Fromm, M., et al. (2005). E-cadherin is essential for in vivo epidermal barrier function by regulating tight junctions. *EMBO Journal, 24*, 1146–1156.

Ussing, H. H. (1949). The distinction by means of tracers between active transport and diffusion. *Acta Physiologica Scand., 19*, 43–56.

Van Itallie, C. M., Holmes, J., Bridges, A., Gookin, J. L., Coccaro, M. R., Proctor, W., et al. (2008). The density of small tight junction pores varies among cell types and is increased by expression of claudin-2. *Journal of Cell Science, 121*, 298–305.

Van Itallie, C., Rahner, C., & Anderson, J. M. (2001). Regulated expression of claudin-4 decreases paracellular conductance through a selective decrease in sodium permeability. *The Journal of Clinical Investigation, 107*, 1319–1327.

Van Itallie, C. M., Rogan, S., Yu, A., Vidal, L. S., Holmes, J., & Anderson, J. M. (2006). Two splice variants of claudin-10 in the kidney create paracellular pores with different ion selectivities. *American Journal of Physiology. Cell Physiology, 291*, F1288–F1299.

Watson, C. J., Rowland, M., & Warhurst, G. (2001). Functional modeling of tight junctions in intestinal cell monolayers using polyethylene glycol oligomers. *American Journal of Physiology. Cell Physiology, 281*, 388–397.

Whittembury, G., Malnic, G., Mello-Aires, M., & Amorena, C. (1988). Solvent drag of sucrose during absorption indicates paracellular water flow in the rat kidney proximal tubule. *Pflügers Archiv, 412*, 541–547.

Wills, N. K., Lewis, S. A., & Eaton, D. C. (1979). Active and passive properties of rabbit descending colon: A microelectrode and nystatin study. *The Journal of Membrane Biology, 45*, 81–108.

Xu, J., Kausalya, P. J., Phua, D. C., Ali, S. M., Hossain, Z., & Hunziker, W. (2008). Early embryonic lethality of mice lacking ZO-2, but not ZO-3, reveals critical and nonredundant roles for individual zonula occludens proteins in mammalian development. *Molecular and Cellular Biology, 28*, 1669–1678.

Yu, A. S. L., Cheng, M. H., Angelow, S., Günzel, D., Kanzawa, S. A., Schneeberger, E. E., et al. (2009). Molecular basis for cation selectivity in claudin-2-based paracellular pores: Identification of an electrostatic interaction site. *The Journal of General Physiology, 133*, 111–127.

Yu, A. S., McCarthy, K. M., Francis, S. A., McCormack, J. M., Lai, J., Rogers, R. A., et al. (2005). Knockdown of occludin expression leads to diverse phenotypic alterations in epithelial cells. *American Journal of Physiology. Cell Physiology, 288*, C1231–C1241.

Zeissig, S., Bürgel, N., Günzel, D., Richter, J. F., Mankertz, J., Wahnschaffe, U., et al. (2007). Changes in expression and distribution of claudin 2, 5 and 8 lead to discontinuous tight junctions and barrier dysfunction in active Crohn's disease. *Gut, 56*, 61–72.

CHAPTER 4

Structure–Function Studies of the Claudin Pore

Alan S. L. Yu
Division of Nephrology, Department of Medicine and
Department of Physiology and Biophysics,
University of Southern California Keck School of Medicine, Los Angeles, California, USA

I. OVERVIEW

Claudins are tight junction membrane proteins that form both the paracellular pore and barrier. Recent studies have begun to elucidate the characteristics of the claudin pore and identify the function of specific protein domains and residues. In this chapter, we review evidence that the first extracellular domains of multiple claudin monomers fold to enclose a charge-selective paracellular ion pore with a diameter of 6.5–8 Å. The location of specific amino acid residues relative to the pore, and their role in determining charge selectivity, is discussed.

Current Topics in Membranes, Volume 65

1063-5823/2010 $35.00
DOI: 10.1016/S1063-5823(10)65004-1

II. INTRODUCTION

Claudins form the major structural protein component of both the paracellular barrier and paracellular pores in epithelial tissues. However, the structure of the paracellular pore, and the molecular basis for the permeability properties of claudin pores, is still poorly understood. In channels that mediate transcellular transport, ions move perpendicular to the plane of the cell membrane, traversing the low-dielectric environment at the interior of the lipid bilayer, with stabilization and selectivity conferred by the transmembrane domains of the pore protein. Paracellular ion permeation through claudins is likely to be quite different since ions move parallel to and extracellular to the plane of the lipid bilayer (Fig. 1). This pathway is potentially highly hydrated and, as discussed below, is thought to be lined predominantly by the extracellular domains of claudins. The aim of this chapter is to review current knowledge about the structure and function of the claudin pore and the types of methods that have been used to address this, and delineate some of the major unanswered questions in this area.

III. METHODOLOGICAL CHALLENGES IN MEASURING PORE FUNCTION OF INDIVIDUAL CLAUDINS

The general approach to elucidating the paracellular properties of claudins is to overexpress it by transfection into a host cell line. The cells are then grown to a confluent monolayer. Ideally, they are mounted in Ussing chambers, and transepithelial permeabilities are determined either from the measurements of diffusion potentials (i.e., dilution or biionic potentials) at steady

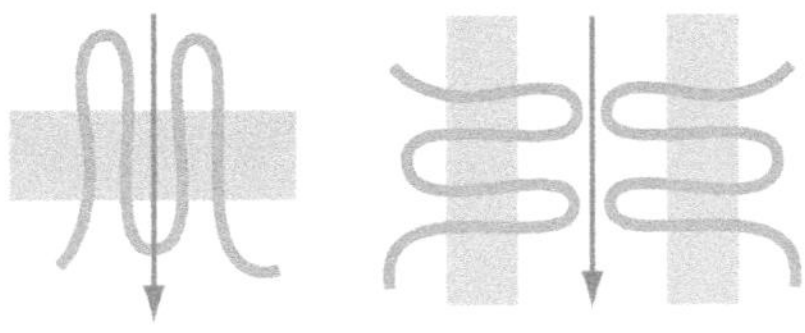

FIGURE 1 Comparison of the different transepithelial transport routes. Transcellular/transmembrane channels (*left*) mediate ion transport (arrow) perpendicular to the plane of the lipid bilayer (gray), with the pore wall formed predominantly by intramembrane domains of the channel polypeptide. Paracellular pores such as claudins (*right*) mediate transport parallel to and extracellular to the lipid bilayer, with the pore walls presumably constituted by the extracellular domains of claudin polypeptides. © Yu et al. (2009). Originally published in *The Journal of General Physiology*, doi:10.1085/jgp.200810154.

state or from the kinetic measurements of conductance or unidirectional radiotracer fluxes. However, several issues complicate the interpretation of such studies:

(a) All epithelia that form confluent monolayers with measurable permeability already express endogenous claudins. Thus, overexpressed claudins could potentially affect the levels of endogenous claudins (Yu, Enck, Lencer, & Schneeberger, 2003) and/or form heteromultimers with them (Furuse, Sasaki, & Tsukita, 1999). Host cell lines that lack claudins and hence tight junctions cannot be induced to form confluent monolayers even when claudins are transfected into them.
(b) Claudins form not only the paracellular pore but also its barrier. Indeed, most claudin isoforms (e.g., claudin-4, -8, -15), when transfected into cell lines, decrease ion permeability (Van Itallie, Rahner, & Anderson, 2001; Yu et al., 2003), thus behaving primarily as "barrier-building" rather than "pore-forming" claudins. This poses a problem because the permeability properties of such claudins cannot be directly measured but can only be indirectly inferred from the change in the overall cell permeability.
(c) For most claudins that behave in a pore-forming manner, the increase in permeability above background is quite small.

Although these limitations cannot currently be completely overcome, one can mitigate them substantially by the judicious selection of claudin isoform, host cell line, and expression system, as described in a recent study (Yu et al., 2009). In this study, three criteria were used to optimize the accuracy of pore permeability measurements. First, a pore-forming claudin isoform was chosen so that transfection into a cell line would increase permeability, thus enabling the quantitation of pore properties. By contrast, barrier-forming claudins decrease the permeability of the cell line into which they are transfected, precluding any direct measure of claudin pore permeability. Second, the combination of pore-forming claudin isoform (claudin-2, which forms cation pores; Amasheh et al., 2002; Furuse, Furuse, Sasaki, & Tsukita, 2001) and host cell line (MDCK I, which has high transepithelial and hence paracellular resistance) was chosen to maximize the signal-to-noise ratio. Third, an inducible stable expression system (TetOff) was chosen (Gossen & Bujard, 1992). This ensured that the claudin expression was uniform in clones expressing wild-type and mutant claudin-2. Furthermore, the permeability of uninduced cells could be subtracted from the permeability of cells induced to express claudin-2, yielding a direct measure of the permeability of the claudin-2 pore. As discussed below, this method allowed the identification of a Na^+-binding site in the claudin-2 pore.

IV. THE FIRST EXTRACELLULAR DOMAIN OF CLAUDINS LINES THE PARACELLULAR PORE

Numerous studies have now shown that overexpression, mutation, knockdown, or knockout of various claudin isoforms in cell lines, in human inherited diseases, and in transgenic animal models, lead to alterations in paracellular permeability (extensively reviewed in Angelow, Ahlstrom, & Yu, 2008; Angelow & Yu, 2007; Furuse & Tsukita, 2006; Schneeberger & Lynch, 2004; Tsukita & Furuse, 2002; Van Itallie & Anderson, 2004, 2006). There are two possible explanations for this: either claudins themselves form the paracellular barrier and pore or they are regulators of the paracellular barrier and pore. The evidence that claudins are in fact the principal component of the paracellular barrier and pore comes largely from the group of James Anderson and Christina van Itallie. Colegio, Van Itallie, Rahner, and Anderson (2003) showed that when claudin-2 is transfected into MDCK II cells, it confers low transepithelial resistance (TER) and paracellular cation selectivity, whereas claudin-4 is associated with high TER and relative anion selectivity (Fig. 2). By generating chimeras between these two isoforms, they showed that the TER and charge selectivity properties are determined solely by the first extracellular domain. In a companion paper (Fig. 3), they showed that mutations that reversed the charge of a basic amino acid in the first extracellular domain of claudin-4, and of acidic amino acids in the first

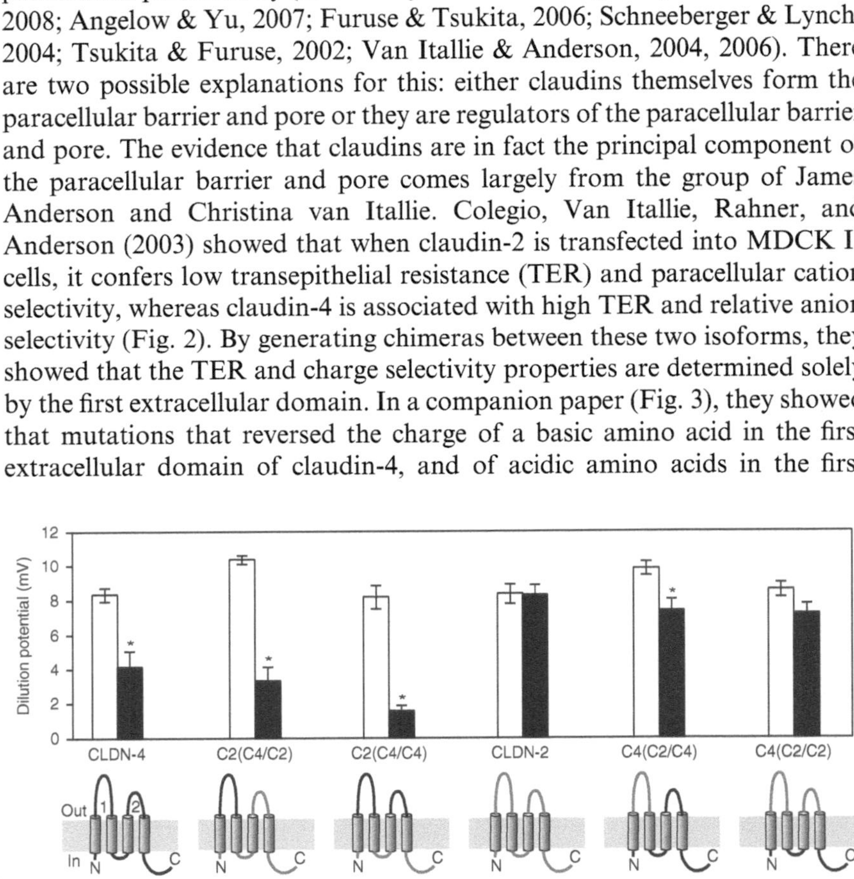

FIGURE 2 The first extracellular domains of claudins are sufficient to determine paracellular charge selectivity. *Lower panel*: Construction and predicted membrane topology of claudin extracellular loop chimeras (black segments, claudin-4; gray/red segments, claudin-2). *Upper panel*: Dilution potentials compared between MDCK II TetOff monolayers that were uninduced (open bars) and induced (filled bars) to express the claudin chimera transgene. Increasing positive dilution potential indicates increasing selectivity for Na^+ over Cl^-. *$p < 0.05$ compared to uninduced. The data show that claudin-4 decreases cation selectivity, while claudin-2 has no effect, and that these effects are largely dependent on the first extracellular loop. Modified with permission from Colegio et al. (2003).

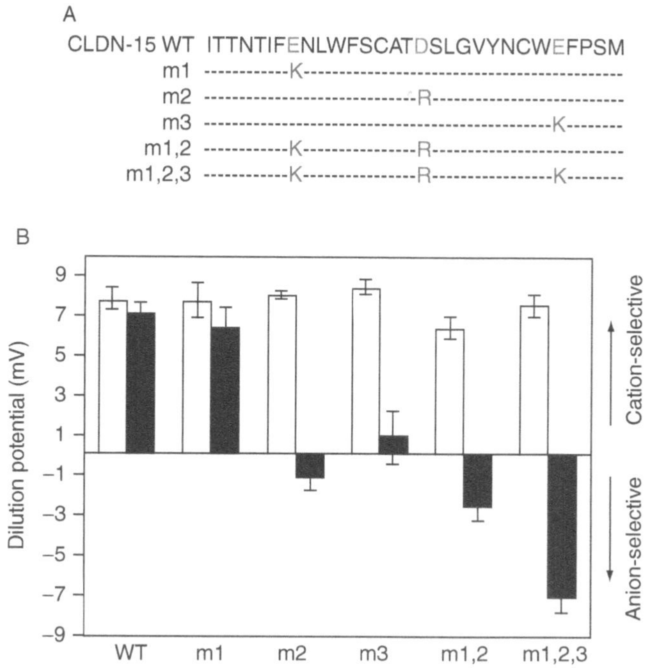

FIGURE 3 Replacing acidic (red/gray) with basic (blue/black) residues on the first extracellular domain of claudin-15 reverses the paracellular selectivity from Na^+ to Cl^- ions. (A) Amino acid sequence alignment of the first extracellular domain of claudin-15 and charge-reversal mutants: m1 (E46K); m2 (D55R); m3 (E64K); m1,2 (E46K, D55R); and m1,2,3 (E46K, D55R, E64K). (B) Results of NaCl dilution potential measurements in clonal cell lines expressing wild-type and mutant claudin-15 in the noninduced (open bars) and induced state (solid bars). Note that progressively replacing the acidic residues, individually or in combination, with basic residues causes a progressive change in charge selectivity of claudin-15 from cation-selective (positive dilution potentials) to anion-selective (negative dilution potential). Modified with permission from Colegio et al. (2002).

extracellular domain of claudin-15, led to predictable changes in paracellular charge selectivity consistent with an electrostatic effect (Colegio, Van Itallie, McCrea, Rahner, & Anderson, 2002). These data suggest that claudins are not merely regulators of paracellular permeability, but that the first extracellular domain of claudin actually lines the lumen of the paracellular pore.

V. MOLECULAR BASIS OF CHARGE SELECTIVITY

The paracellular permeability of native epithelia can often be highly charge selective. Cation-selective epithelia are widespread, for example, rat jejunum (P_{Na}/P_{Cl} = 5–10) (Munck & Schultz, 1974; Wright, 1966) and rabbit gall bladder ($P_{Na}/P_{Cl} = 8$) (Wright & Diamond, 1968), but there are also several examples of anion-selective epithelia, including frog skin ($P_{Na}/P_{Cl} = 0.2$) (Mandel & Curran, 1972), Necturus proximal tubule ($P_{Na}/P_{Cl} = 0.14$) (Edelman & Anagnostopoulos, 1978), and canine trachea ($P_{Na}/P_{Cl} = 0.2$) (Welsh & Widdicombe, 1980). Claudins often alter paracellular charge selectivity when overexpressed in epithelial cells (Table I), suggesting that claudin proteins at the tight junction are the determinants of paracellular charge selectivity. In general, there are three possible mechanisms by which claudins could form charge-selective paracellular pores.

1. *Diffuse electrostatic environment within pore*: Charges within the pore diffusely distributed along the pore wall, generating an electrostatic effect within the pore lumen.
2. *Discrete intrapore binding sites*: Focal charged site(s) on the pore wall that stabilize ions locally within the pore lumen by electrostatic interaction.
3. *Surface charges*: Charges just *outside* the pore opening attract ions, increasing the concentration of ions of one charge that enter the pore.

TABLE I

Charge-Selectivity of Claudins Derived from Overexpression Studies

Permeability effect[a]	Claudin isoform	References
Pore-forming claudins		
$\uparrow P_{Na}$	2	Amasheh et al. (2002), Furuse et al. (2001), Yu et al. (2009)
	10b	Van Itallie et al. (2006)
	16	Hou et al. (2005), Kausalya et al. (2006)
$\uparrow P_{Ca}$[b]	12	Fujita et al. (2008)
$\uparrow P_{Cl}$	10a	Van Itallie et al. (2006)
Barrier-forming claudins		
$\downarrow P_{Na}$	4	Van Itallie et al. (2001)
	8	Yu et al. (2003)
	14	Ben-Yosef et al. (2003)
$\downarrow P_{Cl}$	15	Colegio et al. (2002)

[a]The effect of the claudin depends on the background permeability of the host cell line and the endogenous claudins (Angelow, Schneeberger, & Yu, 2007; Van Itallie et al., 2003). This table is not meant to be comprehensive but lists selected claudins for which we can draw a reasonable consensus as to their overall functional effect. Claudins that have subtle effects or for which studies show conflicting results are not listed.

[b]Selectively increases P_{Ca} but not P_{Na}. Some of the isoforms that are listed as increasing P_{Na} also increase P_{Ca}.

Several studies have been published that appear to support Model 1. As shown in Fig. 3, charge-reversing mutations of the negatively charged residues in claudin-15 progressively switched it from a cation-selective to an anion-selective pore. Importantly, of the three charges, two are in the second half of the first extracellular domain and their effects are additive (Colegio et al., 2002). In studies comparing claudins 2, 4, 11, and 15, Van Itallie, Fanning, and Anderson (2003) found a linear relationship between the net charge of all the residues in the second half of the first extracellular domain and the charge selectivity of the claudin pore. This would suggest a mechanism in which the additive effect of all the lumen-exposed charges creates a general electrostatic environment within the pore that determines charge selectivity. However, charge-reversing mutations are potentially misleading since they introduce new charges that did not exist in the native protein and therefore may reflect artificially created electrostatic effects, rather than inform on the location of the normal selectivity filter.

Hou, Paul, and Goodenough (2005) have studied the effect of multiple charge-neutralizing mutations in the first extracellular loop, using claudin-16 as their model (see also Chapter 7 for a more detailed discussion). They found five acidic residues that affected cation permeability, but these were interspersed with other acidic residues that had no effect. Mutation of each of the five functionally important residues had only a modest effect (11–33% reduction in P_{Na}) and combining the mutations appeared to be additive. One explanation of these results is that the first extracellular domain of claudin-16 is folded into a three-dimensional conformation that brings together the side-groups of multiple noncontiguous acidic residues to form the selectivity filter. However, the replacement of a charged residue, even with a neutral residue, can have multiple consequences, including steric effects and the disruption of folding due to the abolition of salt bridges, all of which could abrogate permeability.

Yu et al. (2009) took a different approach and concluded that, at least for claudin-2, Model 2 (discrete-binding site) is correct. As discussed earlier, claudin-2 was stably expressed in MDCK I cells using the TetOff inducible expression system. By this means, it was shown that wild-type claudin-2 is highly cation-selective ($P_{Na}/P_{Cl} \sim 8$). The three negatively charged residues in the first extracellular domain of claudin-2, E53, D65, and D76, were then neutralized by mutating them, individually or in combination, to their polar, uncharged counterparts (aspartate to asparagine, and glutamate to glutamine) (Yu et al.). Compared to wild-type claudin-2, the D65N mutant had a threefold reduction in conductance and cation selectivity (Fig. 4A). E53Q and D76N were no different from wild-type, and a triple mutant (E53Q/D65N/D76N) was identical to the D65N single mutant. This suggested that

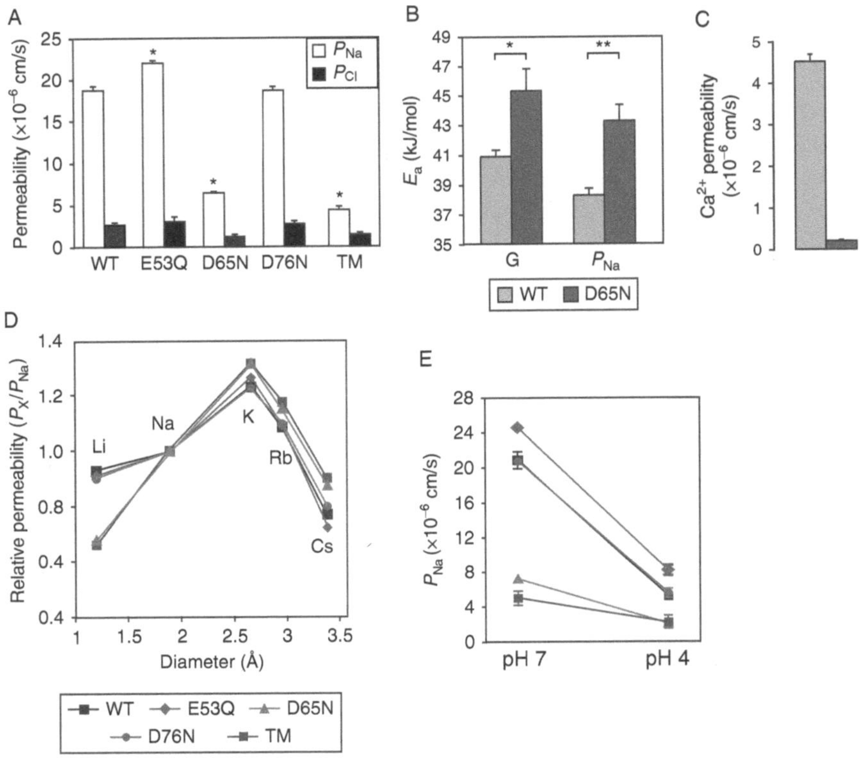

FIGURE 4 Aspartate-65 in claudin-2 is a discrete electrostatic cation interaction site. (A) Na^+ and Cl^- permeability (P_{Na}, P_{Cl}) of claudin-2 wild-type (WT) and mutants. TM, triple mutant (i.e., E53Q/D65N/D76N). *$p < 0.001$ compared to wild-type. (B) Activation energies (E_a) for conductance and permeability to Na^+, as determined from Arrhenius plots. *$p < 0.05$, **$p < 0.01$. (C) Effect of D65N on permeability to Ca^{2+}, as measured by radiotracer flux assay (see part B for explanation of symbols). (D) Relative permeability to alkali metal cations normalized to Na^+ permeability (P_X/P_{Na}) of wild-type and mutant claudin-2. (E) Effect of acidification to pH 4 on P_{Na} of wild-type and mutant claudin-2 (see part D for explanation of symbols). © Yu et al. (2009). Originally published in *The Journal of General Physiology*, doi:10.1085/jgp.200810154.

D65 might form a charged intrapore ion-binding site that determines claudin-2. The following data were presented to show convincingly that the effect of D65N on conductance and charge selectivity is a direct electrostatic effect.

(a) *Activation energy of Na^+ permeation is increased*: If D65 is an energetically favorable binding site for Na^+ that increases its rate of permeation through the pore, then mutating it would increase the activation energy, and this is indeed what was observed (Fig. 4B).

(b) *Eisenman sequence of alkali metal permeation is shifted to lower order*: In Eisenman's theory, the equilibrium ion selectivity of a charged site is determined by the difference between the energy cost of dehydration of the ion and the energy gain from binding to the site, with the latter being dependent on the electric field strength at the surface of the site (Eisenman & Horn, 1983; Eisenman, Sandblom, & Walker, 1967). The higher the field strength, the more selective the site will be for smaller ions (e.g., Li^+) and the lower the field strength, the more selective for larger ions (e.g., Cs^+). The finding that D65N shifts the sequence to a lower order (i.e., favoring larger ions) suggests that D65 is a high field strength binding site (Fig. 4D).

(c) *Calcium permeability is reduced disproportionately to Na^+ permeability*: Reduction or abolition of an intrapore negative charge would be expected to disproportionately reduce the permeability of multivalent cations, like Ca^{2+}, relative to the monovalent cations (Fig. 4C).

(d) *Loss of pH titratability*: Low extracellular pH inhibits Na^+ permeation and reduces charge selectivity of wild-type claudin-2, suggesting that a negative charge that binds Na^+ and is contributed by a carboxylate group has been titrated. Loss of pH inhibition in the D65N mutant suggests that the culprit negative charge is D65 itself (Fig. 4E).

This same study also addressed the issue of potential surface charge effects (Model 3). At low Na^+ concentrations and hence low ionic strength, screening of any surface charges by the bulk solution is minimized and the local Na^+ concentration at the pore entrance would be increased (Green & Andersen, 1991). This would be expected to result in anomalously high conductance at low Na^+ concentrations. Instead, a linear relationship was observed between conductance through claudin-2 and Na^+ concentration within the physiological range. This argues against Model 3.

While D65 is clearly an important cation-binding site in claudin-2, it may not be the only selectivity determinant. Even when this charge was neutralized (D65N), the pore still exhibited residual cation selectivity ($P_{Na}/P_{Cl} \sim 3$) (Yu et al., 2009). One possible explanation is that the replacement asparagine forms a dipole oriented with its amide oxygen atom facing into the lumen, thus exposing a partial negative charge. Indeed, asparagines can mediate cation binding in narrow filters, most notably in the NMDA channel (Burnashev et al., 1992). It is also possible that there are additional determinants of charge selectivity. It is clear that the only other acidic residues in the first extracellular domain, E53 and D76, are not involved because neutralizing them had no effect on charge selectivity (Yu et al., 2009). There are, though, four other acidic residues on the second extracellular domain that could potentially contribute. In addition, other channels are known to use dipoles (e.g., main chain carbonyl and side-chain hydroxyl groups in KcsA; Doyle et al., 1998) to stabilize ions in very close proximity. This mechanism may also be used by claudins, though this is likely to be less important given that the pore is fairly wide. Aromatic residues may contribute through cation–π interactions.

In summary, the weight of the evidence supports the idea that claudin pore ion conductance and charge selectivity are mediated by intraluminal charges contributed by the side-chains of pore-lining amino acids that stabilize the permeating ion by electrostatic interactions. There is evidence for both Model 1 (diffuse charges) and Model 2 (discrete binding sites) and currently, it is unresolved whether these represent real differences between claudin isoforms, or are simply due to differences in methodology between studies.

VI. SIZE OF CLAUDIN PORES

Studies of paracellular permeability have yielded estimates of pore size that vary widely. It has been consistently found that there is both a high capacity, size-restrictive pathway (e.g., ~8 Å diameter in Caco-2 and T84 cells) and a low-capacity, size-independent pathway (Watson, Rowland, & Warhurst, 2001). Two studies have investigated the size of claudin pores. Van Itallie et al. used a highly discriminatory method based on profiling the apparent permeability to a continuous series of noncharged polyethylene glycols (PEGs). They found that most cells and tissues exhibited a size-restrictive pathway of ~8 Å in diameter and a size-independent pathway. Overexpression of claudin-2 caused an increase in the apparent density of the 8 Å pores (Fig. 5). Interestingly, the overexpression of claudins 4, 14, and 18 did not affect pore size or density, suggesting that claudins that are barrier-forming do so without affecting the size-restrictive pores. Yu et al. (2009) also examined the claudin-2 size profile, but using permeability to a series of organic cations of varying sizes. While this method lacks the accuracy and resolution of PEG profiling, it has the advantage that the functional pore size to cations, the major permeant species through claudin-2, can be estimated (Yu et al., 2009). Using this method, claudin-2 was found to have a pore size for the cations of ~6.5–7.0 Å. There are many possible reasons for the slightly differing estimates of pore diameter, but the most obvious is that organic cations are not symmetrical in shape and so any estimate of their size must be regarded as approximate.

The finding that claudin-2 has a pore size of 6.5–8 Å has important implications. The widest diameter of water is 2.8 Å, so one might expect that the pore would be water-filled. However, the diameters of a dehydrated and fully hydrated Na^+ ion are 3.8 and 7.2 Å, respectively (Nightingale, 1959).

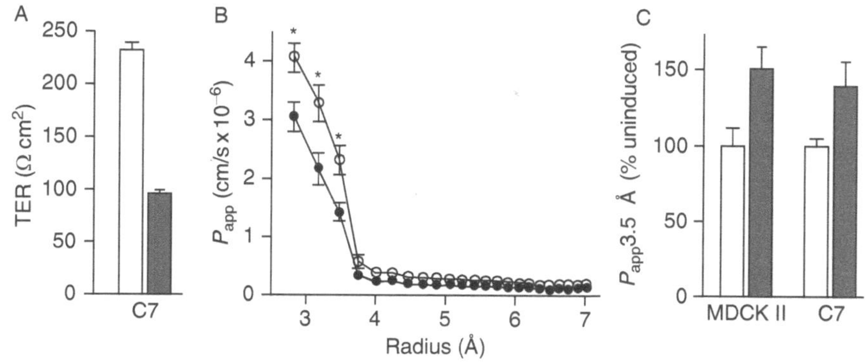

FIGURE 5 Expression of claudin-2 increases the number of size-restrictive pores with radius ~4 Å. (A) Induction (filled bars) of claudin-2 compared with uninduced (unfilled bars) MDCK C7 monolayers results in a large drop in transepithelial resistance (TER). (B) Induction of claudin-2 (unfilled circles) in MDCK II monolayers results in a significant increase in the apparent permeability (P_{app}) specifically for polyethylene glycol (PEG) sizes that are of radius <4 Å, compared with uninduced monolayers (filled circles). (C) Corrected P_{app} of the 3.5 Å PEG species reveals increases in pore number after inducing claudin-2 (filled bars) relative to uninduced (unfilled bars) in both MDCK II and MDCK C7 cells. Data represent mean ± S.E. of four separate clones. *$p < 0.05$ (induced vs. uninduced). Modified with permission from Van Itallie et al. (2008).

It is therefore likely that Na^+ and most other permeating cations would have to shed at least part of their hydration shell to permeate through the narrowest part of the claudin-2 pore. This explains the need for an intrapore Na^+-binding site. Its role is to stabilize Na^+ in a partially dehydrated state within the pore, such that the binding energy offsets the energetic cost of dehydrating the cation.

VII. MAPPING RESIDUES ONTO THE STRUCTURE OF THE CLAUDIN PORE BY CYSTEINE MUTAGENESIS

Having established that the first extracellular domain of claudin forms a pore of 6.5–8 Å with one or more intrapore electrostatic ion-binding sites, the next major question is *how* is the first extracellular domain folded to achieve this? Angelow and Yu (2009) have begun to address this by mutating selected residues in the first extracellular domain of claudin-2 individually to cysteine. This is a potentially very powerful approach to obtain the following structure–function information.

(a) *Which residues are essential for permeation?* If mutation of a residue to cysteine abolishes or qualitatively alters claudin ion permeability, this might indicate that this residue is functionally essential for ion permeation (e.g., ion-binding site). This would allow the identification of novel functional motifs that mediate pore function. It is important to interpret such studies with care, since any mutation could indirectly alter pore properties simply by disrupting protein folding.

(b) *Where is each residue spatially located relative to the pore lumen?* By testing the accessibility of cysteines at each position to the thiol-reactive reagent, *N*-biotinylaminoethyl methanethiosulfonate (MTSEA-biotin), and the ability of pore conductance to be blocked by methanethiosulfonate (MTS) reagents of various sizes and charges, one can deduce where they are located relative to the pore lumen.

(c) *Identify residues that are located in intra- and intermolecular spatial proximity*: Introduction of cysteines may lead to disulfide bonding with other cysteines from the same molecule or from neighboring claudins that are in close proximity. This would shed light on the tertiary and quaternary structure of the claudin-based pore.

In this study, Angelow and Yu (2009) made three significant observations. First, they found that the mutation of D65 (the Na^+-binding site) to cysteine caused dimerization of claudin-2 due to disulfide bonding. This indicates that claudin-2 is multimeric and that D65 must lie near an intermolecular interface (for a disulfide bond to form, the C65 β-carbons are likely to be separated by less than 7 Å; Careaga & Falke, 1992). Furthermore, since the side-chain of D65 is predicted to face into the lumen, it suggests that the pore lumen is itself composed of at least two claudin molecules.

Second, they found that the mutation, I66C, rendered the claudin susceptible to block by MTS reagents. The kinetics of the block suggested that diffusion of these reagents to the reactive site was rate-limiting, indicating that I66 is buried deep in a narrow part of the pore, with its side-chain facing toward the lumen.

Third, they found another mutation, Y35C, that rendered the claudin susceptible to biotinylation by MTSEA-biotin, but not to block by any MTS reagent. This suggested that Y35C is exposed extracellularly outside the mouth of the pore. These findings, which are summarized in Fig. 6, begin to fill in the gaps in information about the location of each residue within the first extracellular fold of claudin.

VIII. STOICHIOMETRY OF CLAUDIN PORES

It is fairly well accepted that claudins are probably multimeric. Claudins have been shown to polymerize laterally (in *cis*) within the tight junction strand of one cell and to interact in a head-to-head manner (in *trans*) with

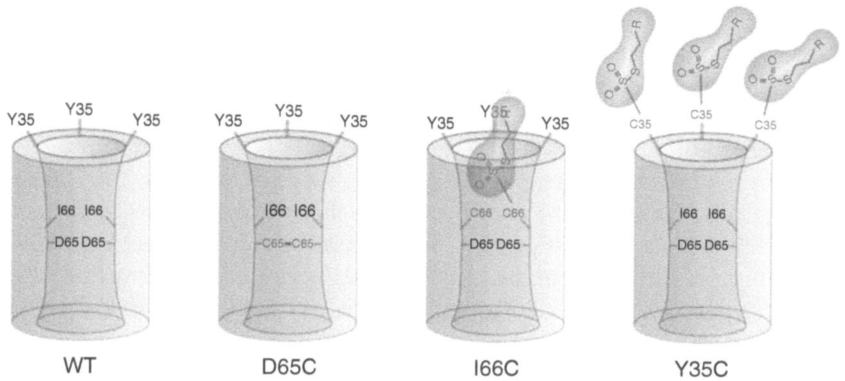

FIGURE 6 Model of claudin-2 showing the putative location of residues in the first extracellular domain derived from cysteine mutagenesis data. The pore is hypothesized to be a homomultimer (residues from two to three subunits are shown). Wild-type (WT) claudin-2 is depicted in the left panel, and the consequences of cysteine mutagenesis in the other panels. D65 is located in the narrowest part of the pore facing the lumen and close to an intersubunit interface, so that the D65C mutation leads to dimerization by disulfide bonding. I66 is also within the pore facing the lumen, but residues from neighboring subunits are further apart, so that disulfide bonding in I66C is precluded. Thiol-reactive MTS reagents (green/gray) enter the pore to react with I66C, partially blocking the pore to ion permeation. Only a single MTS molecule is accommodated in each pore. Y35 is outside the pore facing extracellularly. Thus, the reaction of MTS reagents with Y35C does not block the pore. Furthermore, because there is no steric restriction, every Y35C residue can react with an MTS molecule, so that multiple MTS molecules are associated with each pore multimer. This research was originally published in Angelow and Yu (2009), © The American Society for Biochemistry and Molecular Biology.

claudins on the adjacent cell (Furuse et al., 1999). Limited biochemical evidence suggests that these may occur by direct physical association (discussed in more detail in Chapter 6) (Blasig et al., 2006; Hou et al., 2008). Furthermore, when claudin-4 was solubilized in perfluoro-octanoic acid, multimeric species up to hexamers were observed (Mitic, Unger, & Anderson, 2003). What is unclear is whether claudin pores are also multimeric. In theory, individual pores could be enclosed within a single claudin monomer, formed by multiple claudin monomers interacting in *cis*, or by monomers interacting in *trans*, or both. Two types of studies have provided clues to this issue:

(a) *Functional studies*: Hou et al. have shown that claudin-16 and -19 physically associate within the same cell and interact in a heteromeric manner in *cis* (see Chapter 8 for a detailed discussion). They found that claudin-16 increased P_{Na}, claudin-19 decreased P_{Cl}, and that coexpression of both isoforms caused both an increase in P_{Na} and a decrease in P_{Cl}. What is unclear is whether this represents a synergistic effect on pore function, supporting the hypothesis that the pore is itself heteromeric, or merely an additive effect, as would be expected from the coexistence of two different populations of homomeric pores.

(b) *Structural studies*: As discussed earlier, Angelow and Yu (2009) found that the claudin-2 intrapore residue, D65, when mutated to cysteine, caused dimer formation, suggesting that the claudin-2 pore is at least dimeric. They also made an additional interesting observation when comparing the claudin-2 mutants, Y35C and I66C. The I66C pore was blocked by MTS reagents, but was weakly labeled when exposed to MTSEA-biotin. By contrast, Y35C was not blocked at all by MTS reagents, but MTSEA-biotin labeled it very strongly. A plausible explanation for this is that the pore is multimeric. Because the lumen is narrow, access to residues within the pore is sterically hindered, so only one of the intrapore cysteines in I66C can be biotinylated. However, Y35C is located outside the pore, so every residue of the multimeric complex is biotinylatable (Fig. 6). Thus, it was suggested that the difference in MTSEA-biotin labeling of Y35C compared to I66C could be a clue to the stoichiometry of a multimeric claudin pore.

IX. CONCLUSIONS

In summary, current evidence supports the idea that the first extracellular domains (perhaps predominantly the second half) of multiple claudin monomers fold to enclose a 6.5–8 Å diameter paracellular pore, and present residues with charged side-chains that protrude into the lumen and act as electrostatic ion-binding sites. The structure of the extracellular domain of claudin has yet to be solved at atomic resolution, so many questions remain to be answered.

References

Amasheh, S., Meiri, N., Gitter, A. H., Schoneberg, T., Mankertz, J., Schulzke, J. D., et al. (2002). Claudin-2 expression induces cation-selective channels in tight junctions of epithelial cells. *Journal of Cell Science, 115*(Pt 24), 4969–4976.

Angelow, S., Ahlstrom, R., & Yu, A. S. (2008). Biology of claudins. *American Journal of Physiology. Renal Physiology, 295*(4), F867–F876.

Angelow, S., Schneeberger, E. E., & Yu, A. S. (2007). Claudin-8 expression in renal epithelial cells augments the paracellular barrier by replacing endogenous claudin-2. *The Journal of Membrane Biology, 215*(2–3), 147–159.

Angelow, S., & Yu, A. S. (2007). Claudins and paracellular transport: An update. *Current Opinion in Nephrology and Hypertension, 16*(5), 459–464.

Angelow, S., & Yu, A. S. (2009). Structure-function studies of claudin extracellular domains by cysteine-scanning mutagenesis. *The Journal of Biological Chemistry, 284*(42), 29205–29217 M109.043752 [pii], 10.1074/jbc.M109.043752 [doi].

Ben-Yosef, T., Belyantseva, I. A., Saunders, T. L., Hughes, E. D., Kawamoto, K., Van Itallie, C. M., et al. (2003). Claudin 14 knockout mice, a model for autosomal recessive deafness DFNB29, are deaf due to cochlear hair cell degeneration. *Human Molecular Genetics, 12*(16), 2049–2061.

Blasig, I. E., Winkler, L., Lassowski, B., Mueller, S. L., Zuleger, N., Krause, E., et al. (2006). On the self-association potential of transmembrane tight junction proteins. *Cellular and Molecular Life Sciences, 63*(4), 505–514.

Burnashev, N., Schoepfer, R., Monyer, H., Ruppersberg, J. P., Gunther, W., Seeburg, P. H., et al. (1992). Control by asparagine residues of calcium permeability and magnesium blockade in the NMDA receptor. *Science, 257*(5075), 1415–1419.

Careaga, C. L., & Falke, J. J. (1992). Thermal motions of surface alpha-helices in the D-galactose chemosensory receptor. Detection by disulfide trapping. *Journal of Molecular Biology, 226*(4), 1219–1235.

Colegio, O. R., Van Itallie, C. M., McCrea, H. J., Rahner, C., & Anderson, J. M. (2002). Claudins create charge-selective channels in the paracellular pathway between epithelial cells. *American Journal of Physiology. Cell Physiology, 283*(1), C142–C147.

Colegio, O. R., Van Itallie, C., Rahner, C., & Anderson, J. M. (2003). Claudin extracellular domains determine paracellular charge selectivity and resistance but not tight junction fibril architecture. *American Journal of Physiology. Cell Physiology, 284*(6), C1346–C1354.

Doyle, D. A., Morais Cabral, J., Pfuetzner, R. A., Kuo, A., Gulbis, J. M., Cohen, S. L., et al. (1998). The structure of the potassium channel: Molecular basis of K^+ conduction and selectivity. *Science, 280*(5360), 69–77.

Edelman, A., & Anagnostopoulos, T. (1978). Further studies on ion permeation in proximal tubule of necturus kidney. *American Journal of Physiology, 235*(2), F89–F95.

Eisenman, G., & Horn, R. (1983). Ionic selectivity revisited: The role of kinetic and equilibrium processes in ion permeation through channels. *The Journal of Membrane Biology, 76*(3), 197–225.

Eisenman, G., Sandblom, J. P., & Walker, J. L. Jr., (1967). Membrane structure and ion permeation. Study of ion exchange membrane structure and function is relevant to analysis of biological ion permeation. *Science, 155*(765), 965–974.

Fujita, H., Sugimoto, K., Inatomi, S., Maeda, T., Osanai, M., Uchiyama, Y., et al. (2008). Tight junction proteins claudin-2 and -12 are critical for vitamin D-dependent Ca^{2+} absorption between enterocytes. *Molecular Biology of the Cell, 19*(5), 1912–1921, E07-09-0973 [pii], 10.1091/mbc.E07-09-0973 [doi].

Furuse, M., Furuse, K., Sasaki, H., & Tsukita, S. (2001). Conversion of zonulae occludentes from tight to leaky strand type by introducing claudin-2 into Madin-Darby canine kidney I cells. *The Journal of Cell Biology, 153*(2), 263–272.

Furuse, M., Sasaki, H., & Tsukita, S. (1999). Manner of interaction of heterogeneous claudin species within and between tight junction strands. *The Journal of Cell Biology, 147*(4), 891–903.

Furuse, M., & Tsukita, S. (2006). Claudins in occluding junctions of humans and flies. *Trends in Cell Biology, 16*(4), 181–188.

Gossen, M., & Bujard, H. (1992). Tight control of gene expression in mammalian cells by tetracycline-responsive promoters. *Proceedings of the National Academy of Sciences of the United States of America, 89*(12), 5547–5551.

Green, W. N., & Andersen, O. S. (1991). Surface charges and ion channel function. *Annual Review of Physiology, 53*, 341–359.

Hou, J., Paul, D. L., & Goodenough, D. A. (2005). Paracellin-1 and the modulation of ion selectivity of tight junctions. *Journal of Cell Science, 118*(Pt 21), 5109–5118.

Hou, J., Renigunta, A., Konrad, M., Gomes, A. S., Schneeberger, E. E., Paul, D. L., et al. (2008). Claudin-16 and claudin-19 interact and form a cation-selective tight junction complex. *The Journal of Clinical Investigation, 118*(2), 619–628.

Kausalya, P. J., Amasheh, S., Gunzel, D., Wurps, H., Muller, D., Fromm, M., et al. (2006). Disease-associated mutations affect intracellular traffic and paracellular Mg^{2+} transport function of Claudin-16. *The Journal of Clinical Investigation, 116*(4), 878–891.

Mandel, L. J., & Curran, P. F. (1972). Response of the frog skin to steady-state voltage clamping. I. The shunt pathway. *The Journal of General Physiology, 59*(5), 503–518.

Mitic, L. L., Unger, V. M., & Anderson, J. M. (2003). Expression, solubilization, and biochemical characterization of the tight junction transmembrane protein claudin-4. *Protein Science, 12*(2), 218–227.

Munck, B. G., & Schultz, S. G. (1974). Properties of the passive conductance pathway across in vitro rat jejunum. *The Journal of Membrane Biology, 16*(2), 163–174.

Nightingale, E. R. (1959). Phenomenological theory of ion salvation. Effective radii of hydrated ions. *The Journal of Physical Chemistry, 63*, 1381–1387.

Schneeberger, E. E., & Lynch, R. D. (2004). The tight junction: A multifunctional complex. *American Journal of Physiology. Cell Physiology, 286*(6), C1213–C1228.

Tsukita, S., & Furuse, M. (2002). Claudin-based barrier in simple and stratified cellular sheets. *Current Opinion in Cell Biology, 14*(5), 531–536.

Van Itallie, C. M., & Anderson, J. M. (2004). The molecular physiology of tight junction pores. *Physiology (Bethesda), 19*, 331–338.

Van Itallie, C. M., & Anderson, J. M. (2006). Claudins and epithelial paracellular transport. *Annual Review of Physiology, 68*, 403–429.

Van Itallie, C., Fanning, A. S., & Anderson, J. M. (2003). Reversal of charge selectivity in cation or anion selective epithelial lines by expression of different claudins. *American Journal of Physiology. Cell Physiology, 286*, F1078–F1084.

Van Itallie, C. M., Holmes, J., Bridges, A., Gookin, J. L., Coccaro, M. R., Proctor, W., et al. (2008). The density of small tight junction pores varies among cell types and is increased by expression of claudin-2. *Journal of Cell Science, 121*(Pt 3), 298–305. jcs.021485 [pii], 10.1242/jcs.021485 [doi].

Van Itallie, C., Rahner, C., & Anderson, J. M. (2001). Regulated expression of claudin-4 decreases paracellular conductance through a selective decrease in sodium permeability. *The Journal of Clinical Investigation, 107*(10), 1319–1327.

Van Itallie, C. M., Rogan, S., Yu, A. S., Seminario-Vidal, L., Holmes, J., & Anderson, J. M. (2006). Two splice variants of claudin-10 in the kidney create paracellular pores with different ion selectivities. *American Journal of Physiology. Renal Physiology, 291*, F1288–F1299.

Watson, C. J., Rowland, M., & Warhurst, G. (2001). Functional modeling of tight junctions in intestinal cell monolayers using polyethylene glycol oligomers. *American Journal of Physiology. Cell Physiology, 281*(2), C388–C397.

Welsh, M. J., & Widdicombe, J. H. (1980). Pathways of ion movement in the canine tracheal epithelium. *American Journal of Physiology, 239*(3), F215–F221.

Wright, E. M. (1966). Diffusion potentials across the small intestine. *Nature, 212*(5058), 189–190.

Wright, E. M., & Diamond, J. M. (1968). Effects of pH and polyvalent cations on the selective permeability of gall-bladder epithelium to monovalent ions. *Biochimica et Biophysica Acta, 163*(1), 57–74.

Yu, A. S., Cheng, M. H., Angelow, S., Gunzel, D., Kanzawa, S. A., Schneeberger, E. E., et al. (2009). Molecular basis for cation selectivity in claudin-2-based paracellular pores: Identification of an electrostatic interaction site. *The Journal of General Physiology, 133*(1), 111–127.
Yu, A. S., Enck, A. H., Lencer, W. I., & Schneeberger, E. E. (2003). Claudin-8 expression in MDCK cells augments the paracellular barrier to cation permeation. *The Journal of Biological Chemistry, 278*, 17350–17359.

CHAPTER 5

The Investigation of *cis*- and *trans*-Interactions Between Claudins

Reiner F. Haseloff, Jörg Piontek, and Ingolf E. Blasig
Leibniz Institute for Molecular Pharmacology (FMP), Berlin-Buch, Germany

I. OVERVIEW

Claudins control the paracellular permeability of the tight junctions in epithelial and endothelial cells forming single cell layers. Several claudins may constitute a network of strands, establishing a continuous barrier within the intercellular clefts of the monolayer. Depending on the claudin composition of a given tissue, the intracellular space may be tight for any solute or permeable for compounds of different molecular weight or differently charged ions.

These functions are largely based on intermolecular claudin–claudin interactions. Strand formation between claudins requires two components: a longitudinal association along the plasma membrane of the cell—*cis*-interaction—and an interplay from one cell surface to the next one—*trans*-interaction.

This chapter is aimed at reviewing methods for the analysis of *cis*- and *trans*-interacting claudins.

Current Topics in Membranes, Volume 65 1063-5823/10 $35.00

DOI: 10.1016/S1063-5823(10)65005-3

II. INTRODUCTION

Claudins are tetraspan proteins with intracellular termini. The transmembrane topology results in one intra- and two extracellular loops. Claudins determine the barrier and pore properties of the tight junctions between the cell membranes of two opposing cells, as found in endothelial or epithelial monolayers. The intra- and intermolecular interactions of the extracellular loops very probably specify the intercellular properties of claudins (Krause et al., 2008). The first extracellular loop contains 40–50 amino acids, and it is assumed that this is critical for establishing paracellular tightness and determining the ion selectivity of the pores (Alexandre, Jeansonne, Renegar, Tatum, & Chen, 2007; Amasheh et al., 2002; Colegio, Van Itallie, Rahner, & Anderson, 2003; Hou, Paul, & Goodenough, 2005; Van Itallie et al., 2006; Wen, Watry, Marcondes, & Fox, 2004). The 15–30 amino acids of the second extracellular loop may constrict the paracellular cleft and hold the neighboring plasma membranes together (Blasig et al., 2006; Piontek et al., 2008). There is also evidence that the second loop may have a tightening function, too (Piehl, Piontek, Cording, Wolburg, & Blasig, 2010). In addition, one has to assume that the two loops cooperate with each other, as their transmembrane domains are close and would tend to adhere to each other; moreover, claudins may constitute oligomers. Sequence analysis of claudins has led to differentiation into two groups, designated as *classic* claudins (1–10, 14, 15, 17, 19) and *nonclassic* claudins (11–13, 16, 18, 20–24), according to their degree of sequence similarity. This is also reflected in the sequence–structure–function relationships derived for extracellular loops 1 and 2. The concepts evolved from these findings and the first tentative molecular models for homophilic interactions may explain the different functional contribution of the two extracellular loops at tight junctions.

Claudins may form a network of strands continuously sealing the intercellular clefts of the monolayer (Tsukita & Furuse, 2000). The proteins constituting the tight junctions, in particular the expression pattern of the claudins, determine the barrier properties which can differ considerably between different organs and cell types (Peppi & Ghabriel, 2004; Pfeiffer et al., 2008) and which can change under disease conditions (Yu, Kanzawa, Usorov, Leeuwen, & Peters, 2008). These barrier properties are essentially based on intermolecular claudin–claudin interactions. For strand formation, two processes have to be considered: longitudinal association along the plasma membranes of the adhering cells and an interplay from one cell surface to the next one (Piontek et al., 2008). This orthotropic self-association is known for other junctional proteins, such as the cadherins located at the adherens junctions (Ahrens et al., 2002). We therefore adapted the terms of cadherin–cadherin interaction to that of claudins (Krause et al., 2008). That means that the attachment between a claudin from the surface of one cell and a claudin from the surface of the next cell is called *trans*-interaction. Alternatively, the adhesion of a claudin molecule along the membrane of a cohesive cell is termed *cis*-interaction. Potential interactions between claudin molecules at the tight junctions are depicted in Fig. 1.

The strand-forming nature of claudins demonstrates not only the association between single claudin molecules but also the capability to polymerize. *In vitro* evidence suggests that claudins possess a general potential for oligomerization. In seminative electrophoresis with perfluorooctanoate gels, claudin-5 is detected as a larger oligomer than claudin-1. Claudin-5 preferentially forms hexamers (Coyne, Gambling, Boucher, Carson, & Johnson, 2003) but does not polymerize within one cell. Only *trans*-interaction between oligomers at tight junctions and between two claudin-expressing cells triggers the formation of polymeric strands between the cells (Piontek et al., 2008). As tight junction-carrying cells express different claudins, it is likely that heterophilic claudin–claudin interactions also take place.

From gel electrophoretic and freeze-fracture data, a two-step mechanism of the strand formation can be assumed (Fig. 2). Oligomers of varying size—up to hexamers, but no higher polymers—can be detected (Blasig et al., 2006; Coyne et al., 2003; Mitic, Unger, & Anderson, 2003). Some claudins, such as claudin-4 and -5, seem to occur preferentially as hexamers (Coyne et al., 2003; Mitic et al., 2003). In electron microscopy, discontinuous strands, including particles in grooves, have been observed when claudin-2 (Furuse, Sasaki, Fujimoto, & Tsukita, 1998; Furuse, Sasaki, & Tsukita, 1999), claudin-3 (Morita, Furuse, Fujimoto, & Tsukita, 1999), claudin-5 (Coyne et al., 2003; Morita, Sasaki,

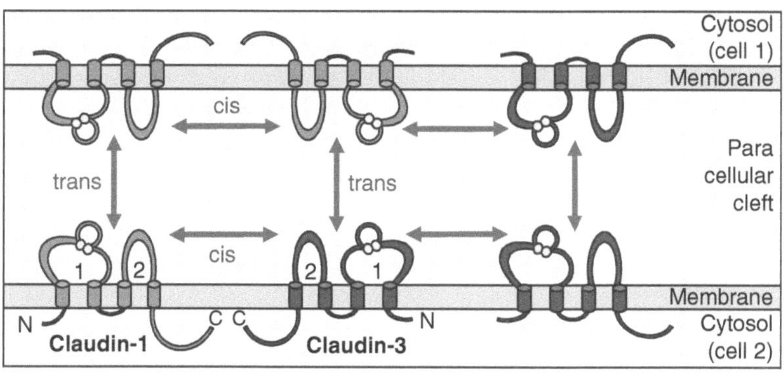

FIGURE 1 Possible interactions between claudins as reported for claudin-1 and -3 (Furuse et al., 1999), including homophilic and heterophilic *cis*- and *trans*-interactions based on a nomenclature described for cadherins (Ahrens et al., 2002), a group of proteins found in adherens junctions. Numbers mark first or second extracellular loop; small circles denote disulfide bounds assumed to be formed within the first loop due to the oxidizing environment prevailing extracellularly.

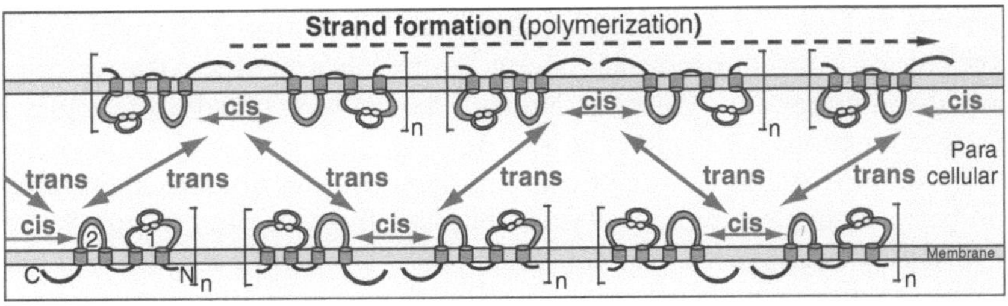

FIGURE 2 Scheme of claudin oligomerization and strand formation (two-step mechanism). (i) Claudins form oligomers of *n* claudin molecules via *cis*-interaction along one plasma membrane, as demonstrated by fluorescence resonance energy transfer (Blasig et al., 2006). The oligomers seem to consist predominantly of hexamers (Coyne et al., 2003). However, no continuous polymer is formed in the cell membrane when *trans*-interaction is impossible or prevented, as proven by freeze-fracture electron microscopy. (ii) *Trans*-interaction, probably between claudin oligomers, is the essential prerequisite for the strand formation. Claudins in the tight junctions of opposing cell membranes are able to form polymers along the two membranes only after *trans*-interaction, resulting in the formation of tight junction strands. This stepwise interaction scheme is consistent with mutagenesis data, homology models, and the presence/absence of a network of strands in freeze-fracture replica (Piontek et al., 2008).

Furuse, & Tsukita, 1999; Piontek et al., 2008), claudin-6 (Morita et al., 2002), or claudin-11 (Morita, Sasaki, Fujimoto, Furuse, & Tsukita, 1999) were transfected in tight junction-free cells. The particles are of limited size, matching that of connexin hexamers as determined by electron microscopy, size exclusion chromatography, or density gradient centrifugation (Perkins, Goodenough, & Sosinsky, 1997; Saez, Berthoud, Branes, Martinez, & Beyer, 2003; Schmitz & Wolburg, 1991). Since claudins are roughly of the same size as connexins and also consist of four transmembrane helices, it is conceivable that they may also form hexamers *in vivo*, even though the configuration would be different. According to the two-step mechanism, claudins first oligomerize via *cis*-interaction, and this can obviously occur already within the cell. However, the claudins do not polymerize into strands within a given membrane. *Trans*-interaction between oligomers in the two opposing plasma membranes is only possible at tight junctions. Then, this *trans*-interaction triggers the formation of polymeric strands. In addition, this mechanism would prevent uncontrolled polymerization in intracellular compartments.

III. ESTIMATION OF INTERACTIONS, WITHOUT DIFFERENTIATION BETWEEN *CIS*- AND *TRANS*-INTERACTION

A. *Conventional Techniques*

Standard procedures to analyze claudin–claudin interactions are based on measurement of the transcellular electrical resistance and determination of the paracellular permeability of compounds known to lack transcellular permeation. Generally, these techniques provide a rough estimate of the intercellular *trans*-interaction. However, they are also afflicted with disadvantages, as they are influenced by the material, pore number and size, or coating (Angelow, El Husseini, Kanzawa, & Yu, 2007; Barker & Simmons, 1981; Cereijido, Robbins, Dolan, Rotunno, & Sabatini, 1978; Jovov et al., 2007; Lo, Keese, & Giaever, 1999; Van Itallie, Fanning, & Anderson, 2003; Wen et al., 2004) of the filter inserts applied. On the basis of the resistance, differentiation is not possible between the contribution by cell adhesion to the support and that by the tight junctions. Impedance measurements can overcome this problem and allow differentiation between paracellular and transcellular resistance (Arndt, Seebach, Psathaki, Galla, & Wegener, 2004; Giaever & Keese, 1991; Lo et al., 1999; McCoy & Wang, 2005; Tiruppathi, Malik, Del Vecchio, Keese, & Giaever, 1992). However, despite the fact that continuous measurements are possible, this technique is less sensitive for the evaluation of modulators of the tight junctions. Determination of permeability is even more labor intensive. Standardization is necessary, as variation arises from the size of the filter, the marker compound used, and the concentration of the marker; moreover, one has to work in the linear part of the permeability–time curve. The permeability coefficient needs to be calculated to obtain a lab-invariant parameter (Gaillard et al., 2001). A more detailed discussion of permeability measurements is given in Chapter 3 of this book, entitled "Biophysical methods to study claudin permeability".

Freeze-fracture electron microscopy has been the method of choice to visualize tight junction strands for three decades (Staehelin, 1973, 1974). Homopolymeric strands have been demonstrated for claudins 1–6, 9, 11, 14, and 19 after transfection in HEK-293 cells, murine L-fibroblasts, or COS-7 cells (Furuse et al., 1998, 1999; Miyamoto et al., 2005; Morita et al., 1999; Nunes et al., 2006; Piontek et al., 2008; Sonoda et al., 1999; Turksen & Troy, 2001; Wattenhofer et al., 2005). Heteropolymeric strands have been described for the combinations claudin-1/-2, -1/-3, -2/-3, and -1/-3/-5 (Coyne et al., 2003; Furuse et al., 1999). In addition to strand morphology, paracellular tightness and hence the *trans*-interaction can be evaluated by means of immunoelectron microscopy (Sheikov, McDannold, Sharma, & Hynynen, 2008). This has been accomplished by administration on one side of the tight junction barrier of electron dense agents, such as lanthanum-containing compounds permeable neither for membranes nor tight junctions. However, depending on the permeation of the barrier by these agents, only a qualitative decision is possible on the leakiness or tightness of the junctions.

Another well-known group of techniques to demonstrate protein–protein interactions is based on affinity purification (cf. Section III.B). Alternatively, the binding between two proteins can be shown by fluorescence resonance energy transfer (FRET), as displayed in Fig. 3, if the proteins have been

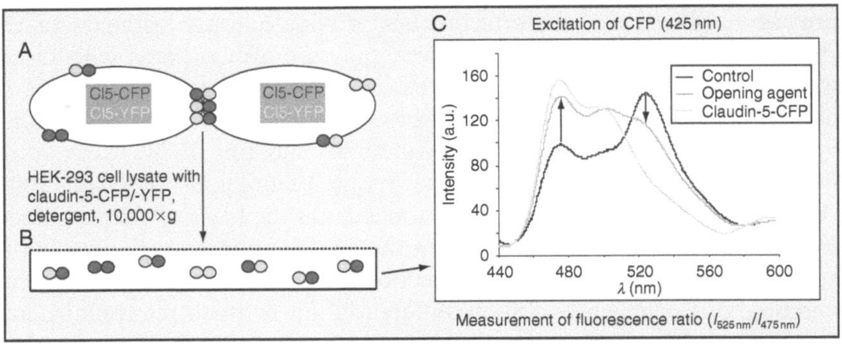

FIGURE 3 Demonstration of claudin–claudin interactions using fluorescence resonance energy transfer (FRET). (A) Scheme of the enrichment of claudin-5 (Cl5) and, consequently, the tightening of the paracellular cleft between opposing cells in the area of the tight junctions in cells cotransfected with claudin-5-CFP/-YFP (circles). (B) Extraction of the Cl5 constructs and (C) measurement of the FRET between the fluorescent fusion proteins of Cl5 as a measure of the claudin self-association (black spectrum) and the influence of test substances on the FRET intensity (middle spectrum, arrows). CFP/YFP, cyano-/yellow-fluorescence protein. Arrows, FRET reduction by a junction opening agent.

labeled with fluorescence tags. Colocalization studies are often used as an exploratory analysis before starting a protein–protein interaction study. These experiments do not prove a direct association, but without colocalization it is possible to exclude direct adherence and abandon further investigations.

B. Proteomic Approaches for Identifying Interaction Partners of Claudins

Proteomic approaches applying state-of-the-art mass spectrometry (MS) provide a powerful tool for the analysis of claudin–protein interactions even within large networks of interacting proteins (Gingras, Gstaiger, Raught, & Aebersold, 2007). The most prevalent technique of preparation of claudin complexes is based on affinity purification directed at a protein which is known to be part of the protein complex; commonly used tools are antibodies or recombinant proteins/protein domains (bait protein). As a second step, prefractionation on the protein level is often required, since the number of proteins copurified with the bait protein is usually very high. These procedures are followed by a proteolysis of the complete set of proteins, in most cases, by the use of trypsin. The resulting peptides are extracted and subjected to mass spectrometry; tandem MS techniques which deliver sequence information are now a prerequisite. Database searches yield a list of proteins consisting of the bait protein, its interaction partners, and contaminating proteins. The critical steps within this workflow are sample preparation, affinity-based purification and, dependent on these preceding steps, prefractionation of the resulting complex.

The preparation of the sample is a crucial point in the search for protein interaction partners, since claudin–protein associations must be left intact. For experiments based on cell cultures, this requires very gentle lysis conditions, and the use of detergents should be avoided whenever possible. The lipophilicity of the claudins is a considerable obstacle for the identification of their interaction partners by proteomic methods; techniques based on tandem affinity purification are highly valuable for soluble proteins (Gloeckner, Boldt, Schumacher, Roepman, & Ueffing, 2007; Rigaut et al., 1999) but are less efficient for integral membrane proteins. In general, prefractionation of the sample prior to affinity purification (e.g., by the removal of high abundant proteins) can help to diminish the amount of nonspecifically binding proteins.

The ideal bait protein for a purification binds with high affinity exclusively to the target protein—and does not exist. Low-affinity association of a high-abundant protein to the bait can decrease the efficiency of the purification considerably. MS of an eluate obtained from a preliminary coimmunoprecipitation experiment using cell lysates and a monoclonal antibody against claudin-4 (which is not suspected of being cross-reactive to other claudins) identified hundreds of proteins, many of them of nuclear or mitochondrial origin and therefore unlikely interaction partners of claudins (unpublished data). One has to consider that nonspecific binding can occur with the matrix or with the tag (when using recombinant proteins) and also with the bait protein itself; it therefore needs to be diminished or at least identified (Chang, 2006; Fremont, Wang, & King, 2005; Nguyen & Goodrich, 2006). Quantitative proteomics approaches, for example, based on isotopic labeling techniques, allow the discrimination of nonspecifically binding proteins if a suitable control sample is available. This is illustrated by an example of a workflow of an affinity-based purification shown in Fig. 4, which applies stable isotope labeling by amino acids in cell culture (SILAC; Ong et al., 2002).

For tight junction proteins, in particular claudins, one has to take into account that it is impossible to detect homologous interactions of proteins by these proteomics approaches. Moreover, MS identification of claudin peptides is restricted as their number is relatively small due to the relatively low molecular weight of the claudins. Nevertheless, there are a few reports available which exploit proteomic approaches for identifying the interaction partners of claudins: A tyrosine-phosphorylated, GST-fusion construct of the entire cytoplasmic region of macrophage colony-stimulating factor receptor 1

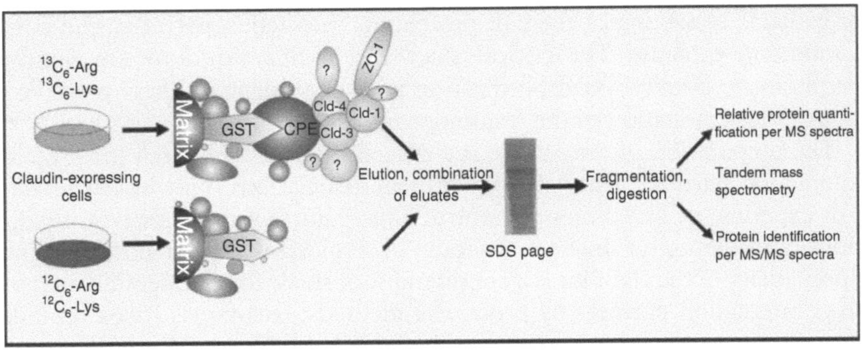

FIGURE 4 Workflow of an affinity-based purification of claudins and potential interaction partners (modified from Lohrberg et al., 2009). Cell lysates are obtained from claudin-expressing cells cultivated in medium supplemented with either ^{12}C-Arg/Lys or ^{13}C-Arg/Lys; affinity purification is accomplished using a recombinant claudin-binding truncated glutathion-*S*-transferase (GST) fusion protein; *Clostridium perfringens* enterotoxin ($CPE_{116-319}$); GST coupled to the matrix serves as the control. Eluates are combined, electrophoretically separated and digested in-gel; the extracted peptides are subjected to LC–MS/MS. Protein identification and relative quantification are based on MS/MS- and MS spectra, respectively (LC, liquid chromatography, MS, mass spectrometry).

(CSF-1R) was used to characterize proteins putatively associating with the activated CSF-1R (Cross, Nguyen, Bogdanoska, Reynolds, & Hamilton, 2005), and claudin-11 was identified among other proteins. Purification of the tight junction complex in T84 epithelial cells was accomplished by a coimmunoprecipitation approach (based on the binding of tight junction proteins to atypical protein kinase zeta) (Tang, 2006), and claudin-16 was identified by peptide mass fingerprinting as the only claudin present. A recent study, with the objective of single-step enrichment of claudins, provided evidence for the heterologous interaction of different claudins in two epithelial cell lines (Lohrberg et al., 2009); claudin-1, -2, -3, -4, -6, and -7 were identified by MS. An interaction of claudin-9 with the tetraspanin protein CD9 was found in a pull-down assay, applying a cysteine-targeted cross-linking step (Kovalenko, Yang, & Hemler, 2007). However, the CD9–claudin-1 complexes did not reside in tight junctions.

IV. ANALYSIS OF *CIS*-INTERACTIONS ALONG THE PLASMA MEMBRANE

FRET detected in a confocal microscope is an elegant way to analyze *cis*-interactions along a membrane. In principle, all membranes can be analyzed. However, the investigation of intracellular membranes is difficult due to their mobility and, in part, small dimensions. For FRET studies, the cell membrane is well qualified if one can demonstrate that the claudin of interest is localized in this membrane. This can be accomplished by colocalization studies applying membrane markers, e.g., transmembranal proteins, which are ubiquitously distributed throughout the plasma membrane, or by applying membrane-specific dyes. Aquaporin-1 or corticotropin-releasing factor receptor 1 (Blasig et al., 2006) have been used in this context as membrane markers. Dyes, such as trypan blue (Wuller et al., 2004), usually require optimization of the experimental protocols.

For a FRET assay, claudin molecules need to be tagged with suitable fluorescence labels, such as cyan-fluorescent protein (CFP) or yellow-fluorescent protein (YFP). The differently marked claudin molecules are cotransfected at comparable expression levels in tight junction-free cells or in cells carrying endogenous claudins. After excitation of CFP, not only the emission of the CFP fluorescence but, in the case of FRET, also YFP emission, is observed due to resonance energy transfer (Fig. 5). However, FRET is only possible when the two chromophores are closely associated in two adjacent claudin molecules. FRET typically occurs if the distance of the two fluorophores is ≤ 6 nm (Sekar & Periasamy, 2003). Thus, the FRET signal demonstrates close proximity between the fluorescent proteins. Consequently, a physical interaction is assumed between two claudins in the plasma membrane of a living cell. A *trans*-interaction between the tagged claudins of two opposing plasma membranes is excluded because, in this case, the distance between the terminal CFP and YFP tags is too large to result in a FRET signal.

When FRET occurs, the fluorescence light emission of the photodonor (e.g., CFP) is weakened and that of the photoacceptor is intensified. This FRET signal was found to be considerably higher in HEK-293 cells which do not express claudins 1–5, as compared to MDCK cells with tight junction strands and endogenous claudins 1, 2, 4, or 7 (Furuse, Furuse, Sasaki, & Tsukita, 2001). In the latter case, the FRET intensity is much weaker, presumably due to competing interactions with endogenous claudins. A protein consisting of CFP covalently bound to YFP (tandem fluorescence protein, without any claudin) or HEK cells cotransfected with claudin-5-CFP and claudin-5-YFP can be used as positive controls. Cells cotransfected with the claudin-free fluorescence proteins, CFP and YFP, may serve as negative controls. Unfortunately, both YFP and CFP localize to the cytoplasm. The use of YFP-aquaporin-1 or CRFR-YFP is more suitable, as these are membrane proteins which do not interact with claudins (Piontek et al., 2008).

The position of the fluorescence proteins within the claudin sequence is of technical relevance for the detectability of any FRET but may also have functional consequences. In our experiments, FRET signals can be obtained

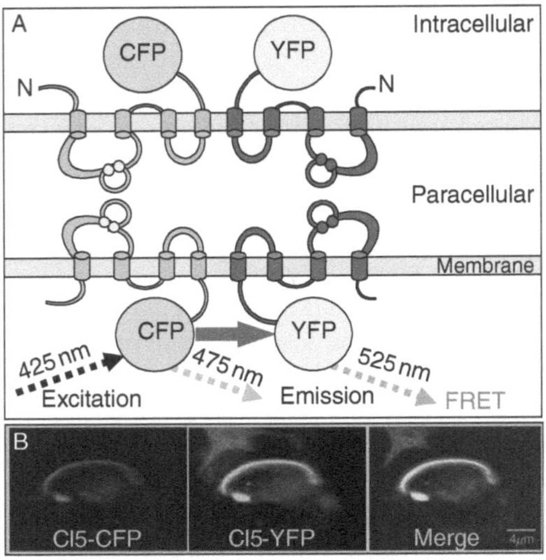

FIGURE 5 Analysis of *cis*-interaction between claudins, measured along the plasma membrane by means of fluorescence resonance energy transfer (FRET). FRET is shown between cyan-fluorescence protein (CFP) and yellow-fluorescence protein (YFP), C-terminally fused to different claudin-5 molecules (Cld5). (A) Scheme of the FRET system. (B) Living HEK-293 cells cotransfected with equal amounts of Cld5-CFP and Cld5-YFP show colocalization of claudin-5-CFP (blue) and claudin-5-YFP (yellow) at the plasma membrane of the same cell in a confluent culture; bar, 10 μm. Cl5-CFP and Cl5-YFP exhibit fluorescence signals after excitation. The tight junction area was marked (red) and the respective CFP fluorescence in this area can be determined before and after YFP photobleaching. FRET from Cl5-CFP to Cl5-YFP, cotransfected in HEK-293 cells, reveals claudin-5 homoassociation within the cell membrane of the same cell. (See Color Insert.)

if the fluorescence proteins were placed at the C-terminal end of claudins 1–6. For CFP and YFP fused to the N-terminus of claudin-5 molecules (Blasig et al., 2006), FRET detection is more complicated. The reason still needs to be elucidated. The C-termini of claudins may associate with PDZ domains of scaffolding proteins of the tight junctions, such as ZO-1 (Zhang et al., 2006), which is assumed to play a role in the proper organization of the tight junctions. This binding function is impeded by the fusion of fluorescence proteins to claudins at their C-terminal end. Hence, heterophilic claudin–ZO-1 interaction cannot be analyzed. On the other hand, the C-terminal position excludes the influence of ZO-1 and allows to study the claudin–claudin interaction independent of proteins containing PDZ domains. FRET analysis of heterologous *cis*-interactions between different members of the claudin protein family or between claudins and other tight junction proteins have recently been published (Harris et al., 2010).

FRET can be detected in different modes. One possibility is described in Fig. 3, although this gives highly variable values and interference between the CFP and YFP detection channels. The latter means that CFP emission is measured in the YFP channel and YFP emission in the CFP channel of the detector. An alternative to circumvent this cross talk is based on the application of acceptor photobleaching, where only the CFP emission intensity is considered. For FRET acceptor photobleaching (Blasig et al., 2006), CFP and YFP are excited at 458 and 514 nm and detected from 463–495 nm and 527–634 nm, respectively. Photobleaching of YFP in the area of the tight junctions can be achieved by a sufficient number of pulses of the 514 nm line of an argon ion laser at high emission power. FRET efficiency (E_F) is calculated as $E_F = (I_A - I_B) \times 100/I_A$, where I_B and I_A refer to the CFP intensity before and after photobleaching. In each experiment, the pair claudin-CFP and claudin-YFP was the internal standard. The relative E_F was calculated as the E_F of a distinct claudin-CFP/YFP-protein (claudin mutant or nonbinding membrane protein) pair divided by E_F of claudin-CFP/claudin-YFP. For quantitation, the laser and detector settings were kept constant, and only cells with CFP and YFP intensities similar to the intensities of the internal standard (claudin-CFP/claudin-YFP) were considered. Using this quantitative mode, 20–40 measurements were found sufficient to demonstrate statistically significant differences.

V. DETERMINATION OF *TRANS*-INTERACTIONS BETWEEN OPPOSING CELL MEMBRANES

To demonstrate *trans*-interactions between claudins from one cell membrane to the cell membrane of an opposing cell, a cell scan assay has been developed (Fig. 6). This allows to quantify the enrichment of a claudin at contacts between two claudin-expressing cells. The plasma membrane of living HEK-293 cells, transfected with a claudin construct carrying a fluorescence tag, is visualized by trypan blue. The amount of the claudin in the plasma membrane is quantified by confocal microscopy to determine the fluorescence intensity that colocalized with the trypan blue fluorescence peaks in intensity profiles. The intensity ratio is determined as half the fluorescence intensity of the claudin at contacts between two claudin-expressing cells divided by the intensity of the claudin fluorescence at contacts between the claudin-expressing- and nonexpressing cells. As two cells contribute to the fluorescence in a tight junction, the resulting value has to be divided by 2. For claudin-5, the enrichment factor was determined to be approximately 3 (Piontek et al., 2008). Other claudins seem to exhibit lower enrichment. However, systematic studies have not been published yet.

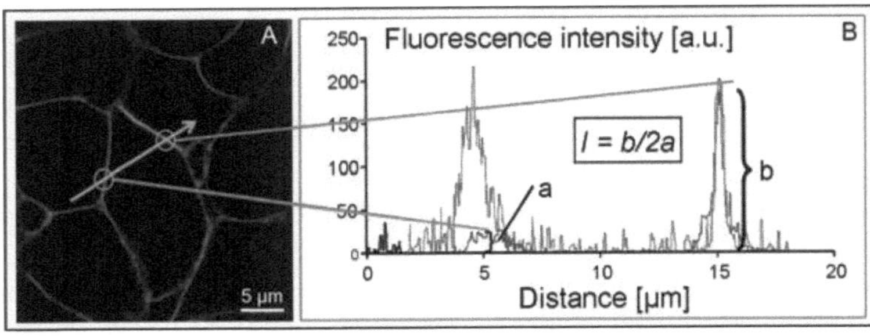

FIGURE 6 Detection of the *trans*-interaction between claudins localized in the tight junction formed by the plasma membranes of two opposing cells, using a live-cell scan. (A) Confluent HEK-293 cells after transfection with claudin-5 C-terminally fused to yellow-fluorescent protein (green fluorescence). The cell membranes are visualized by trypan blue (red fluorescence). Note that the claudin is highly enriched in the membrane contact area between two transfected cells only, seen in the center of the picture. (B) Live-cell scan through a claudin-transfected cell from a cell membrane adhering to a nontransfected cell (a, left) to the cell membrane contacting a transfected cell (b, right). In a, only a small amount of the claudin can be traced, whereas in b, much higher intensity is registered, demonstrating enrichment of claudin molecules in the tight junction from the two transfected cells and, hence, *trans*-interaction. The inset gives the formula to quantify the intensity of the *trans*-interaction. (See Color Insert.)

An important aspect of the investigations of claudin–claudin associations is their verification by performing qualified control experiments. If a possible interaction is detected, this verification can be achieved by demonstrating the loss of interaction, as a result of a specific manipulation of the claudin, such as the exchange of amino acids involved in the respective association. An example is given for the second extracellular loop of claudin-5. The replacement of an aromatic amino acid within this loop completely prevented the enrichment—resulting in intensity ratios of about 1 (Piontek et al., 2008). Another possibility is the application of agents specifically associating with the binding area of the respective claudin molecules. *Clostridium perfringens* enterotoxin, the ligand of the second extracellular loop of claudin-3 or -4 (and other claudins), can serve as a suitable means to reduce the *trans*-interaction ratio (Winkler et al., 2009).

References

Ahrens, T., Pertz, O., Haussinger, D., Fauser, C., Schulthess, T., & Engel, J. (2002). Analysis of heterophilic and homophilic interactions of cadherins using the c-Jun/c-Fos dimerization domains. *Journal of Biological Chemistry, 277*, 19455–19460.

Alexandre, M. D., Jeansonne, B. G., Renegar, R. H., Tatum, R., & Chen, Y. H. (2007). The first extracellular domain of claudin-7 affects paracellular Cl^- permeability. *Biochemical and Biophysical Research Communications, 357*, 87–91.

Amasheh, S., Meiri, N., Gitter, A. H., Schoneberg, T., Mankertz, J., Schulzke, J. D., et al. (2002). Claudin-2 expression induces cation-selective channels in tight junctions of epithelial cells. *Journal of Cell Science, 115*, 4969–4976.

Angelow, S., El Husseini, R., Kanzawa, S. A., & Yu, A. S. (2007). Renal localization and function of the tight junction protein, claudin-19. *American Journal of Physiology. Renal Physiology, 293*(1), F166–F177.

Arndt, S., Seebach, J., Psathaki, K., Galla, H. J., & Wegener, J. (2004). Bioelectrical impedance assay to monitor changes in cell shape during apoptosis. *Biosensors and Bioelectronics, 19*(6), 583–594.

Barker, G., & Simmons, N. L. (1981). Identification of two strains of cultured canine renal epithelial cells (MDCK cells) which display entirely different physiological properties. *Quarterly Journal of Experimental Physiology, 66*(1), 61–72.

Blasig, I. E., Winkler, L., Lassowski, B., Mueller, S. L., Zuleger, N., Krause, E., et al. (2006). On the self-association potential of transmembrane tight junction proteins. *Cellular and Molecular Life Sciences, 63*, 505–514.

Cereijido, M., Robbins, E. S., Dolan, W. J., Rotunno, C. A., & Sabatini, D. D. (1978). Polarized monolayers formed by epithelial cells on a permeable and translucent support. *Journal of Cell Biology, 77*(3), 853–880.

Chang, I. F. (2006). Mass spectrometry-based proteomic analysis of the epitope-tag affinity purified protein complexes in eukaryotes. *Proteomics, 6*, 6158–6166.

Colegio, O. R., Van Itallie, C. M., Rahner, C., & Anderson, J. M. (2003). Claudin extracellular domains determine paracellular charge selectivity and resistance but not tight junction fibril architecture. *American Journal of Physiology. Cell Physiology, 284*, C1346–C1354.

Coyne, C. B., Gambling, T. M., Boucher, R. C., Carson, J. L., & Johnson, L. G. (2003). Role of claudin interactions in airway tight junctional permeability. *American Journal of Physiology. Lung Cellular and Molecular Physiology, 285*, L1166–L1178.

Cross, M., Nguyen, T., Bogdanoska, V., Reynolds, E., & Hamilton, J. A. (2005). A proteomics strategy for the enrichment of receptor-associated complexes. *Proteomics, 5*, 4754–4763.

Fremont, J. J., Wang, R. W., & King, C. D. (2005). Coimmunoprecipitation of UDP-glucuronosyltransferase isoforms and cytochrome P450 3A4. *Molecular Pharmacology, 67*, 260–262.

Furuse, M., Furuse, K., Sasaki, H., & Tsukita, S. (2001). Conversion of zonulae occludentes from tight to leaky strand type by introducing claudin-2 into Madin–Darby canine kidney I cells. *Journal of Cell Biology, 153*, 263–272.

Furuse, M., Sasaki, H., Fujimoto, K., & Tsukita, S. (1998). A single gene product, claudin-1 or -2, reconstitutes tight junction strands and recruits occludin in fibroblasts. *Journal of Cell Biology, 143*(2), 391–401.

Furuse, M., Sasaki, H., & Tsukita, S. (1999). Manner of interaction of heterogeneous claudin species within and between tight junction strands. *Journal of Cell Biology, 147*, 891–903.

Gaillard, P. J., Voorwinden, L. H., Nielsen, J. L., Ivanov, A., Atsumi, R., Engman, H., et al. (2001). Establishment and functional characterization of an in vitro model of the blood-brain barrier, comprising a co-culture of brain capillary endothelial cells and astrocytes. *European Journal of Pharmaceutical Sciences, 12*(3), 215–222.

Giaever, I., & Keese, C. R. (1991). Micromotion of mammalian cells measured electrically. *Proceedings of the National Academy of Sciences of the United States of America, 88*(17), 7896–7900.

Gingras, A. C., Gstaiger, M., Raught, B., & Aebersold, R. (2007). Analysis of protein complexes using mass spectrometry. *Nature Reviews Molecular Cell Biology, 8*, 645–654.

Gloeckner, C. J., Boldt, K., Schumacher, A., Roepman, R., & Ueffing, M. (2007). A novel tandem affinity purification strategy for the efficient isolation and characterisation of native protein complexes. *Proteomics, 7*, 4228–4234.

Harris, H. J., Davis, C., Mullins, J. G., Hu, K., Goodall, M., Farquhar, M. J. et al., (2010). Claudin association with CD81 defines hepatitis C virus entry. *Journal of Biological Chemistry* 2010 Apr. 7, doi: 10.1074/jbc.M110.104836.

Hou, J., Paul, D. L., & Goodenough, D. A. (2005). Paracellin-1 and the modulation of ion selectivity of tight junctions. *Journal of Cell Science, 118*, 5109–5118.

Jovov, B., Van Itallie, C. M., Shaheen, N. J., Carson, J. L., Gambling, T. M., Anderson, J. M., et al. (2007). Claudin-18: A dominant tight junction protein in Barrett's esophagus and likely contributor to its acid resistance. *American Journal of Physiology. Gastrointestinal and Liver Physiology, 293*(6), G1106–G1113.

Kovalenko, O. V., Yang, X. W. H., & Hemler, M. E. (2007). A novel cysteine cross-linking method reveals a direct association between claudin-1 and tetraspanin CD9. *Molecular and Cellular Proteomics, 6*, 1855–1867.

Krause, G., Winkler, L., Mueller, S. L., Haseloff, R. F., Piontek, J., & Blasig, I. E. (2008). Structure and function of claudins. *Biochimica et Biophysica Acta—Biomembranes, 1778*, 631–645.

Lo, C. M., Keese, C. R., & Giaever, I. (1999). Cell-substrate contact: Another factor may influence transepithelial electrical resistance of cell layers cultured on permeable filters. *Experimental Cell Research, 250*(2), 576–580.

Lohrberg, D., Krause, E., Schümann, M., Piontek, J., Winkler, L., Blasig, I. E., et al. (2009). A strategy for enrichment of claudins based on their affinity to *Clostridium perfringens* enterotoxin. *BMC Molecular Biology, 10*, 61.

McCoy, M. H., & Wang, E. (2005). Use of electric cell-substrate impedance sensing as a tool for quantifying cytopathic effect in influenza A virus infected MDCK cells in real-time. *Journal of Virological Methods, 130*(1–2), 157–161.

Mitic, L. L., Unger, V. M., & Anderson, J. M. (2003). Expression, solubilization, and biochemical characterization of the tight junction transmembrane protein claudin-4. *Protein Science, 12*, 218–227.

Miyamoto, T., Morita, K., Takemoto, D., Takeuchi, K., Kitano, Y., Miyakawa, T., et al. (2005). Tight junctions in Schwann cells of peripheral myelinated axons: A lesson from claudin-19-deficient mice. *Journal of Cell Biology, 169*(3), 527–538.

Morita, K., Sasaki, H., Furuse, M., & Tsukita, S. (1999). Endothelial claudin: Claudin-5/TMVCF constitutes tight junction strands in endothelial cells. *Journal of Cell Biology, 147*(1), 185–194.

Morita, K., Furuse, M., Fujimoto, K., & Tsukita, S. (1999). Claudin multigene family encoding four-transmembrane domain protein components of tight junction strands. *Proceedings of the National Academy of Sciences of the United States of America, 96*, 511–516.

Morita, K., Sasaki, H., Fujimoto, K., Furuse, M., & Tsukita, S. (1999). Claudin-11/OSP-based tight junctions of myelin sheaths in brain and Sertoli cells in testis. *Journal of Cell Biology, 145*, 579–588.

Morita, K., Furuse, M., Yoshida, Y., Itoh, M., Sasaki, H., Tsukita, S., et al. (2002). Molecular architecture of tight junctions of periderm differs from that of the maculae occludentes of epidermis. *Journal of Investigative Dermatology, 118*, 1073–1079.

Nguyen, T. N., & Goodrich, J. A. (2006). Protein-protein interaction assays: Eliminating false positive interactions. *Nature Methods, 3*, 135–139.

Nunes, F. D., Lopez, L. N., Lin, H. W., Davies, C., Azevedo, R. B., Gow, A., et al. (2006). Distinct subdomain organization and molecular composition of a tight junction with adherens junction features. *Journal of Cell Science, 119*(Pt 23), 4819–4827.

Ong, S. E., Blagoev, B., Kratchmarova, I., Kristensen, D. B., Steen, H., Pandey, A., et al. (2002). Stable isotope labeling by amino acids in cell culture, SILAC, as a simple and accurate approach to expression proteomics. *Molecular and Cellular Proteomics, 1*, 376–386.

Peppi, M., & Ghabriel, M. N. (2004). Tissue-specific expression of the tight junction proteins claudins and occludin in the rat salivary glands. *Journal of Anatomy, 205*, 257–266.

Perkins, G., Goodenough, D., & Sosinsky, G. (1997). Three-dimensional structure of the gap junction connexon. *Biophysical Journal, 72*, 533–544.

Pfeiffer, F., Kumar, V., Butz, S., Vestweber, D., Imhof, B. A., Stein, J. V., et al. (2008). Distinct molecular composition of blood and lymphatic vascular endothelial cell junctions establishes specific functional barriers within the peripheral lymph node. *European Journal of Immunology, 38*, 2142–2155.

Piehl, C., Piontek, J., Cording, J., Wolburg, H., & Blasig, I. E. (2010). Participation of the second extracellular loop of claudin-5 in paracellular tightening against ions, small and large molecules. *Cellular and Molecular Life Sciences*, Mar 24. doi: 10.1007/s00018-010-0332-8.

Piontek, J., Winkler, L., Wolburg, H., Mueller, S. L., Zuleger, N., Piehl, C., et al. (2008). Formation of tight junction: Determinants of homophilic interaction between classic claudins. *FASEB Journal, 22*, 146–158.

Rigaut, G., Shevchenko, A., Rutz, B., Wilm, M., Mann, M., & Seraphin, B. (1999). A generic protein purification method for protein complex characterization and proteome exploration. *Nature Biotechnology, 17*, 1030–1032.

Saez, J. C., Berthoud, V. M., Branes, M. C., Martinez, A. D., & Beyer, E. C. (2003). Plasma membrane channels formed by connexins: Their regulation and functions. *Physiological Reviews, 83*, 1359–1400.

Schmitz, Y., & Wolburg, H. (1991). Gap junction morphology of retinal horizontal cells is sensitive to pH alterations in vitro. *Cell and Tissue Research, 263*, 303–310.

Sekar, R. B., & Periasamy, A. (2003). Fluorescence resonance energy transfer (FRET) microscopy imaging of live cell protein localizations. *Journal of Cell Biology, 160*, 629–633.

Sheikov, N., McDannold, N., Sharma, S., & Hynynen, K. (2008). Effect of focused ultrasound applied with an ultrasound contrast agent on the tight junctional integrity of the brain microvascular endothelium. *Ultrasound in Medicine and Biology, 34*, 1093–1104.

Sonoda, N., Furuse, M., Sasaki, H., Yonemura, S., Katahira, J., Horiguchi, Y., et al. (1999). *Clostridium perfringens* enterotoxin fragment removes specific claudins from tight junction strands: Evidence for direct involvement of claudins in tight junction barrier. *Journal of Cell Biology, 147*(1), 195–204.

Staehelin, L. A. (1973). Further observations on the fine structure of freeze-cleaved tight junctions. *Journal of Cell Science, 13*(3), 763–786.

Staehelin, L. A. (1974). Structure and function of intercellular junctions. *International Review of Cytology, 39*, 191–283.

Tang, V. W. (2006). Proteomic and bioinformatic analysis of epithelial tight junction reveals an unexpected cluster of synaptic molecules. *Biology Direct, 1*, 37.

Tiruppathi, C., Malik, A. B., Del Vecchio, P. J., Keese, C. R., & Giaever, I. (1992). Electrical method for detection of endothelial cell shape change in real time: Assessment of endothelial barrier function. *Proceedings of the National Academy of Sciences of the United States of America, 89*(17), 7919–7923.

Tsukita, S., & Furuse, M. (2000). The structure and function of claudins, cell adhesion molecules at tight junctions. *Annals of the New York Academy of Sciences, 915*, 129–135.

Turksen, K., & Troy, T. C. (2001). Claudin-6: A novel tight junction molecule is developmentally regulated in mouse embryonic epithelium. *Developmental Dynamics, 222*(2), 292–300.

Van Itallie, C. M., Fanning, A. S., & Anderson, J. M. (2003). Reversal of charge selectivity in cation or anion-selective epithelial lines by expression of different claudins. *American Journal of Physiology. Renal Physiology, 285*(6), F1078–F1084.

Van Itallie, C. M., Rogan, S., Yu, A., Vidal, L. S., Holmes, J., & Anderson, J. M. (2006). Two splice variants of claudin-10 in the kidney create paracellular pores with different ion selectivities. *American Journal of Physiology. Renal Physiology, 291*, F1288–F1299.

Wattenhofer, M., Reymond, A., Falciola, V., Charollais, A., Caille, D., Borel, C., et al. (2005). Different mechanisms preclude mutant CLDN14 proteins from forming tight junctions in vitro. *Human Mutation, 25*(6), 543–549.

Wen, H., Watry, D. D., Marcondes, M. C., & Fox, H. S. (2004). Selective decrease in paracellular conductance of tight junctions: Role of the first extracellular domain of claudin-5. *Molecular and Cellular Biology, 24*(19), 8408–8417.

Winkler, L., Gehring, C., Wenzel, A., Mueller, S. L., Piehl, C., Krause, G., et al. (2009). Molecular determinants of the interaction between *Clostridium perfringens* enterotoxin and claudins. *Journal of Biological Chemistry, 284*, 18863–18872.

Wuller, S., Wiesner, B., Loffler, A., Furkert, J., Krause, G., Hermosilla, R., et al. (2004). Pharmacochaperones post-translationally enhance cell surface expression by increasing conformational stability of wild-type and mutant vasopressin V2 receptors. *Journal of Biological Chemistry, 279*, 47254–47263.

Yu, A. S. L., Kanzawa, S. A., Usorov, A., Leeuwen, I S L V, & Peters, D. J. M. (2008). Tight junction composition is altered in the epithelium of polycystic kidneys. *Journal of Pathology, 216*, 120–128.

Zhang, Y. N., Yeh, S., Appleton, B. A., Held, H. A., Kausalya, P. J., Phua, D. C. Y., et al. (2006). Convergent and divergent ligand specificity among PDZ domains of the LAP and zonula occludens (ZO) families. *Journal of Biological Chemistry, 281*, 22299–22311.

CHAPTER 6

Regulation of Claudins by Posttranslational Modifications and Cell-Signaling Cascades

Lorenza González-Mariscal, Erika Garay, and Miguel Quirós
Department of Physiology, Biophysics and Neuroscience, Center for Research and Advanced Studies (Cinvestav), México D.F., México

Current Topics in Membranes, Volume 65 1063-5823/10 $35.00

DOI: 10.1016/S1063-5823(10)65006-5

I. OVERVIEW

The tight junction (TJ) is a dynamic structure that changes according to the environment and physiological or pathological conditions of the organism. Claudins are the main constituents of TJ filaments, and their function is regulated by several posttranslational modifications including phosphorylation, palmitoylation, and signaling through cytokines, prostaglandins, and hormones. Here, we describe the putative phosphorylation sites present in the claudin family of proteins and analyze those phosphorylations that trigger a specific cellular response. We also describe how cytokines, transforming growth factor β, prostaglandin E_2, and progesterone promote the development of leakier TJs in epithelia by reducing the expression of claudins that act as cation barriers and increasing the appearance of those that function as cation pores. In contrast, epithelial growth factor (EGF), the follicle-stimulating hormones, testosterone and aldosterone, induce the generation of tighter TJs by increasing the expression of claudins that block the paracellular passage of cations.

II. INTRODUCTION

Claudins are transmembrane proteins that form a network of strands that surrounds epithelial cells at the tight junction (TJ) (Furuse, Fujita, Hiiragi, Fujimoto, & Tsukita, 1998). Claudins constitute aqueous pores that regulate the transit of ions and molecules through the paracellular pathway in a size and charge sensitive manner.

In this chapter, we will describe the phosphorylation sites present in claudins and how phosphorylation, palmitoylation, and signaling through cytokines, prostaglandins, and hormones regulate claudin function.

III. CLAUDIN PHOSPHORYLATION

Phylogenetic tree analysis of full-length sequences of claudins has revealed a very great sequence homology between claudins 1–10, 14, 15, 17, and 19. These claudins have hence been defined as classic claudins, while the remaining 11–13, 16, 18, 20–24 are considered as nonclassic (Krause et al., 2008). Using the softwares Group-based Prediction System 2.0 (cut-off of high threshold), NetPhosK 1.0 (0.5 threshold), and PredPhospho version 2 (97% specificity), we have identified the putative phosphorylation sites for protein kinase C (PKC), protein kinase A (PKA), mitogen-activated protein kinase (MAPK), serine/threonine kinase with no K in the kinase domain (WNK), myosin light chain kinase (MLCK), sarcoma tyrosine kinase (Src), Rho kinase (RhoK), and ephrin (Eph) receptor family, present in the intracellular carboxyl segment of classic and nonclassic claudins (Table I). The number of putative phosphorylation sites varies in accordance with the length of the carboxyl tail of each claudin, thus mouse claudin-13 that exhibits a tail with 26 residues has three phosphorylation sites, whereas human claudin-2 with a tail of 45 amino acids has 10 sites. A search of common phosphorylation sites among claudins reveals that the tyrosine present at position -1 of all claudins, except 11, 12, 13, 16, 22, and 23, is a conserved putative Eph phosphorylation site. Besides, from this observation the analysis of common phosphorylation sites could only be made in the classic claudin group as sequence variation of the carboxyl tail between the nonclassic claudins is to strong to allow a correct alignment. Figure 1 shows that only four putative phosphorylation sites are conserved among certain classic claudins: (1) the Ser/Thr located at position -7 to -9 of claudins 1–9, 14–15, 17, and 19; (2) the Ser/Thr located at position -24 to -26 of claudins 1–3, 6, 8, and 19; (3) Ser206 of human claudins 1 and 7; and (4) Tyr 199 of human claudin-1 which is conserved as Tyr 201 of human claudin-7.

A. Phosphorylation of Claudins that Promotes their Assembly at Tight Junctions

1. Protein Kinase A

a. PKA induces claudin-5 phosphorylation and expression in brain endothelia. In brain endothelial cells, it has long been known that cyclic AMP (cAMP) promotes TJ sealing and complexity (Adamson, Liu, Fry, Rubin, & Curry, 1998; Wolburg et al., 1994). Yet, only recently it was found that cAMP induces the phosphorylation of claudin-5 at serine residues and increases claudin-5 immunoreactivity along cell boundaries in blood–brain barrier (BBB) endothelial cells via a PKA-dependent pathway (Ishizaki et al., 2003). In this respect, it is interesting to observe in Table I that Ser286 of human claudin-5 constitutes a putative PKA phosphorylation site conserved in mouse and rat sequences.

b. Phosphorylation of claudin-16 at Ser217 by PKA protects against internalization and degradation. In MDCK cells, transfected claudin-16 is phosphorylated at Ser217 by PKA (Ikari et al., 2008). Inhibitors of PKA and adenylate cyclase decrease transepithelial Mg^{2+} transport and TER. This effect is explained by observing that while phosphorylated claudin-16 localizes at TJs, the dephosphorylated protein or the S217A mutant is

TABLE I

Phosphorylation Sites Present in the Carboxyl Segment of Claudins

Cln	Enzyme	Putative site (Sp)	Species with concensus	Phosphorylation demostrated by	P-effect
1	PKC	S185 (H)	R and B		
	WNK	T190 (H)	M, R, and B		
	PKA	T190 (H)	B	French et al. (2009)	↑Cytoplasmic and ↓nuclear
	cPKC	T190 (M)		French et al. (2009)	↑Cytoplasmic and ↓nuclear
	Eph	Y193 (H)	M, R, and B		
	p38	T195 (H)	M, R, and B	French et al. (2009) by PKA/PKC	↑Cytoplasmic and ↓nuclear
	EphB2	Y199 (H)	M, R, and B		
	Erk5/JNK/p38	T203 (M)	R	Fujibe et al. (2004)	TJ assembly
	NP	S205 (M)	H, R, and B	Villén et al. (2007)	
	PKC	S206 (H)	M, R (T), and B	Simard, Di Pietro, Young, Plaza, and Ryan (2006) in chicken	Direction of heart looping
				Villén et al. (2007)	ND
	Eph	Y210 (H)	M, R, and B		
2	PKC	S186 (H)			
	MAPK	S186 (M)	R and B		
	cPKC/nPKC	S187 (H)			
	MLCK/PKA	S192 (H)			

	EphB1	Y194 (H)	M, R, and B		
	cSrc	Y195 (H)	B		
	Erk5	T206 (H)	M, R, and B		
	cPKC/nPKC/p38/ Erk1/2/JNK	S208 (H)	M, R, and B		
	Erk5/PKC	S211 (M)	R		
	nPKC	S219 (M)	R		
	PKA	S223 (H)	M, R, and B		
	PKC	T227 (H)	M, R, and B		
	Eph	Y229 (H)	M, R and B		
3	RhoK	T192 (H)		D'Souza, Agarwal and Morin (2005) by PKA	TJ disruption
	PKC	T194 (H)	M, R, and B		
	EphB2	Y198 (H)	B		
	PKC	S199 (H)	S198 (M, R, B)		
	RhoK	T203 (B)			
	WNK	T204 (H)	T203 (M, R, B)		
	WNK/PKA	S209 (H)			
	PKC/WNK	T209 (B)			
	PKC	T212 (H)	T211 (M, R, B)		
	Eph	Y219 (H)	M, R, and B		
4	PKC	T189 (H)		D'Souza et al. (2007)	TJ disruption
	cSRC	Y194 (R)			
	nPKC	S194 (H)	S195 (R), B	D'Souza et al. (2007)	TJ disruption
	PKC	S203 (M)	R		
	PKA	S206 (H)	S207 (M, R)		

(*continued*)

TABLE I *(continued)*

Cln	Enzyme	Putative site (Sp)	Species with concensus	Phosphorylation demostrated by	P-effect
	ERK5	S206 (B)			
	Eph	Y208 (H)	Y209 (M, R), B	Tanaka, Kamata, and Sakai (2005)	TJ disruption
5	PKA	S286 (H)	S201 (M, R)		
	PKA	S203 (B)			
	WNK/PKA	T206 (R)			
	PKA/Rhok	T292 (H)	T207 (M, R, B)	Soma et al. (2004), Yamamoto et al. (2008)	TJ disruption
	PKC/WNK	T294 (H)	T209 (B)		
	Eph	Y302 (H)	Y217 (M, R, B)		
6	PKC	S187 (B)			
	PKC	T190 (M)	R and B		
	cPKC	S194 (H)	B		
	MLCK	S194 (B)			
	PKA	S201 (H)	B		
	PKC	S201 (B)			
	p38/PKA/PKC	S202 (R)			
	PKC	S203 (H)	M		
	PKA	S212 (H)	S211 (M, R) S213(B)		
	Eph	Y214 (H)	Y213 (M, R), B		
	Eph	Y219 (H)	Y218 (M, R) Y220 (B)		

7	PKC	S200 (H)	M, R, and B		
	Eph	Y201 (H)	M, R, and B		
	PKC	S204 (H)	M and R		
	PKC	S206 (H)	M, R, and B	Tatum et al. (2007) for WNK in pig	TJ assembly
	Eph	Y210 (H)	M, R, and B		
8	PKC	T189 (M)	R		
	PKC	S193 (B)			
	cPKC/nPKC	S194 (H)	M, R, and B		
	PKC	S198 (M)	R		
	PKA	S198 (B)			
	nPKC	S201 (H)	M, R, and B		
	Rhok	T204 (H)	M, R, and B		
	PKC/Rhok	T205 (H)	M, R, and B		
	nPKC	S208 (H)	M, R, and B		
	PKC	T301 (H)			
	ERK1/2/JNK/p38	S215 (H)	M, R, and B		
	nPKC	S217 (H)	M, R, and B		
	Eph	Y219 (M)	R and B		
	PKC	S220 (B)			
	Eph	Y224 (H)	M, R, and B		
9	cPKC	S188 (M)	R		
	WNK	S206 (H)	M, R, and B		
	WNK	S209 (H)	M, R, and B		
	Eph	Y216 (H)	M, R, and B		

(*continued*)

TABLE I *(continued)*

Cln	Enzyme	Putative site (Sp)	Species with concensus	Phosphorylation demostrated by	P-effect
10	MAPK	T187 (H)	M and T244 (R)		
	Eph	Y194 (M)	Y251 (R)		
	PKC	S200 (H)	S202 (M), S259 (R), S204 (B)		
	p38	S200 (H)	S204 (B)		
	PKC	S201 (H)	S203 (M), S260 (R), S205 (B)		
	p38	S203 (M)	S260 (R)		
	PKC	T205 (M)	T207 (B)		
	PKC	T216 (M)	T273 (R)		
	p38/JNK	T218 (B)			
	PKC	T214 (H)	T217 (M), T274 (R)		
	Eph	Y225 (H)	Y228 (M),Y285 (R), Y230 (B)		
11	PKC	T193 (H)	M, R, and B		
	PKC	S194 (M)	R and B		
	PKC	S196 (H)			
	WNK	S197 (H)	M, R, and B	Munton et al. (2007)	
	ERK1/2/JNK/p38	S198 (H)	M, R, and B	Munton et al. (2007)	
	PKC	T200 (H)	M, R, and B		
	ERK5	S204 (H)	M, R, and B		
12	PKC	T198 (H)			
	PKA	S201 (H)	M and R		
	MLCK	S203 (M)	R		
	p38	S204 (H)	M, R, and B		

	JNK	S204 (M)	R	
	EphB2	Y211 (H)	R and B	
	EphA3/B1/B2	Y220 (H)	M, R, and B	
	cPKC	S221 (H)		
	PKC	S225 (H)	M, R, and B	
		S228 (M)	H, R, and B	Villén et al. (2007)
	PKA	S231 (H)	M, R, and B	
13	JNK/p38	T196 (M)		
	cPKC/nPKC	S198 (M)		
	PKA	S202 (M)		
14	cSRC	Y191 (R)		
	WNK	T201 (H)		
	MLCK/PKA/PKC	T202 (H)		
	PKC	T207 (H)	M and R	
	Eph	Y211 (H)	M, R, Y209 (B)	
	EphA3/EphB1	Y217 (H)	M, R, Y215 (B)	
	PKC/PKA	S224 (H)	M, R, Y222 (B)	Villén et al. (2007)
	nPKC	S227 (B)		
		S227 (M)		Villén et al. (2007)
		S231 (M)		Villén et al. (2007)
	Eph	Y238 (H)	M, R, Y236 (B)	
15	cPKC	S181 (M)	R	
	PKC	S186 (M)		
	cPKC	S187 (H)	M and R	
	cPKC/nPKC	S194 (H)		
	nPKC	T206 (R)		

(*continued*)

TABLE I (*continued*)

Cln	Enzyme	Putative site (Sp)	Species with concensus	Phosphorylation demostrated by	P-effect
	WNK	T210 (M)	R		
	cPKC	S211 (H)			
	PKA	S211 (M)	R		
	PKC	S218 (H)	S217 (R)		
	Eph	Y227 (H)	Y226 (M and R)		
16	nPKC	S272 (H)	S221 (B)		
	PKC	T222 (B)			
	PKC	S227 (B)			
	PKC	T209 (M)	R and T228(B)		
	PKC	S232 (B)			
	PKC	S287 (H)	S217 (M and R)	Ikari, Matsumoto, Harada, Takagi, Hayashi, et al. (2006) by PKA	TJ assembly
	Eph	Y288 (H)	Y218 (M and R)		
	PKC	T293 (H)	T223 (M and R)		
	MLCK	T295 (H)	T225 (M, R), T244 (B)		
	Rhok	T295 (H)	T225 (M and R)		
	PKC	T244 (B)			
17	PKC	T207 (H)			
	nPKC	S209 (R)			
	cPKC/PKA/ RhoK	T213 (H)			
	PKC/PKA	T214 (H)	M and R		

	RhoK	T214 (M)	R	
	PKC	S217 (H)		
	nPKC	S219 (M)		
	WNK/PKA	S219 (M)	T219 (R)	
	RhoK	T219 (R)		
	PKA	S220 (H)	M, R	
	MLCK/nPKC/ RhoK	S220 (M)	R	
	PKC/RhoK	T221 (H)	R	
	cPKC	S222 (H)		
	Eph	Y223 (H)	M and R	
18	MAPK	T203 (M)	T200 (R)	
	c-SRC	Y206 (H)	B	
	PKC	S210 (H)	S213 (M), R, B	
	PKC	S217 (H)		
	WNK	S229 (B)		
	PKC	S233 (H)	S230 (M), R, B	
	PKC	T235 (H)		
	NP	T250 (M)	H, R, and B	Dai et al. (2007)
	PKC	S253 (H)	Astringencia	Dai et al. (2007) in mouse
	ERK2/JNK/p38	S253 (B)		
	Eph	Y254 (H)		
	cPKC	S256 (B)	Astringencia	Dai et al. (2007) in mouse
	Eph	Y260 (H)	Y263 (M), R, B	

(*continued*)

TABLE I *(continued)*

Cln	Enzyme	Putative site (Sp)	Species with concensus	Phosphorylation demostrated by	P-effect
19	cPKC	T185 (B)			
	PKA	S194 (H)	M and R		
	Erk5	S194 (H)			
	JNK/p38	S195 (H)	B		
	WNK	S195 (H)			
	WNK	S201 (M)	R		
	WNK	S204 (H)	B		
	PKC	S204 (M)	R		
	Eph	Y210 (H)	M, R, and B		
20	PKC	S202 (M)	R		
	PKA	S202 (M)			
	MLCK	T202 (R)			
	nPKC	T203 (M)	R		
	PKA	S203 (R)			
	PKC	T204 (R)			
	Eph	Y218 (M)	R		
21	nPKC	S189 (H)			
	PKC	S190 (H)			
	ERK5/p38	S190 (M)			
	WNK	S196 (H)	R		
	cSRC	Y199 (H)	R		
22	cPKC	S197 (H)			
	WNK/PKA	S203 (H)	M and R		

cPKC/nPKC/PKA	S204 (H)	M and R
PKC	T207 (H)	M and R
RhoK	T207 (R)	
EphA3	Y209 (M)	R
ERK1/2/ERK5/JNK/p38	T218 (M)	R
PKC	T240 (M)	T239 (R)
PKC/MLCK	S241 (M)	
EphA3	Y257 (H)	M, Y256 (R)
EphB2	Y257 (M)	Y256 (R)
MLCK	T258 (H)	M
PKA	S260 (H)	
WNK	T271 (M)	T270 (R)
nPKC	T273 (M)	S272 (R)
RhoK	T272 (R)	
nPKC	S274 (M)	S273 (R)
cPKC	S278 (M)	S277 (R)
cPKC/nPKC	S279 (M)	S278 (R)
WNK/RhoK	S278 (R)	
cPKC	S280 (M)	S279 (R)
cPKC/nPKC	S282 (M)	S281 (R)
PKC	T279 (H)	T283 (M), T282 (R)
RhoK	S279 (R)	
RhoK	S281 (R)	
RhoK	T282 (R)	

M, mouse; R, rat; H, human; B, bull; NP, not predicted; ND, not determined; cPKC, classical PKC; nPKC, novel PCK; aPKC, atypical PKC.

translocated to the lysosome. Therefore, phosphorylation of claudin-16 at Ser217 is proposed to inhibit ubiquitylation and further degradation (Ikari, Matsumoto, Harada, Takagi, Hayashi, et al., 2006). Interestingly, Table I shows that this residue identified in the human claudin-16 sequence as Ser287 is conserved in mouse and rat and is identified *in silico* as a putative PKC phosphorylation site.

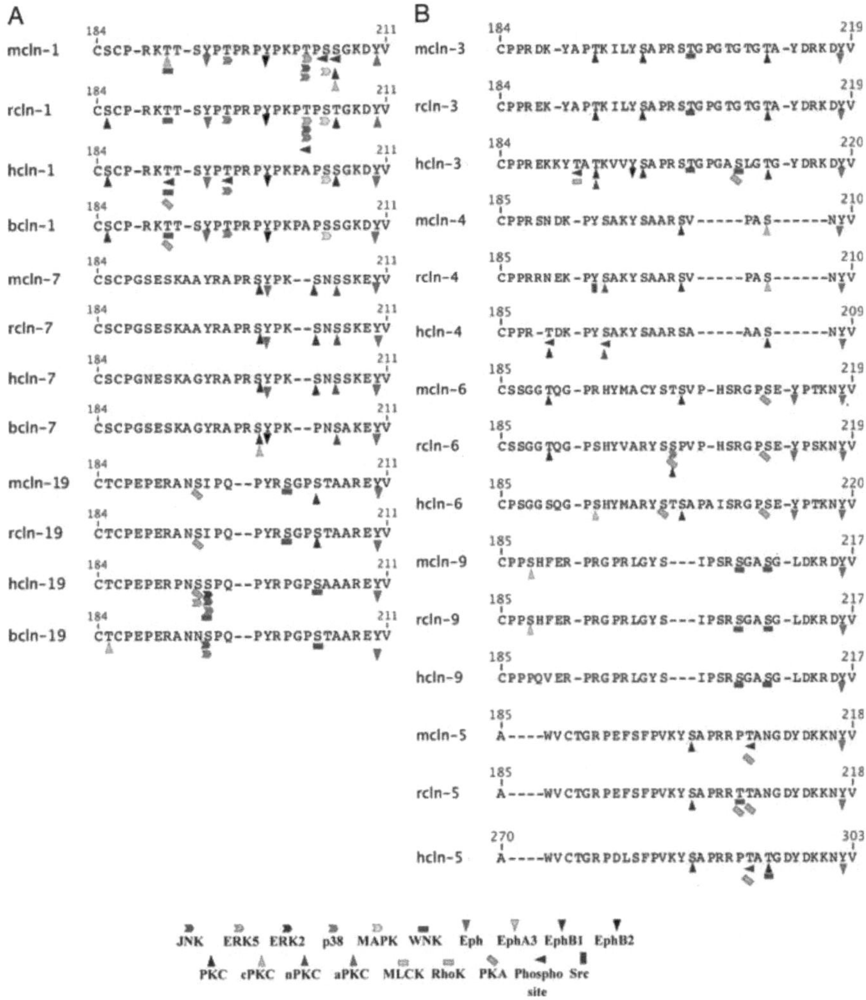

FIGURE 1 (Continued)

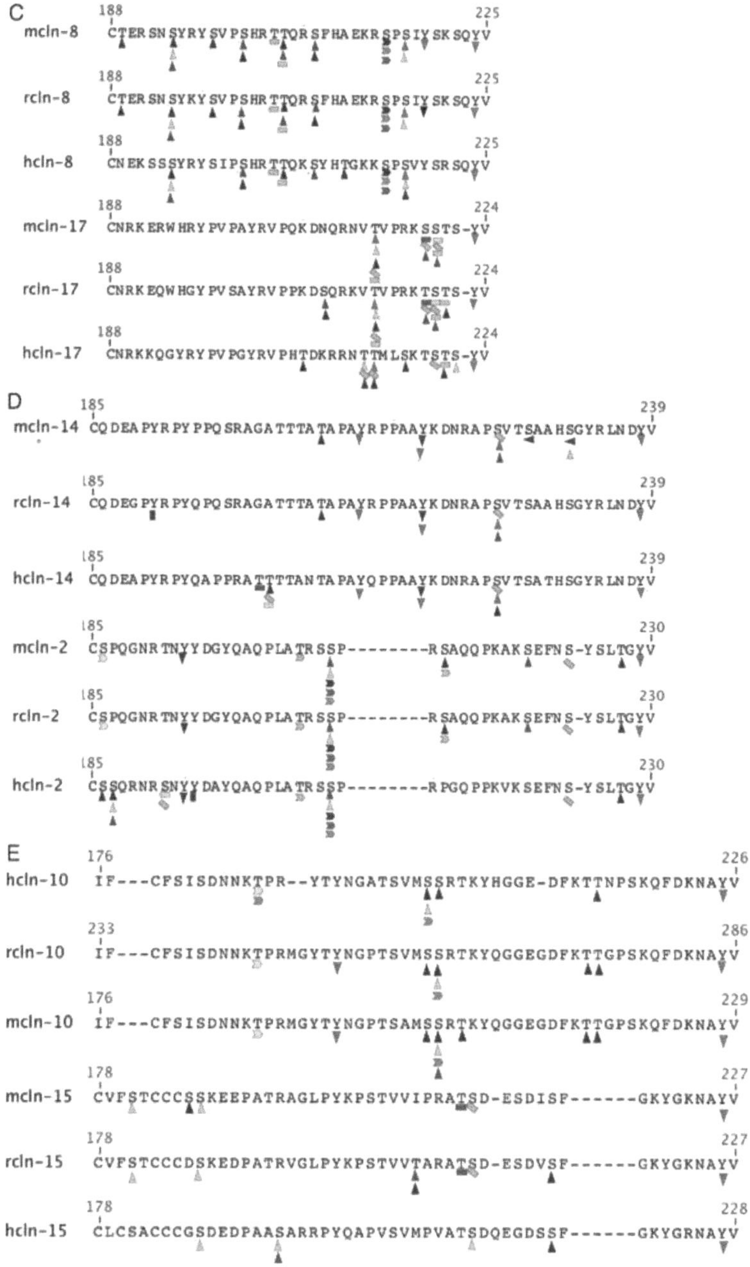

FIGURE 1 Putative phosphorylation sites of classic claudins 1–10, 14, 15, 17, and 19. The phosphorylation sites were predicted employing the Group-based Prediction System 2.0, NetPhosK 1.0, and PredPhospho version 2. Claudins were grouped according to their phylogenetic proximity (Krause et al., 2008) in the following manner: (A) claudins 1, 7, and 19; (B) claudins 3–6, and 9; (C) claudins 8 and 17; (D) claudins 8 and 14; and (E) claudins 10 and 15; m, mouse; r, rat; h, human; b, bull.

2. Protein Kinase C

a. Atypical PKC regulates tight junction formation through claudin-4 phosphorylation at Ser194. In the human epidermal keratinocyte cell line HaCat, TJs form upon treatment with the c-Jun N-terminal kinase (JNK) inhibitor SP600125. In this process, atypical PKC (aPKC) phosphorylates claudin-4 at serine 194. Overexpression of a mutant claudin-4 lacking this residue perturbs TJ formation, indicating that aPKC regulates TJ formation through the phosphorylation of claudin-4 (Aono & Hirai, 2008). It is noteworthy that serine 194 of claudin-4 is conserved as a putative PKC phosphorylation site in human, rat, and bull sequences (Table I).

b. Novel PKCε promotes the association of claudin-1 to the detergent insoluble fraction. In epithelial intestinal T84 cells, bryostatin, a selective activator of conventional PKCα, and novel PKCε, and δ, increases TER in a manner associated with a shift in the subcellular distribution of claudin-1 from a detergent soluble into a detergent insoluble fraction (Yoo et al., 2003), hence suggesting that PKC activity promotes claudin-1 incorporation into TJs. Novel PKCε appears to be the isozyme responsible. Table I shows that serines 185 and 206 of human claudin-1 are putative PKC phosphorylation sites conserved in rat and bull, or mouse, rat and bull sequences, respectively, and could therefore constitute the targets of nPKCε. Furthermore, Ser185 is maintained at position −24 to −26 of claudins 1–3, 6, 8, and 19 as a putative phosphorylation site for PKC, PKA, and MAPK (Fig. 1). Hence, these observations suggest that the phosphorylation of Ser185 might be critical for the integration of several claudins to intercellular junctions.

c. Novel PKCθ activity is required for claudins 1 and 4 expression and function in intestinal cells. In Caco-2 cells, novel PKCθ (nPKCθ) is associated with claudins 1 and 4. Diverse strategies that lead to nPKCθ underexpression reduce claudin-1 and -4 assembly and serine/threonine phosphorylation. Interestingly, nPKCθ overexpression disrupts claudin assembly. Hence, these results suggest that a stable level of nPKCθ activity is required for claudin-1 and -4 expression at the TJ and for the preservation of the intestinal epithelial barrier (Banan et al., 2005). Table I identifies Ser194 of human claudin-4 as a target of nPKC conserved in rat and bull; however, the phosphorylation of ovarian cancer cells on this residue depends on the activity of nPKCε and not on that of nPKCθ (D'Souza, Indig, & Morin, 2007).

d. Inhibition of phosphatase PP2A induces phosphorylation of claudin-1 and TJ formation in a process mediated by aPKC. Protein phosphatase 2A (PP2A) is a serine/threonine phosphatase that associates with the apical junctional complex and inhibits the activity and distribution of aPKC at the cell borders during calcium-induced junction biogenesis. Enhanced expression of PP2A induces dephosphorylation of TJ proteins, including claudin-1, whereas the inhibition of PP2A with okadaic acid promotes the phosphorylation and recruitment of claudin-1 and other TJ proteins to the cell borders (Nunbhakdi-Craig et al., 2002).

e. Activation of novel PKCδ and θ with TPA increases TJ barrier function and claudin-1 expression. In primary human nasal epithelial cells, treatment with 12-o-tetradecanoylphorbol-13-acetate (TPA) increases TER and upregulates TJ proteins occludin, ZO-1, ZO-2, and claudin-1. These changes are prevented by treatment with novel PKCδ and θ inhibitors. According to Table I, residues Ser185 and S206 of human claudin-1 could hypothetically constitute the phosphorylation targets of the aforementioned enzymes.

3. MAPK Phosphorylation of Claudin-1 at Thr203 Promotes Barrier Function in Lung Endothelia

Threonine 203 of claudin-1 is a putative MAPK phosphorylation site. The importance of this site for TJ barrier function in lung endothelial cells is highlighted by the observation of the T203A mutation decreases TER and increases paracellular permeability (Fujibe et al., 2004). Thr203 is conserved in mouse and rat and constitutes a putative phosphorylation site for Erk5, JNK, and p38. In addition, Thr203 localizes at position −8 of rat claudin-1, where Ser/Thr targets of various kinases are found in the classic human claudins 1–9, 14–15, 17, and 19.

B. Phosphorylation of Claudins that Modulates their Function as Tight Junction Pores

1. PKA Promotes Mg^{2+} Reabsorption by Claudin-16 in Renal Cells

In the body, Mg^{2+} content is regulated by reabsorption in renal epithelia, in particular, at the paracellular pathway of the thick ascending limb (TAL) of Henle (Quamme & de Rouffignac, 2000). Claudins are charge-selective pores that promote specific ion permeability, and claudin-16 found almost exclusively at the TAL is critical for the control of the paracellular transport of Mg^{2+} and Ca^{2+}. In renal MDCK cells expressing claudin-16, the inhibition of PKA reduces serine phosphorylation of claudin-16 and transepithelial Mg^{2+} transport, thus suggesting that serine phosphorylation is required for claudin-16 function as a Mg^{2+} selective pore (Ikari, Matsumoto, Harada, Takagi, Degawa, et al., 2006).

2. WNK1 and WNK4 Induce Phosphorylation of Claudins and Generate a Chloride Shunt in Kidney Cells

With no K, kinases 1 and 4 (WNK1 and WNK4) are Ser/Thr kinases involved in the development of an autosomal dominant disorder named pseudohypoaldosteronism type II (PHAII) or Gordon syndrome, characterized for hypertension, high potassium levels in serum (hyperkalemia), and metabolic acidosis due to reduced renal H^+ secretion (Wilson et al., 2001). The pathogenesis of PHAII is thought to involve an altered transcellular ion flux due to defective WNK kinases (Mayan et al., 2002), as well as an increased chloride paracellular permeability that explains increased NaCl reabsorption and decreased K^+ and H^+ secretion (Schambelan, Sebastian, & Rector, 1981; Take, Ikeda, Kurasawa, & Kurokawa, 1991).

Mutations in WNK1 gene causing PHAII are deletions in the first intron that increase WNK1 mRNA expression. Interestingly, when WNK1 is overexpressed in MDCK cells, chloride permeability augments significantly, together with claudin-4 phosphorylation at Thr residues. The mechanism leading to this phosphorylation remains unclear as claudin-4 and WNK1 do not coimmunoprecipitate and WNK1 localizes to the cytosol rather than TJs (Ohta et al., 2006).

Missense mutations on WNK4 gene are also responsible for PHAII. WNK4 localizes at TJs (Wilson et al., 2001), and when the disease-causing mutants are expressed in MDCK cells, they increase the paracellular chloride permeability and enhance the phosphorylation of claudins 1, 2, 3, 4, and 7 (Tatum et al., 2007; Yamauchi et al., 2004). *In silico* analysis of claudins identifies the following putative WNK phosphorylation sites: Thr190 in human claudin-1 and conserved in mouse, rat, and bull; Thr204 in human claudin-3 and conserved in mouse, rat, and bull; Ser209 in human claudin-2; and Thr209 in bull claudin-3.

Ser206 of claudin-7 has been identified as a WNK4 phosphorylation site through an *in vitro* kinase assay (Tatum et al., 2007), although our *in silico* analysis identifies this site as a PKC phosphorylation site conserved in human, mouse, rat, and bull (Table I).

Aldosterone increases the abundance of WNK4 in the kidney, and in MDCK cells, promotes the apical to basal transport of Cl^- and a transient phosphorylation of claudin-4 in Thr residues (Le Moellic et al., 2005).

C. Phosphorylation of Claudins that Inhibits their Assembly at Tight Junctions

1. PKA and ROCK Phosphorylation of Thr207 in Claudin-5 Produces Leaky TJs

In rat lung endothelial cells, the transfection of wild-type claudin-5 produces a paracellular barrier permeable to mannitol (182 Da) but not to the bigger molecule inulin (5 kDa). In contrast, the transfection of claudin-5 with

T207A mutation forms a barrier against both molecules, thus suggesting that the phosphorylation of Thr207 in claudin-5 by PKA produces a size-selective loosening in endothelial TJs (Soma et al., 2004).

ROCK, the serine/threonine-specific protein kinase activated by GTP-bound RhoA, also phosphorylates claudin-5 at Thr207. This phosphorylation diminishes barrier tightness and enhances monocyte migration across the BBB in mouse and human brains affected with human immunodeficiency virus-1 encephalitis (HIVE) (Yamamoto et al., 2008).

Thr207 of rat claudin-5 that corresponds to human Thr292 is a PKA and RhoK putative phosphorylation site conserved in all the analyzed species (Table I).

2. EphA2 Phosphorylation of Claudin-4 PDZ-Binding Motif Perturbs the Barrier Function

Ephrin receptors are frequently overexpressed in cancerous tissues. EphA2 and claudin-4 interact via their extracellular domains. This association stimulates the tyrosine kinase activity of EphA2, leading to claudin-4 phosphorylation at Tyr208. This residue localizes within the PDZ-binding motif NYV of claudin-4. Tyr208 phosphorylation diminishes claudin-4 interaction with the PDZ-containing protein ZO-1, and as a result, claudin-4 can no longer integrate in an efficient manner to the cell borders, thus delaying *de novo* formation of TJs (Tanaka, Kamata, & Sakai, 2005). Interestingly, Tyr208 located at position − 1 is conserved in claudins 1–10, 14, 15, 17–20 as an Eph phosphorylation site.

3. MLCK Phosphorylation of Claudin-5 Promotes Blood–Brain Barrier Disfunction

Ethanol metabolites, acetaldehyde and reactive oxygen species (ROS) stimulate in brain microvascular endothelial cells the 1,4,5-triphosphate receptor (IP3R) gated intracellular Ca^{2+} release that leads to the activation of MLCK and possibly of PTK. The activation of these kinases triggers the phosphorylation of myosin light chain (MLC) and of TJ proteins, including that of claudin-5 on serine (Haorah et al., 2005) and tyrosine residues (Haorah, Knipe, Gorantla, Zheng, & Persidsky, 2007), and causes cytoskeletal rearrangements and disruption of the TJ manifested by a decrease in TER and an increase in monocyte migration across the BBB.

D. *Kinases that Promote other Claudin Functions*

1. PKA Modulates Claudin-1-Mediated Entry of Hepatitis Virus C into Epithelial Cells

Hepatitis C virus (HCV) entry into epithelial cells is dependent on the host cell expression of CD81 and scavenger receptor B1, as well as on the junctional proteins occludin, claudin-1, -6, or -9. Inhibition of PKA activity in HPC hepatoma permissive cells Huh7.5 reduces virus entry, induces a relocalization of claudin-1 from the plasma membrane to an intracellular vesicular location, and disrupts the interaction between claudin-1 and CD81 (Farquhar et al., 2008). Although HCV infection of Huh-7.5 cells increases cAMP levels and the number and intensity of bands representing phosphorylated PKA substrates, no evidence of claudin-1 phosphorylation by PKA is found in cells incubated with HCV. The identity of the PKA target that induces the relocation of claudin-1 upon HCV infection remains unknown.

2. PKC Phosphorylation of Claudin-1 at Thr206 Randomizes the Direction of Heart Looping

Claudin-1 is normally expressed on the left side of the chick embryo, and the maintenance of this asymmetric expression appears to be required for normal rightward heart looping because when claudin-1 is overexpressed on the right side, the direction of heart looping is randomized. This change is abolished when Thr206, a putative PKC phosphorylation site of claudin-1 conserved in human, mouse, rat, and bull (Table I), is mutated for alanine (Simard, Di Pietro, Young, Plaza, & Ryan, 2006).

It is noteworthy that Thr206 is located at position −5 of claudin-1 and hence its phosphorylation might regulate the interaction with PDZ domains present in associated molecules.

E. Phosphorylation of Claudins in Cancerous Tissues

Loss of TJs and barrier function in transformed tissues has long been recognized (Gonzalez-Mariscal, Lechuga, & Garay, 2007). Hence, the observation that certain cancers like breast, colon, esophagus, ovary, pancreas, and prostate are characterized for overexpressing certain TJ proteins, including claudin-3 and -4, throws up an interesting enigma. The answer to this riddle might be in the posttranslational modifications that these proteins exhibit in the transformed tissues that allow them to be expressed, but not to exert their barrier function or to localize at the TJ.

1. PKA Phosphorylation of Claudin-3 at Thr192 Decreases TJ Strength in Ovarian Cancer Cells

Thr192 of human claudin-3 is a putative target of RhoK (Table I) found by site-directed mutagenesis and an *in vitro* kinase assay to be phosphorylated by PKA. In ovarian cancer cells, the transfection of a claudin-3 with a mutation that mimics the phosphorylation of this residue (T192D) decreases the strength of the TJ. This observation is important since, in ovarian cancer cells, PKA is frequently activated and claudin-3 is present in a phosphorylated state (D'Souza, Agarwal, & Morin, 2005).

2. Novel PKCε Phosphorylation of Claudin-4 at Thr189 and Ser194 Disrupts the TJ Barrier Function of Ovarian Cancer Cells

In ovarian cancer cells, claudin-4 is overexpressed and phosphorylated by PKC. Residues threonine 189 and serine 194 of claudin-4 are the targets of phosphorylation by PKC stimulation with TPA, and the overexpression of claudin-4 with phosphomimetic mutations T189D/S194D results in a disruption of the barrier function. nPKCε identified as the isoform responsible for these effects is found to colocalize with claudin-4 at the TJ (D'Souza et al., 2007).

3. PKA and PKC Phosphorylation Regulates the Subcellular Localization of Claudin-1 and the Metastatic Potential of Melanoma Cells

In metastatic melanoma cells, claudin-1 is overexpressed, but the subcellular localization of the protein is dysregulated, moving from the cell borders to the cytoplasm (Leotlela et al., 2007). This change correlates with an increased migration and secretion of MMP-2. Instead, in benign lesions and less aggressive melanomas, claudin-1 is expressed in the nucleus. When melanoma cells are transfected with claudin-1 coupled to a NLS (NLS–claudin-1), the protein exhibits nuclear and cytoplasmic localization. However, if the cells are treated with phorbol esters as PKC stimulators, NLS–claudin-1 is excluded from the nucleus. Phosphomimetic mutations on PKC and PKA phosphorylation sites (PKC/PKA: Thr190 and Thr195; PKA: Ser192) result in cytoplasmic claudin-1 expression. Cells transfected with the S69A claudin-1 mutant, which is nonphosphorytable by PKA, have a largely nuclear expression of the protein and show a lower degree of invasion than cells overexpressing wild-type claudin-1 that have plenty of cytoplasmic claudin. These data indicate that the phosphorylation of claudin-1 regulates its subcellular localization, and this in turn has effects on the metastatic capacity of the cell (French et al., 2009).

IV. INHIBITION OF PALMITOYLATION IMPAIRS THE EFFICIENT LOCALIZATION OF CLAUDIN-14 AT TIGHT JUNCTIONS

Palmitoylation is a posttranslational modification that consists of the reversible addition of palmitate and other long chain fatty acids to proteins at cysteine residues via a thio-ester linkage (for review, see Nadolski & Linder, 2007). Although palmitoylation has no single sequence requirement outside of the presence of a cysteine residue, palmitoylation sites are often found in proteins with transmembrane domains, in pairs or in longer stretches of cysteine residues at the interface of the cytoplasm and membrane, or in the cytoplasmic C-terminal tails.

Claudins contain two pairs of membrane-proximal cysteines following both the second and the fourth transmembrane domains. In claudin-14, the mutation of any of these sets of cysteines to serines decreases the incorporation of [^{3}H]-palmitic acid and the mutation of all four cysteines eliminates palmitoylation. Inhibition of palmitoylation decreases the localization of claudin-14 at TJs, diminishes its association with detergent-resistant membranes, and impairs its ability to form a paracellular barrier. Preventing palmitoylation does not affect claudin-14 half-life or the ability to form TJ fibrils in transfected fibroblasts (Van Itallie, Gambling, Carson, & Anderson, 2005).

V. CHANGES IN CLAUDIN EXPRESSION TRIGGERED BY DIVERSE ELEMENTS AND SIGNALING PATHWAYS

TJs are dynamic structures whose degree of sealing is finely tuned by the physiological and pathological conditions of the tissue. Among the factors that regulate the permeability of the paracellular pathway by modulating the expression level of different claudins are enzymes of the MAPK pathway, GTPases of the Rho family, as well as cytokines, interleukins, growth factors, prostaglandins, and hormones (Fig. 2).

A. MAP Kinase

MAP kinases phosphorylate Ser/Thr residues and are activated by growth factors and stress. Cell lines treated with aspirin (Oshima, Miwa, & Joh, 2008a) or HIV Tat protein (Bai et al., 2008) that activate the phosphorylation of Erk1/2, p38, or JNK, or transformed with activated Ras (Chen, Lu, Schneeberger, & Goodenough, 2000) or Raf (Lan, Kojima, Osanai, Chiba, & Sawada, 2004), exhibit disruption of their TJs and disappearance of TJ components from the cell borders. The nature of the affected claudins varies according to the tissue. Thus, a decreased expression of claudin-7 is observed in gastric cells (Oshima et al., 2008a), of claudins 1 and 2 in hepatocytes (Lan et al., 2004), of claudin-1 in renal cultures (Chen et al., 2000), and of claudins 1, 2, 3, and 4 in retinal pigment cells, together with an increased expression of claudin-2 (Bai et al., 2008).

In LLC-PK1 renal cells, ras transfection generates a $>40\%$ increase in TER and at the same time produces leaky TJs to nonelectrolytes. This difficult-to-interpret result is accompanied by the appearance of multilayered foci of cells, a decreased expression of claudin-2, and an increased abundance of claudins 1, 4, and 7 (Mullin et al., 2005).

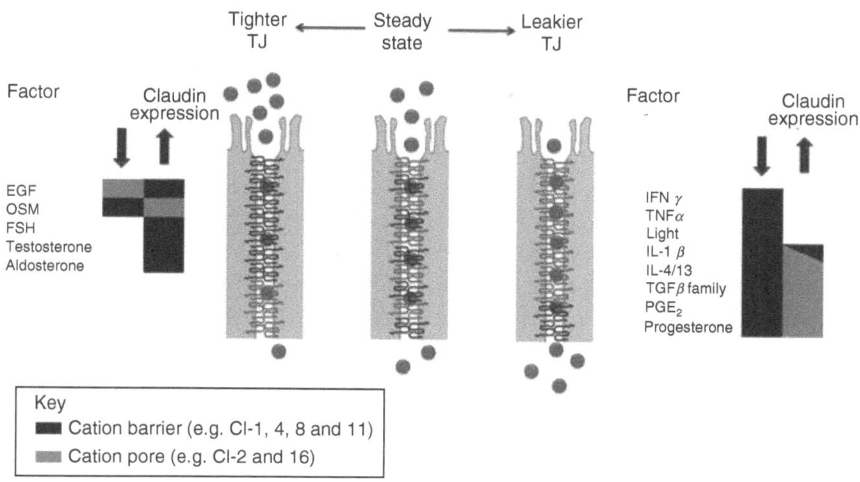

FIGURE 2 Factors that regulate tight junction sealing by changing the expression profile of claudins in the tissue. Proinflammatory cytokines, interleukins, TGFβ, PGE_2, and progesterone promote a leakier state of TJs by decreasing the expression of claudins that function as cation barriers (e.g., Cl-1, 4, 8, and 11) and by promoting, at the same time, the expression of claudins that work as cation pores (e.g., Cl-2). Instead, EGF and the hormones FSH, testosterone, and aldosterone induce the establishment of tighter TJs by increasing the expression of claudins which act as barriers to cations. (See Color Insert.)

In low-resistance type II MDCK cells, the activation of the MAPK pathway by hepatocyte growth factor (HGF) decreases the expression of claudin-2. Since this claudin behaves as a cationic pore, its reduced expression triggers a significant increase in TER. Moreover, claudin-2 expression and a concomitant reduction in TER can be induced in high-resistance MDCK I cells by treatment with U0126, an inhibitor of MEK1/2, thus demonstrating that claudin-2 expression is negatively regulated by MAPK in renal epithelial cells (Lipschutz, Li, Arisco, & Balkovetz, 2005).

B. Rho Rack and CDC42 Pathways

The function of the TJ is regulated by its association to the F-actin cytoskeleton. Therefore, it is no surprise to find that the Rho family of small GTPases RhoA, Rac1, and Cdc42 that modulate the cytoskeleton have a profound effect on the epithelial barrier and the expression of junctional proteins. In renal epithelial MDCK cells, the use of the constitutively active forms of RhoA, Rac1, and Cdc42, and of the dominant-negative Rac1, perturbs the epithelial barrier function and decreases the expression of claudins 1 and 2 (Bruewer, Hopkins, Hobert, Nusrat, & Madara, 2004).

RhoA, Rac1, and Cdc42 constitute eukaryotic targets for several pathogenic organisms. In accordance, infection of intestinal cells with *Citrobacter rodentium*, a causative agent of murine colonic hyperplasia, that produces attaching and effacing lesions that are indistinguishable from those induced by clinical enteropathogenic *Escherichia coli* (EPEC) strains, and is hence often used as a model of EPEC infection, disrupts the TJ expression of claudins 4 and 5 through a process that can be partially prevented by the inhibition of RhoK with Y27632 (Flynn & Buret, 2008).

C. Cytokines

Proinflammatory cytokines are elevated in epithelia suffering chronic inflammation or autoimmune diseases, like the intestinal mucosa in inflammatory bowel disease (IBD) (Bouma & Strober, 2003), the thyroid of patients with Graves' or Hashimoto syndrome (Heuer, Aust, Ode-Hakim, & Scherbaum, 1996), and the salivary and lacrimal glands in Sjögren's syndrome (Boumba, Skopouli, & Moutsopoulos, 1995; Fox, Kang, Ando, Abrams, & Pisa, 1994; Halse, Tengner, Wahren-Herlenius, Haga, & Jonsson, 1999; Ohyama et al., 1996; Oxholm, Daniels, & Bendtzen, 1992).

In some of these diseases, a change in the expression pattern of claudins has been observed. For example, in IBD, claudin-2 (Oshima, Miwa, & Joh, 2008b; Prasad et al., 2005; Zeissig et al., 2007) is strongly expressed in the inflamed crypt epithelium, while it is absent or barely detectable in normal colon (Prasad et al., 2005), and claudin-18 mRNA is upregulated in patients with ulcerative colitis (Zwiers et al., 2008). In contrast, claudins 3, 4, 5, and 7 (Oshima et al., 2008b; Prasad et al., 2005; Zeissig et al., 2007) that are present throughout the normal colon epithelium are reduced or redistributed in the inflamed tissue.

1. Interferon

Interferons (IFNs) are a multigene family of inducible cytokines with antiviral activity. Type II IFN, also known as immune or IFNγ, is induced by immune and inflammatory stimuli and is synthesized only by certain cells of the immune system including natural killer cells, CD4 Th1 cells, and CD8 cytotoxic suppressor cells (Samuel, 2001). IFNγ signaling starts by its association to the IFNγ receptor and proceeds by the activation of the JAK–STAT pathway (for review, see Bach, Aguet, & Schreiber, 1997).

a. IFNγ disrupts epithelial barrier function and alters claudin expression. IFNγ added to the basolateral surface of epithelial cell lines triggers a disruption of the epithelial barrier function. In thyrocyte (Tedelind, Ericson, Karlsson, & Nilsson, 2003) and parotid (Baker et al., 2008) cell lines, treatment with IFNγ decreases the expression of claudin-1, whereas in the intestinal cell line T-84, claudins 1 and 4 redistribute away from TJs (Bruewer et al., 2003) and the expression of claudin-2 diminishes (Willemsen, Hoetjes, van Deventer, & van Tol, 2005). These results are in contrast with the increased expression of claudin-2 present in the colon of IBD patients. The difference might be due to the presence of additional cytokines in the colon that together regulate claudin-2 expression in a more complex fashion.

It is noteworthy that the addition of IFNγ to the apical surface of colonic cells decreases the expression of claudin-2 and increases TER (Wisner, Harris, Green, & Poritz, 2008). The fact that a different response is obtained depending upon the apical versus basolateral administration of TNFγ, suggests that TNFγ receptors on each surface might initiate a different signaling cascade.

2. Tumor Necrosis Factor Alpha

Tumor necrosis factor alpha (TNFα), initially discovered as a result of its antitumor activity, is one of the major mediators of inflammation capable of regulating essential biological functions such as cell differentiation, proliferation, survival, apoptosis, and a variety of cellular responses to stress and injury (for review, see Sethi, Sung, & Aggarwal, 2008). TNFα is primarily produced by monocytes and macrophages in response to inflammatory stimuli and is also secreted by many other cell types.

a. TNFα disrupts the gate function of tight junction affecting claudin expression. In inflammation, the overproduction of TNFα leads to the induction of inflammatory genes, cell death, and the recruitment and activation of immune cells (Shen & Pervaiz, 2006; Warren, Ward, & Johnson, 1988). In addition, it has been demonstrated that TNFα plays a role in the control of epithelial permeability and the downregulation of TJ proteins expression. Thus, in brain microvascular endothelial cells hCMEC/D3, the basolateral addition of TNFα decreases TER and the expression of claudin-5 (Forster et al., 2008) through a mechanism that can be prevented by pretreatment with the glucocorticoid hydrocortisone. In epithelia, the basolateral addition of TNFα potentiates the disruption of the TJ gate function provoked by IFNγ in the parotid cell line Par-C10 (Baker et al., 2008) and the intestinal cells T-84 (Bruewer et al., 2003). In the latter, the process is independent of the proapoptotic abilities of TNFα and IFNγ (Bruewer et al., 2003). Treatment with these cytokines produces in T-84 cells an internalization of several TJ integral proteins, including claudins 1 and 4, and in Par-C10 cells, decreases anion secretion and the expression of claudin-1 (Baker et al., 2008). The effects induced in the salivary epithelium by TNFα and not IFNγ are reversible.

Genetic and pharmacological inhibition of TNFα signaling by the respective use of a TNFα receptor (TNF-αR) 1 knockout mice (TNF-αR1KO) and Etanecerpt (ETC), a dimeric soluble form of the TNF-αR2 that by binding to TNFα molecules, blocks their interaction with cell surface TNFRs, indicates that TNFα inhibition reduces TJ permeability and alterations of TJ and AJ proteins, including claudins 2, 4, and 5, in lung tissue subjected to acute lung inflammation induced by carrageenan (Mazzon & Cuzzocrea, 2007).

TNFα appears to be primarily responsible for the defects in enterocytes TJs present in experimental colitis induced by dinitrobenzenesulfonic acid (DNBS). Thus, when this agent is administered to TNF-αR1KO mice or to wild-type mice treated with infliximab (IFX), a humanized monoclonal antibody against TNFα, or ETC, deleterious changes occurring in TJs like the disappearance of occludin and ZO-1 and the appearance of claudin-2 in the ileal epithelium are prevented (Fries, Muja, Crisafulli, Cuzzocrea, & Mazzon, 2008).

b. LIGHT produces endocytosis of claudin-1. LIGHT is a member of the TNF family that regulates T cell activation, triggers in mice an experimental IBD similar to Crohn's, and is overexpressed in the intestinal biopsies of patients with active Crohn's (Wang et al., 2005). LIGHT activation of intestinal epithelial lymphotoxin β receptor (LTβR) induces MLCK activation, phosphorylation of MLC, and caveolar endocytosis of occludin and claudin-1. This chain of events leads to the disruption of the TJ barrier function (Schwarz et al., 2007).

3. Interleukins

The name interleukin is derived from "inter," meaning a way of communication, and "leukin" that highlights that many of these proteins are produced by leukocytes and act on them. Interleukins signaling proceeds by the activation of the JAK–STAT pathway.

a. Interleukin-1β. Interleukin-1β (IL-1β) is a highly inflammatory cytokine whose production is tightly regulated (for review, see Dinarello, 1996). IL-1β is a systemic hormone-like mediator that can trigger responses on cells both near and distant from the site of its synthesis.

In epithelial cells and astrocytes, the expression of claudins and the gap junction proteins connexins (Cx) is regulated by IL-1β. Thus, in retinal pigment epithelial cells (ARPE-19), claudin-1 expression increases, whereas that of claudin-11 and -12 diminishes, producing as a result an increase in epithelial permeability (Abe, Sugano, Saigo, & Tamai, 2003). In hepatocytes, IL-1β triggers an increase in claudin-2 and a decrease in Cx32 that is mediated by the MAPK pathway (Yamamoto et al., 2004), and interestingly, in astrocytes, the augmented expression of claudin-1 and the decreased expression of Cx43 are accompanied by the development of strand-like arrays of intramembranous particles resembling rudimentary TJs (Duffy, John, Lee, Brosnan, & Spray, 2000).

b. Interleukins 4 and 13. Interleukin-4 (IL-4) plays a key regulatory role in allergic responses and has antitumor and antiinflammatory effects (for review, see Chomarat & Banchereau, 1997). IL-4 acts upon B lymphocytes, monocytes, dendritic cells, and fibroblasts, and its expression and production is restricted to activated T lymphocytes, mast cells, and basophils. IL-4 transducing signals are mediated through JAKs and STAT pathways.

Interleukin-13 (IL-13), a key mediator in the pathogenesis of allergic inflammation, is secreted predominantly by activated T_H2 cells (for review, see Hershey, 2003). IL-13 mediates its effects by interacting with IL-4Rα and two binding proteins: IL-13Rα1 and IL-13Rα2. Signaling is mediated predominantly through IL-4Rα.

Treatment of intestinal epithelial cells T-84 with IL-4 (Wisner et al., 2008) or IL-13 increases the expression of claudin-2 and paracellular permeability through a process mediated by the PI3K pathway (Prasad et al., 2005).

c. Oncostatin M. Oncostatin M (OSM) cytokine belongs to the IL-6 subfamily. Although OSM is very closely related to the leukemia inhibitor factor (LIF), it exhibits unique biological activities in inflammation, hematopoiesis, and development (for review, see Tanaka & Miyajima, 2003). In inflammation, OSM regulates the production of cytokines and chemokines. In primary hepatocytes, treatment with OSM decreases claudin-1 expression and increases that of claudin-2 through a process mediated by PKC, MAPK, and PI3K (Imamura et al., 2007). In contrast to results obtained by others with IL-1β (Prasad et al., 2005) and IL-4 (Wisner et al., 2008), this change in claudin expression decreases paracellular permeability and increases TER.

d. Interleukin-10. The major source of interleukin-10 (IL-10) *in vivo* seems to be macrophages, although the cytokine is also produced by T helper 2 cells, monocytes, B cells, eosinophils, and mast cells (for review, see Asadullah, Sterry, & Volk, 2003). Macrophages are stimulated to produce IL-10 by several endogenous and exogenous factors such as endotoxin, TNFα, catecholamines, and cAMP elevating drugs. Interaction of IL-10 with its receptor activates the JAK and STAT pathways and results in the inhibition of the immune function by suppressing the expression of proinflammatory cytokines, chemokines, adhesion molecules, as well as antigen presenting costimulatory molecules in monocytes, macrophages, neutrophils, and T cells. Much of these effects are regulated by IL-10 inhibition of NF-κB activity.

In IL-10 KO mice, treatment with dinitrobenzene sulfonic acid (DNBS) to experimentally induce colitis, produces hepatobiliary abnormalities that resemble those described in patients with chronic IBD (Mazzon, Puzzolo, Caputi, & Cuzzocrea, 2002). Liver analysis of KO versus wild-type animals reveals a significant increase in the proinflammatory cytokines TNF, IL-1β, and IL-6, an augmented TJ permeability and a significant alteration in the localization of claudin-1 and ZO-1. Although these results suggest that the antiinflammatory cytokine IL-10 plays a crucial role controlling hepatobiliary injuries in IBD, the clinical results of IL-10 therapy in Crohn's disease have been unsatisfactory (Asadullah et al., 2003).

D. Prostaglandins

Prostaglandins are lipid-derived autacoids produced by the sequential metabolism of arachidonic acid by the cyclooxygenase and prostaglandin synthase enzymes. Prostaglandins are ubiquitously produced and act locally to elicit a diverse set of pharmacological effects that modulate many systems in the body. Prostaglandin is a potent proinflammatory mediator implicated in a broad array of diseases (for review, see Hata & Breyer, 2004).

In epithelial cells, treatment with PGE_2 decreases TER and alters claudin expression. In intestinal Caco-2 cells, a discontinuous staining of claudin-1 is observed (Tanaka, Diaz, De Souza, & Morgado-Diaz, 2008), while in renal MDCK cells, PGE_2 treatment blocks the enhanced expression of claudin-4 and the dislocalization of claudin-2 triggered by EGF pretreatment (Flores-Benitez et al., 2009).

E. Growth Factors

1. Epithelial Growth Factor Reinforces the Barrier Function of Tight Junctions

In epithelial cells, EGF increases the TER and changes the pattern of expression and the localization of different TJ proteins. Thus, EGF triggers an increased expression of claudins 4 and 7 in alveolar epithelial cells (Chen et al., 2005) and of claudin-4 in MDCK cells (Flores-Benitez et al., 2007; Ikari et al., 2009). In the latter, this process involves the activation of the MAPK pathway (Flores-Benitez et al., 2009) and the elevated expression of

Sp1 (Ikari et al., 2009), a transcription factor whose knock down decreases claudin-4 in ovarian cancer cells (Honda, Pazin, Ji, Wernyj, & Morin, 2006). These changes are accompanied by a decreased expression of claudin-3 and -5 in alveolar cells (Chen et al., 2005) and of claudin-2 in MDCK cells not mediated by MAPK (Flores-Benitez et al., 2009). In a similar way, the transfection of MDCK cells with a noncleavable, membrane anchored, heparin-binding EGF-like factor, capable of juxtacrine activation of EGF receptors, increases TER and decreases claudin-2 expression (Singh, Sugimoto, Dhawan, & Harris, 2007).

2. Transforming Growth Factor Perturbs the Function of Tight Junctions

Transforming growth factor (TGF) β has three isoforms: TGF-β1, TGF-β2, and TGF-β3. TGF-β1 participates in the control of the immune system and is secreted by every leukocyte lineage (Letterio & Roberts, 1998), whereas TGF-β3 is involved in cell differentiation, embryogenesis, and development, and without it, mammals develop cleft palate (Taya, O'Kane, & Ferguson, 1999).

In MDCK cells, TGF-β1 promotes epithelial mesenchymal transition (EMT) by: (1) activating the Ras-Raf-MEK-ERK-AP1 cascade that upregulates snail expression; (2) activating the PI3 kinase signaling that leads to the inactivation of GSK-3β, an enzyme responsible for leading β-catenin and snail to degradation in the proteosome; and (3) inducing Smad signaling that promotes the transcription of LEF transcription factor, upon endocytosis of the TGF-β1 receptor complex. The formation of β-catenin/LEF-1 complexes promotes the transcription of genes that induce EMT and inhibits the expression of claudins 1 and 2 (Medici, Hay, & Goodenough, 2006). In addition, TGF-β1 in hepatocytes decreases the expression of claudin-1, increases claudin-2, alters the fence function of TJs, reduces the number of TJ strands, and fragments them (Kojima et al., 2008). In Sertoli cells, other members of the TGF family like TGF-β3 (Lui, Lee, & Cheng, 2001) and growth differentiation factor (GDF) 9 (Nicholls, Harrison, Gilchrist, Farnworth, & Stanton, 2009), reduce the expression of claudin-11 and decrease TER.

F. Hormones

1. Sexual Hormones

a. Regulation by hormones of claudins present at the blood–testis barrier. The blood–testis barrier (BTB) is an inter-Sertoli cell junctional complex that localizes around the basal aspects of the seminiferous epithelium (for review, see Pelletier, 2001). TJs in Sertoli cells form a permeability barrier that creates a specialized microenvironment required for germ cell meiosis and maturation that is biochemically and immunologically distinct from the remainder of the testis. Disruption of the BTB leads to a cessation of spermatogenesis and to the exposure of the sequestered antigens to immunological attack.

The development of the BTB is associated with the endocrine status of the organism. Thus, Sertoli cells TJs first appear in puberty as follicle-stimulating hormone (FSH) and luteinizing hormone (LH) increase. Treatment of the primary cultures of Sertoli cells with testosterone increases TER, claudin-11 mRNA, and protein and promotes claudin-11 localization at intercellular contacts (Gye, 2003; Kaitu'u-Lino, Sluka, Foo, & Stanton, 2007). In transgenic mice with a complete absence of androgen receptors in Sertoli cells, claudin-3 (Meng, Holdcraft, Shima, Griswold, & Braun, 2005) and claudin-11 (Tan et al., 2005) mRNAs are downregulated, and fetal exposure to flutamide, a nonsteroidal antiandrogen drug, inhibits claudin-11 expression in rat Sertoli cells during the establishment of the BTB and in adulthood (Florin et al., 2005). Interestingly, the coculture of Sertoli cells with spermatids lowers claudin-11 mRNA levels. Taken together, these results indicate that postmeiotic germ cells and testosterone, respectively, modulate the opening and closing of the BTB during germ cell translocation from the basal to the adluminal compartment in the course of spermatogenesis (Florin et al., 2005).

Treatment of Sertoli cells with FSH produces contradictory results. Thus, while one study shows an increased expression of claudin-11 mRNA (Tarulli, Meachem, Schlatt, & Stanton, 2008), another reports the opposite effect (Hellani et al., 2000). In addition, the exposure of hamsters to short day length (8 h of light and 16 h of darkness) suppresses pituitary FSH and LH, and consequently, testicular testosterone. This results in the disruption of the BTB and spermatogonial development. In this model, claudin-11 and occludin exhibit a cytoplasmic distribution, claudin-3 localizes at the apical regions of Sertoli cells, and JAM-A reactivity is observed around germinal cells residing on the basement membrane. In addition, the mRNA levels of occludin and claudins 11 and 3 are significantly increased. FSH replacement produces a rapid relocalization of these TJ proteins to the basal portion of the seminiferous tubules (Tarulli et al., 2008).

b. Regulation by hormones of claudins present at the uterus and cervix. The distribution of claudins and other TJ proteins changes in the rat uterus during the estrous cycle. On proestrus, when estradiol and progesterone reach their highest serum levels and mating takes place, ZO-1, occludin, and claudins 1 and 5 are found at TJs, while claudins 3 and 7 display a basolateral distribution. In contrast, on metestrus day, when estradiol and progesterone serum levels are low and no sexual mating occurs, and the uterine lumen is devoid of secretions, none of these proteins is detected in the TJ region (Mendoza-Rodriguez, Gonzalez-Mariscal, & Cerbon, 2005).

During pregnancy, the expression of claudin-1 in the ovine uterus is negatively regulated by progesterone. Thus, daily injections of progesterone to ewes, produces a sharp decrease in the expression of claudin-1 in the uterine epithelium on days 9 and 12 of pregnancy, whereas the administration of progesterone receptor antagonist RU486 reverses this effect (Satterfield et al., 2007). In contrast, the administration of the progesterone derivative medroxyprogesterone acetate (MPA) to mice, significantly increases in a microarray analysis, the expression of claudin-2 in the cervix, and pretreatment of dams with MPA before lipopolysaccharide (LPS) infusion in the uterus, prevents the LPS-induced decrease of claudins 1 and 2 expression (Xu, Gonzalez, Ofori, & Elovitz, 2008). LPS treatment is used as model of intrauterine inflammation that results in a high rate of preterm birth without maternal death.

2. Suprarenal Hormones

a. Aldosterone. Aldosterone is a mineralocorticoid produced by the adrenal cortex that has long been known to modulate Na^+ reabsorption through the transcellular route of tight epithelia such as the renal collecting duct and the distal colon. More recently, aldosterone has been shown to prevent Na^+ back leakage by promoting the sealing of the paracellular pathway. In human intestinal HT-29 cells, for example, aldosterone increases TER and the expression of Cl-8 (Amasheh et al., 2009), whereas in rat cortical collecting duct cells, the hormone promotes a transient phosphorylation of claudin-4 on threonine residues and reduces paracellular permeability. Later on, aldosterone produces a fall in TER (Le Moellic et al., 2005).

b. Glucocorticoids. Glucocorticoids increase the barrier properties of endothelial cells, therefore, they have been employed for the treatment of edema accompanying brain tumors (Kaal & Vecht, 2004). However their ophthalmic use can induce glaucoma by increasing the resistance encountered by aqueous humor as it exits the eye to return to the circulation (Armaly, 1963). In retinal endothelial cells, glucocorticoid treatment increases the barrier properties of the monolayer and the content of occludin and claudin-5 through a process that increases the transcriptional activity of their respective promoters (Felinski, Cox, Phillips, & Antonetti, 2008).

References

Abe, T., Sugano, E., Saigo, Y., & Tamai, M. (2003). Interleukin-1beta and barrier function of retinal pigment epithelial cells (ARPE-19): Aberrant expression of junctional complex molecules. *Investigative Ophthalmology and Visual Science, 44*, 4097–4104.

Adamson, R. H., Liu, B., Fry, G. N., Rubin, L. L., & Curry, F. E. (1998). Microvascular permeability and number of tight junctions are modulated by cAMP. *American Journal of Physiology, 274*, H1885–H1894.

Amasheh, S., Milatz, S., Krug, S. M., Bergs, M., Amasheh, M., Schulzke, J. D., et al. (2009). Na^+ absorption defends from paracellular back-leakage by claudin-8 upregulation. *Biochemical and Biophysical Research Communications, 378*, 45–50.

Aono, S., & Hirai, Y. (2008). Phosphorylation of claudin-4 is required for tight junction formation in a human keratinocyte cell line. *Experimental Cell Research, 314*, 3326–3339.

Armaly, M. F. (1963). Effect of corticosteroids on intraocular pressure and fluid dynamics. II. The effect of dexamethasone in the glaucomatous eye. *Archives of Ophthalmology, 70*, 492–499.

Asadullah, K., Sterry, W., & Volk, H. D. (2003). Interleukin-10 therapy—Review of a new approach. *Pharmacological Reviews, 55*, 241–269.

Bach, E. A., Aguet, M., & Schreiber, R. D. (1997). The IFN gamma receptor: A paradigm for cytokine receptor signaling. *Annual Review of Immunology, 15*, 563–591.

Bai, L., Zhang, Z., Zhang, H., Li, X., Yu, Q., Lin, H., et al. (2008). HIV-1 Tat protein alter the tight junction integrity and function of retinal pigment epithelium: An in vitro study. *BMC Infectious Diseases, 8*, 77.

Baker, O. J., Camden, J. M., Redman, R. S., Jones, J. E., Seye, C. I., Erb, L., et al. (2008). Proinflammatory cytokines tumor necrosis factor-alpha and interferon-gamma alter tight junction structure and function in the rat parotid gland Par-C10 cell line. *American Journal of Physiology. Cell Physiology, 295*, C1191–C1201.

Banan, A., Zhang, L. J., Shaikh, M., Fields, J. Z., Choudhary, S., Forsyth, C. B., et al. (2005). theta Isoform of protein kinase C alters barrier function in intestinal epithelium through modulation of distinct claudin isotypes: A novel mechanism for regulation of permeability. *Journal of Pharmacology and Experimental Therapeutics, 313*, 962–982.

Bouma, G., & Strober, W. (2003). The immunological and genetic basis of inflammatory bowel disease. *Nature Reviews. Immunology, 3*, 521–533.

Boumba, D., Skopouli, F. N., & Moutsopoulos, H. M. (1995). Cytokine mRNA expression in the labial salivary gland tissues from patients with primary Sjogren's syndrome. *British Journal of Rheumatology, 34*, 326–333.

Bruewer, M., Hopkins, A. M., Hobert, M. E., Nusrat, A., & Madara, J. L. (2004). RhoA, Rac1, and Cdc42 exert distinct effects on epithelial barrier via selective structural and biochemical modulation of junctional proteins and F-actin. *American Journal of Physiology. Cell Physiology, 287*, C327–C335.

Bruewer, M., Luegering, A., Kucharzik, T., Parkos, C. A., Madara, J. L., Hopkins, A. M., et al. (2003). Proinflammatory cytokines disrupt epithelial barrier function by apoptosis-independent mechanisms. *Journal of Immunology, 171*, 6164–6172.

Chen, S. P., Zhou, B., Willis, B. C., Sandoval, A. J., Liebler, J. M., Kim, K. J., et al. (2005). Effects of transdifferentiation and EGF on claudin isoform expression in alveolar epithelial cells. *Journal of Applied Physiology, 98*, 322–328.

Chen, Y., Lu, Q., Schneeberger, E. E., & Goodenough, D. A. (2000). Restoration of tight junction structure and barrier function by down-regulation of the mitogen-activated protein kinase pathway in ras-transformed Madin-Darby canine kidney cells. *Molecular Biology of the Cell, 11*, 849–862.

Chomarat, P., & Banchereau, J. (1997). An update on interleukin-4 and its receptor. *European Cytokine Network, 8*, 333–344.

Dai, J., Jin, W. H., Sheng, Q. H., Shieh, C. H., Wu, J. R., & Zeng, R. (2007). Protein phosphorylation and expression profiling by Yin-yang multidimentional liquid chromatography (Yin-Yang MDLC) mass spectrometry. *Journal of Proteome Research, 6*, 250–262.

Dinarello, C. A. (1996). Biologic basis for interleukin-1 in disease. *Blood, 87*, 2095–2147.

D'Souza, T., Agarwal, R., & Morin, P. J. (2005). Phosphorylation of claudin-3 at threonine 192 by cAMP-dependent protein kinase regulates tight junction barrier function in ovarian cancer cells. *Journal of Biological Chemistry, 280*, 26233–26240.

D'Souza, T., Indig, F. E., & Morin, P. J. (2007). Phosphorylation of claudin-4 by PKCepsilon regulates tight junction barrier function in ovarian cancer cells. *Experimental Cell Research, 313*, 3364–3375.

Duffy, H. S., John, G. R., Lee, S. C., Brosnan, C. F., & Spray, D. C. (2000). Reciprocal regulation of the junctional proteins claudin-1 and connexin43 by interleukin-1beta in primary human fetal astrocytes. *Journal of Neuroscience, 20*, RC114.

Farquhar, M. J., Harris, H. J., Diskar, M., Jones, S., Mee, C. J., Nielsen, S. U., et al. (2008). Protein kinase A-dependent step(s) in hepatitis C virus entry and infectivity. *Journal of Virology, 82*, 8797–8811.

Felinski, E. A., Cox, A. E., Phillips, B. E., & Antonetti, D. A. (2008). Glucocorticoids induce transactivation of tight junction genes occludin and claudin-5 in retinal endothelial cells via a novel cis-element. *Experimental Eye Research, 86*, 867–878.

Flores-Benitez, D., Rincon-Heredia, R., Razgado, L. F., Larre, I., Cereijido, M., & Contreras, R. G. (2009). Control of tight junctions sealing: Roles of epidermal growth factor and prostaglandin E2. *American Journal of Physiology. Cell Physiology, 297*, C611–C620.

Flores-Benitez, D., Ruiz-Cabrera, A., Flores-Maldonado, C., Shoshani, L., Cereijido, M., & Contreras, R. G. (2007). Control of tight junctional sealing: Role of epidermal growth factor. *American Journal of Physiology. Renal Physiology, 292*, F828–F836.

Florin, A., Maire, M., Bozec, A., Hellani, A., Chater, S., Bars, R., et al. (2005). Androgens and postmeiotic germ cells regulate claudin-11 expression in rat Sertoli cells. *Endocrinology, 146*, 1532–1540.

Flynn, A. N., & Buret, A. G. (2008). Tight junctional disruption and apoptosis in an in vitro model of Citrobacter rodentium infection. *Microbial Pathogenesis, 45*, 98–104.

Forster, C., Burek, M., Romero, I. A., Weksler, B., Couraud, P. O., & Drenckhahn, D. (2008). Differential effects of hydrocortisone and TNFalpha on tight junction proteins in an in vitro model of the human blood-brain barrier. *Journal of Physiology, 586*, 1937–1949.

Fox, R. I., Kang, H. I., Ando, D., Abrams, J., & Pisa, E. (1994). Cytokine mRNA expression in salivary gland biopsies of Sjogren's syndrome. *Journal of Immunology, 152*, 5532–5539.

French, A. D., Fiori, J. L., Camilli, T. C., Leotlela, P. D., O'Connell, M. P., Frank, B. P., et al. (2009). PKC and PKA phosphorylation affect the subcellular localization of claudin-1 in melanoma cells. *International Journal of Medical Sciences, 6*, 93–101.

Fries, W., Muja, C., Crisafulli, C., Cuzzocrea, S., & Mazzon, E. (2008). Dynamics of enterocyte tight junctions: Effect of experimental colitis and two different anti-TNF strategies. *American Journal of Physiology. Gastrointestinal and Liver Physiology, 294*, G938–G947.

Fujibe, M., Chiba, H., Kojima, T., Soma, T., Wada, T., Yamashita, T., et al. (2004). Thr203 of claudin-1, a putative phosphorylation site for MAP kinase, is required to promote the barrier function of tight junctions. *Experimental Cell Research, 295*, 36–47.

Furuse, M., Fujita, K., Hiiragi, T., Fujimoto, K., & Tsukita, S. (1998). Claudin-1 and -2: Novel integral membrane proteins localizing at tight junctions with no sequence similarity to occludin. *Journal of Cell Biology, 141*, 1539–1550.

Gonzalez-Mariscal, L., Lechuga, S., & Garay, E. (2007). Role of tight junctions in cell proliferation and cancer. *Progress in Histochemistry and Cytochemistry, 42*, 1–57.

Gye, M. C. (2003). Changes in the expression of claudins and transepithelial electrical resistance of mouse Sertoli cells by Leydig cell coculture. *International Journal of Andrology, 26*, 271–278.

Halse, A., Tengner, P., Wahren-Herlenius, M., Haga, H., & Jonsson, R. (1999). Increased frequency of cells secreting interleukin-6 and interleukin-10 in peripheral blood of patients with primary Sjogren's syndrome. *Scandinavian Journal of Immunology, 49*, 533–538.

Haorah, J., Heilman, D., Knipe, B., Chrastil, J., Leibhart, J., Ghorpade, A., et al. (2005). Ethanol-induced activation of myosin light chain kinase leads to dysfunction of tight junctions and blood-brain barrier compromise. *Alcoholism, Clinical and Experimental Research, 29*, 999–1009.

Haorah, J., Knipe, B., Gorantla, S., Zheng, J., & Persidsky, Y. (2007). Alcohol-induced blood-brain barrier dysfunction is mediated via inositol 1, 4, 5-triphosphate receptor (IP3R)-gated intracellular calcium release. *Journal of Neurochemistry, 100*, 324–336.

Hata, A. N., & Breyer, R. M. (2004). Pharmacology and signaling of prostaglandin receptors: Multiple roles in inflammation and immune modulation. *Pharmacology and Therapeutics, 103*, 147–166.

Hellani, A., Ji, J., Mauduit, C., Deschildre, C., Tabone, E., & Benahmed, M. (2000). Developmental and hormonal regulation of the expression of oligodendrocyte-specific protein/claudin 11 in mouse testis. *Endocrinology, 141*, 3012–3019.

Hershey, G. K. (2003). IL-13 receptors and signaling pathways: An evolving web. *Journal of Allergy and Clinical Immunology, 111*, 677–690.

Heuer, M., Aust, G., Ode-Hakim, S., & Scherbaum, W. A. (1996). Different cytokine mRNA profiles in Graves' disease, Hashimoto's thyroiditis, and nonautoimmune thyroid disorders determined by quantitative reverse transcriptase polymerase chain reaction (RT-PCR). *Thyroid, 6*, 97–106.

Honda, H., Pazin, M. J., Ji, H., Wernyj, R. P., & Morin, P. J. (2006). Crucial roles of Sp1 and epigenetic modifications in the regulation of the CLDN4 promoter in ovarian cancer cells. *Journal of Biological Chemistry, 281*, 21433–21444.

Ikari, A., Atomi, K., Takiguchi, A., Yamazaki, Y., Miwa, M., & Sugatani, J. (2009). Epidermal growth factor increases claudin-4 expression mediated by Sp1 elevation in MDCK cells. *Biochemical and Biophysical Research Communications, 384*, 306–310.

Ikari, A., Ito, M., Okude, C., Sawada, H., Harada, H., Degawa, M., et al. (2008). Claudin-16 is directly phosphorylated by protein kinase A independently of a vasodilator-stimulated phosphoprotein-mediated pathway. *Journal of Cellular Physiology, 214*, 221–229.

Ikari, A., Matsumoto, S., Harada, H., Takagi, K., Degawa, M., Takahashi, T., et al. (2006). Dysfunction of paracellin-1 by dephosphorylation in Dahl salt-sensitive hypertensive rats. *Journal of Physiological Sciences, 56*, 379–383.

Ikari, A., Matsumoto, S., Harada, H., Takagi, K., Hayashi, H., Suzuki, Y., et al. (2006). Phosphorylation of paracellin-1 at Ser217 by protein kinase A is essential for localization in tight junctions. *Journal of Cell Science, 119*, 1781–1789.

Imamura, M., Kojima, T., Lan, M., Son, S., Murata, M., Osanai, M., et al. (2007). Oncostatin M induces upregulation of claudin-2 in rodent hepatocytes coinciding with changes in morphology and function of tight junctions. *Experimental Cell Research, 313*, 1951–1962.

Ishizaki, T., Chiba, H., Kojima, T., Fujibe, M., Soma, T., Miyajima, H., et al. (2003). Cyclic AMP induces phosphorylation of claudin-5 immunoprecipitates and expression of claudin-5 gene in blood-brain-barrier endothelial cells via protein kinase A-dependent and -independent pathways. *Experimental Cell Research, 290*, 275–288.

Kaal, E. C., & Vecht, C. J. (2004). The management of brain edema in brain tumors. *Current Opinion in Oncology, 16*, 593–600.

Kaitu'u-Lino, T. J., Sluka, P., Foo, C. F., & Stanton, P. G. (2007). Claudin-11 expression and localisation is regulated by androgens in rat Sertoli cells in vitro. *Reproduction, 133*, 1169–1179.

Kojima, T., Takano, K., Yamamoto, T., Murata, M., Son, S., Imamura, M., et al. (2008). Transforming growth factor-beta induces epithelial to mesenchymal transition by down-regulation of claudin-1 expression and the fence function in adult rat hepatocytes. *Liver International, 28*, 534–545.

Krause, G., Winkler, L., Mueller, S. L., Haseloff, R. F., Piontek, J., & Blasig, I. E. (2008). Structure and function of claudins. *Biochimica et Biophysica Acta, 1778*, 631–645.

Lan, M., Kojima, T., Osanai, M., Chiba, H., & Sawada, N. (2004). Oncogenic Raf-1 regulates epithelial to mesenchymal transition via distinct signal transduction pathways in an immortalized mouse hepatic cell line. *Carcinogenesis, 25*, 2385–2395.

Le Moellic, C., Boulkroun, S., Gonzalez-Nunez, D., Dublineau, I., Cluzeaud, F., Fay, M., et al. (2005). Aldosterone and tight junctions: Modulation of claudin-4 phosphorylation in renal collecting duct cells. *American Journal of Physiology. Cell Physiology, 289*, C1513–C1521.

Leotlela, P. D., Wade, M. S., Duray, P. H., Rhode, M. J., Brown, H. F., Rosenthal, D. T., et al. (2007). Claudin-1 overexpression in melanoma is regulated by PKC and contributes to melanoma cell motility. *Oncogene, 26*, 3846–3856.

Letterio, J. J., & Roberts, A. B. (1998). Regulation of immune responses by TGF-beta. *Annual Review of Immunology, 16*, 137–161.

Lipschutz, J. H., Li, S., Arisco, A., & Balkovetz, D. F. (2005). Extracellular signal-regulated kinases 1/2 control claudin-2 expression in Madin-Darby canine kidney strain I and II cells. *Journal of Biological Chemistry, 280*, 3780–3788.

Lui, W. Y., Lee, W. M., & Cheng, C. Y. (2001). Transforming growth factor-beta3 perturbs the inter-Sertoli tight junction permeability barrier in vitro possibly mediated via its effects on occludin, zonula occludens-1, and claudin-11. *Endocrinology, 142*, 1865–1877.

Mayan, H., Vered, I., Mouallem, M., Tzadok-Witkon, M., Pauzner, R., & Farfel, Z. (2002). Pseudohypoaldosteronism type II: Marked sensitivity to thiazides, hypercalciuria, normomagnesemia, and low bone mineral density. *Journal of Clinical Endocrinology and Metabolism, 87*, 3248–3254.

Mazzon, E., & Cuzzocrea, S. (2007). Role of TNF-alpha in lung tight junction alteration in mouse model of acute lung inflammation. *Respiratory Research, 8*, 75.

Mazzon, E., Puzzolo, D., Caputi, A. P., & Cuzzocrea, S. (2002). Role of IL-10 in hepatocyte tight junction alteration in mouse model of experimental colitis. *Molecular Medicine, 8*, 353–366.

Medici, D., Hay, E. D., & Goodenough, D. A. (2006). Cooperation between snail and LEF-1 transcription factors is essential for TGF-{beta}1-induced epithelial-mesenchymal transition. *Molecular Biology of the Cell, 17*, 1871–1879.

Mendoza-Rodriguez, C. A., Gonzalez-Mariscal, L., & Cerbon, M. (2005). Changes in the distribution of ZO-1, occludin, and claudins in the rat uterine epithelium during the estrous cycle. *Cell and Tissue Research, 319*, 315–330.

Meng, J., Holdcraft, R. W., Shima, J. E., Griswold, M. D., & Braun, R. E. (2005). Androgens regulate the permeability of the blood-testis barrier. *Proceedings of the National Academy of Sciences of the United States of America, 102*, 16696–16700.

Mullin, J. M., Leatherman, J. M., Valenzano, M. C., Huerta, E. R., Verrechio, J., Smith, D. M., et al. (2005). Ras mutation impairs epithelial barrier function to a wide range of nonelectrolytes. *Molecular Biology of the Cell, 16*, 5538–5550.

Munton, R. P., Tweedie-Cullen, R., Livingstone-Zatchej, M., Weinandy, F., Waidelich, M., Longo, D., et al. (2007). Qualitative and quantitative analysis of protein phosphorylation in naive and stimulated mouse synaptosomal preparations. *Molecular and Cellular Proteomics, 6*, 283–293.

Nadolski, M. J., & Linder, M. E. (2007). Protein lipidation. *FEBS Journal, 274*, 5202–5210.

Nicholls, P. K., Harrison, C. A., Gilchrist, R. B., Farnworth, P. G., & Stanton, P. G. (2009). Growth differentiation factor 9 is a germ cell regulator of Sertoli cell function. *Endocrinology, 150*, 2481–2490.

Nunbhakdi-Craig, V., Machleidt, T., Ogris, E., Bellotto, D., White, C. L., III., Sontag, E. (2002). Protein phosphatase 2A associates with and regulates atypical PKC and the epithelial tight junction complex. *Journal of Cell Biology, 158*, 967–978.

Ohta, A., Yang, S. S., Rai, T., Chiga, M., Sasaki, S., & Uchida, S. (2006). Overexpression of human WNK1 increases paracellular chloride permeability and phosphorylation of claudin-4 in MDCKII cells. *Biochemical and Biophysical Research Communications, 349*, 804–808.

Ohyama, Y., Nakamura, S., Matsuzaki, G., Shinohara, M., Hiroki, A., Fujimura, T., et al. (1996). Cytokine messenger RNA expression in the labial salivary glands of patients with Sjogren's syndrome. *Arthritis and Rheumatism, 39*, 1376–1384.

Oshima, T., Miwa, H., & Joh, T. (2008a). Aspirin induces gastric epithelial barrier dysfunction by activating p38 MAPK via claudin-7. *American Journal of Physiology. Cell Physiology, 295*, C800–C806.

Oshima, T., Miwa, H., & Joh, T. (2008b). Changes in the expression of claudins in active ulcerative colitis. *Journal of Gastroenterology and Hepatology, 23*(Suppl. 2), S146–S150.

Oxholm, P., Daniels, T. E., & Bendtzen, K. (1992). Cytokine expression in labial salivary glands from patients with primary Sjogren's syndrome. *Autoimmunity, 12*, 185–191.

Pelletier, R. M. (2001). The tight junctions in the testis, epididymis and vas deferens. In M. Cereijido & J. D. Anderson (Eds.), *Tight junctions* (pp. 599–628). (2nd ed.). Boca Raton, FL: CRC Press.

Prasad, S., Mingrino, R., Kaukinen, K., Hayes, K. L., Powell, R. M., MacDonald, T. T., et al. (2005). Inflammatory processes have differential effects on claudins 2, 3 and 4 in colonic epithelial cells. *Laboratory Investigation, 85*, 1139–1162.

Quamme, G. A., & de Rouffignac, C. (2000). Epithelial magnesium transport and regulation by the kidney. *Frontiers in Bioscience, 5*, D694–D711.

Samuel, C. E. (2001). Antiviral actions of interferons. *Clinical Microbiology Reviews, 14*, 778–809.

Satterfield, M. C., Dunlap, K. A., Hayashi, K., Burghardt, R. C., Spencer, T. E., & Bazer, F. W. (2007). Tight and adherens junctions in the ovine uterus: Differential regulation by pregnancy and progesterone. *Endocrinology, 148*, 3922–3931.

Schambelan, M., Sebastian, A., & Rector, F. C. Jr., (1981). Mineralocorticoid-resistant renal hyperkalemia without salt wasting (type II pseudohypoaldosteronism): Role of increased renal chloride reabsorption. *Kidney International, 19*, 716–727.

Schwarz, B. T., Wang, F., Shen, L., Clayburgh, D. R., Su, L., Wang, Y., et al. (2007). LIGHT signals directly to intestinal epithelia to cause barrier dysfunction via cytoskeletal and endocytic mechanisms. *Gastroenterology, 132*, 2383–2394.

Sethi, G., Sung, B., & Aggarwal, B. B. (2008). TNF: A master switch for inflammation to cancer. *Frontiers in Bioscience, 13*, 5094–5107.

Shen, H. M., & Pervaiz, S. (2006). TNF receptor superfamily-induced cell death: Redox-dependent execution. *FASEB Journal, 20*, 1589–1598.

Simard, A., Di Pietro, E., Young, C. R., Plaza, S., & Ryan, A. K. (2006). Alterations in heart looping induced by overexpression of the tight junction protein claudin-1 are dependent on its C-terminal cytoplasmic tail. *Mechanisms of Development, 123*, 210–227.

Singh, A. B., Sugimoto, K., Dhawan, P., & Harris, R. C. (2007). Juxtacrine activation of EGFR regulates claudin expression and increases transepithelial resistance. *American Journal of Physiology. Cell Physiology, 293*, C1660–C1668.

Soma, T., Chiba, H., Kato-Mori, Y., Wada, T., Yamashita, T., Kojima, T., et al. (2004). Thr(207) of claudin-5 is involved in size-selective loosening of the endothelial barrier by cyclic AMP. *Experimental Cell Research, 300*, 202–212.

Take, C., Ikeda, K., Kurasawa, T., & Kurokawa, K. (1991). Increased chloride reabsorption as an inherited renal tubular defect in familial type II pseudohypoaldosteronism. *New England Journal of Medicine, 324*, 472–476.

Tan, K. A., De Gendt, K., Atanassova, N., Walker, M., Sharpe, R. M., Saunders, P. T., et al. (2005). The role of androgens in Sertoli cell proliferation and functional maturation: Studies in mice with total or Sertoli cell-selective ablation of the androgen receptor. *Endocrinology, 146*, 2674–2683.

Tanaka, M. N., Diaz, B. L., De Souza, W., & Morgado-Diaz, J. A. (2008). Prostaglandin E2-EP1 and EP2 receptor signaling promotes apical junctional complex disassembly of Caco-2 human colorectal cancer cells. *BMC Cell Biology, 9*, 63.

Tanaka, M., Kamata, R., & Sakai, R. (2005). EphA2 phosphorylates the cytoplasmic tail of claudin-4 and mediates paracellular permeability. *Journal of Biological Chemistry, 280*, 42375–42382.

Tanaka, M., & Miyajima, A. (2003). Oncostatin M, a multifunctional cytokine. *Reviews of Physiology Biochemistry and Pharmacology, 149*, 39–52.

Tarulli, G. A., Meachem, S. J., Schlatt, S., & Stanton, P. G. (2008). Regulation of testicular tight junctions by gonadotrophins in the adult Djungarian hamster in vivo. *Reproduction, 135*, 867–877.

Tatum, R., Zhang, Y., Lu, Q., Kim, K., Jeansonne, B. G., & Chen, Y. H. (2007). WNK4 phosphorylates ser(206) of claudin-7 and promotes paracellular Cl(–) permeability. *FEBS Letters, 581*, 3887–3891.

Taya, Y., O'Kane, S., & Ferguson, M. W. (1999). Pathogenesis of cleft palate in TGF-beta3 knockout mice. *Development, 126*, 3869–3879.

Tedelind, S., Ericson, L. E., Karlsson, J. O., & Nilsson, M. (2003). Interferon-gamma down-regulates claudin-1 and impairs the epithelial barrier function in primary cultured human thyrocytes. *European Journal of Endocrinology, 149*, 215–221.

Van Itallie, C. M., Gambling, T. M., Carson, J. L., & Anderson, J. M. (2005). Palmitoylation of claudins is required for efficient tight-junction localization. *Journal of Cell Science, 118*, 1427–1436.

Villén, J., Beausoleil, S. A., Gerber, S. A., & Gygi, S. P. (2007). Large-scale phosphorylation analyses of mouse liver. *Proceedings of the National Academy of Sciences, 104*, 1488–1493.

Wang, J., Anders, R. A., Wang, Y., Turner, J. R., Abraham, C., Pfeffer, K., et al. (2005). The critical role of LIGHT in promoting intestinal inflammation and Crohn's disease. *Journal of Immunology, 174*, 8173–8182.

Warren, J. S., Ward, P. A., & Johnson, K. J. (1988). Tumor necrosis factor: A plurifunctional mediator of acute inflammation. *Modern Pathology, 1*, 242–247.

Willemsen, L. E., Hoetjes, J. P., van Deventer, S. J., & van Tol, E. A. (2005). Abrogation of IFN-gamma mediated epithelial barrier disruption by serine protease inhibition. *Clinical and Experimental Immunology, 142*, 275–284.

Wilson, F. H., Disse-Nicodeme, S., Choate, K. A., Ishikawa, K., Nelson-Williams, C., Desitter, I., et al. (2001). Human hypertension caused by mutations in WNK kinases. *Science, 293*, 1107–1112.

Wisner, D. M., Harris, L. R., III, Green, C. L., & Poritz, L. S. (2008). Opposing regulation of the tight junction protein claudin-2 by interferon-gamma and interleukin-4. *Journal of Surgical Research, 144*, 1–7.

Wolburg, H., Neuhaus, J., Kniesel, U., Krauss, B., Schmid, E. M., Ocalan, M., et al. (1994). Modulation of tight junction structure in blood-brain barrier endothelial cells. Effects of tissue culture, second messengers and cocultured astrocytes. *Journal of Cell Science, 107*(Pt 5), 1347–1357.

Xu, H., Gonzalez, J. M., Ofori, E., & Elovitz, M. A. (2008). Preventing cervical ripening: The primary mechanism by which progestational agents prevent preterm birth? *American Journal of Obstetrics and Gynecology, 198*, 314–318.

Yamamoto, M., Ramirez, S. H., Sato, S., Kiyota, T., Cerny, R. L., Kaibuchi, K., et al. (2008). Phosphorylation of claudin-5 and occludin by rho kinase in brain endothelial cells. *American Journal of Pathology, 172*, 521–533.

Yamamoto, T., Kojima, T., Murata, M., Takano, K., Go, M., Chiba, H., et al. (2004). IL-1beta regulates expression of Cx32, occludin, and claudin-2 of rat hepatocytes via distinct signal transduction pathways. *Experimental Cell Research, 299*, 427–441.

Yamauchi, K., Rai, T., Kobayashi, K., Sohara, E., Suzuki, T., Itoh, T., et al. (2004). Disease-causing mutant WNK4 increases paracellular chloride permeability and phosphorylates claudins. *Proceedings of the National Academy of Sciences of the United States of America, 101*, 4690–4694.

Yoo, J., Nichols, A., Mammen, J., Calvo, I., Song, J. C., Worrell, R. T., et al. (2003). Bryostatin-1 enhances barrier function in T84 epithelia through PKC-dependent regulation of tight junction proteins. *American Journal of Physiology. Cell Physiology, 285*, C300–C309.

Zeissig, S., Burgel, N., Gunzel, D., Richter, J., Mankertz, J., Wahnschaffe, U., et al. (2007). Changes in expression and distribution of claudin 2, 5 and 8 lead to discontinuous tight junctions and barrier dysfunction in active Crohn's disease. *Gut, 56*, 61–72.

Zwiers, A., Fuss, I. J., Leijen, S., Mulder, C. J., Kraal, G., & Bouma, G. (2008). Increased expression of the tight junction molecule claudin-18 A1 in both experimental colitis and ulcerative colitis. *Inflammatory Bowel Diseases, 14*, 1652–1659.

CHAPTER 7

Claudins and Renal Magnesium Handling

Jianghui Hou* **and Martin Konrad**[†]
*Department of Internal Medicine—Renal Division,
Washington University School of Medicine, St. Louis, Missouri, USA
[†]Department of Pediatrics, University of Münster, Münster, Germany

I. OVERVIEW

Claudins are tight junction (TJ) integral membrane proteins that are key regulators of the paracellular pathway. Defects in claudin-16 (CLDN16) and claudin-19 (CLDN19) function result in the inherited human renal disorder familial hypomagnesemia with hypercalciuria and nephrocalcinosis (FHHNC). Significant advances have been made toward understanding the mechanisms underlying the roles of these claudins in mediating paracellular ion reabsorption

Current Topics in Membranes, Volume 65
1063-5823/10 $35.00
DOI: 10.1016/S1063-5823(10)65007-7

in the kidney. Here, we review the biosynthesis, trafficking, and interaction of CLDN16 and CLDN19 molecules; the biophysical properties of CLDN16 and CLDN19 channels; and the pathogenic mechanisms for the role of mutant forms of CLDN16 and CLDN19 in the development of FHHNC.

II. INTRODUCTION

A. Claudins form Paracellular Channels in the Kidney

The *tight junction* (*zonula occludens*) is the most apical member of the junctional complex found in vertebrate epithelia (Farquhar & Palade, 1963). The TJ is deployed as a belt around the circumference of each cell so that in the aggregate it forms a continuous, fish-net structure separating apical and basolateral extracellular compartments. The known integral membrane proteins of the TJ include occludin (a 65-kDa membrane protein bearing four transmembrane domains and two uncharged extracellular loops (ECL)) (Furuse et al., 1993), which has splice variants (Ghassemifar et al., 2002; Muresan, Paul, & Goodenough, 2000), the junctional adhesion molecules (JAMs), a four-member group of glycosylated proteins (Ebnet, Suzuki, Ohno, & Vestweber, 2004), and the claudins.

Claudins (CLDN) are tetraspan proteins consisting of a family of at least 24 members (Furuse, Fujita, Hiiragi, Fujimoto, & Tsukita, 1998; Morita, Furuse, Fujimoto, & Tsukita, 1999; Tsukita, Furuse, & Itoh, 2001). They range in molecular mass from 20 to 28 kDa with charged ECL. The cytoplasmic C-terminus of most claudins ends with a PDZ (*p*ostsynaptic density 95/*d*isks large/*z*onula occludens-1) binding domain that is critical for interaction with the submembrane scaffold and correct localization in the TJ (Hamazaki, Itoh, Sasaki, Furuse, & Tsukita, 2001; Itoh et al., 1999). It is clear that occludin and claudins can coassemble into the TJ freeze-fracture fibrils (Furuse et al., 1998). However, there are few data available permitting an understanding of the molecular interactions between occludin and the claudins or among the claudins themselves. One study has shown that the heterotypic interactions (between cells) of CLDN1 and CLDN3 are permitted, but that interactions between CLDN1 and CLDN2 are not observed (Furuse, Sasaki, & Tsukita, 1999). Coculture of HeLa cells expressing different claudin genes revealed that while CLDN3 and CLDN4 were functionally compatible when expressed in the same cell, they did not heterotypically interact (Daugherty, Ward, Smith, Ritzenthaler, & Koval, 2007). CLDN1 and CLDN5 were heterotypically interacting with CLDN3, but would not heterotypically bind to CLDN4. In contrast, chimeras exchanging the second extracellular domain of CLDN4 with CLDN3 conferred the ability to bind to CLDN1 and CLDN5 (Daugherty et al.,

2007). A point mutation in CLDN3 disrupted its heterotypic binding with other claudins, demonstrating considerable selectivity in heterotypic claudin–claudin interactions. While different claudins can assemble into the same TJ strand, current limitations of resolution do not permit a clear understanding of what heteromeric interactions (within cells) are favored. Efforts have been made to demonstrate the oligomerization properties of CLDN4 in cultured insect cells with ambiguous results (Mitic, Unger, & Anderson, 2003). FRAP (fluorescence recovery after photobleaching) studies suggest that claudin molecules assembled in TJs have limited mobility (Sasaki et al., 2003), consistent with their known interactions with proteins in the TJ scaffold.

Functionally, claudins have been shown to confer ion selectivity to the paracellular pathway, resulting in differences in transepithelial resistance (TER) and paracellular permeabilities, which vary widely between epithelia. Studies have shown that CLDN4, -5, -8, -11, and -14 selectively decrease the permeability of cations through TJs (Ben-Yosef et al., 2003; Colegio, Van Itallie, McCrea, Rahner, & Anderson, 2002; Van Itallie, Rahner, & Anderson, 2001; Wen, Watry, Marcondes, & Fox, 2004; Yu, Enck, Lencer, & Schneeberger, 2003), specifically to Na^+, K^+, H^+, and ammonium. CLDN2 and CLDN15 increase cation permeability (Amasheh et al., 2002; Van Itallie, Fanning, & Anderson, 2003). These properties have been attributed to charged amino acids in the first extracellular domain (Colegio, Van Itallie, Rahner, & Anderson, 2003). These and other studies (Tang & Goodenough, 2003) have led to models of the claudins forming *paracellular channels*, a novel class of channels oriented perpendicular to the membrane plane and serving to join two extracellular compartments (Tsukita & Furuse, 2000; Watson, Rowland, & Warhurst, 2001). Measurement of paracellular permeability using cell membrane-impermeant tracers indicates that there are 4–7 Å channels in the TJ (Tang & Goodenough, 2003; Van Itallie et al., 2008). The paracellular channels in the TJ have properties of ion selectivity, pH dependence, and anomalous mole fraction effects, similar to more conventional transmembrane channels (Tang & Goodenough, 2003).

B. The Paracellular Reabsorption of Magnesium in the Kidney

Magnesium is predominantly stored in bone and the intracellular compartments of muscle and soft tissues. Less than 1% of total body magnesium is circulating in the blood. In healthy individuals, serum magnesium levels are kept in a narrow range (0.7–1.1 mmol/l). Magnesium homeostasis depends on the balance between intestinal absorption and renal excretion. Within physiological ranges, diminished magnesium intake is balanced by enhanced magnesium absorption in the intestine and reduced renal excretion. These transport processes are regulated by metabolic and hormonal influences (Kerstan & Quamme, 2002; Quamme, 1997).

Magnesium reabsorption in the kidney differs in quantity depending on the different nephron segments: 15–20% of filtered Mg^{2+} is reabsorbed in the proximal tubule (PCT); ~70% is reabsorbed in the loop of Henle, primarily through the thick ascending limb (TAL); and only 5–10% is reabsorbed in the distal convoluted tubule (DCT), the part of the nephron where the fine adjustment of renal excretion is accomplished (Quamme & Dirks, 1980). A number of elegant *in vitro* studies using perfused TAL tubules have examined the relationship between the flux of Mg^{2+} and the transepithelial voltage (V_{te}) (Di Stefano, Roinel, De Rouffignac, & Wittner, 1993; Hebert & Andreoli, 1986; Mandon, Siga, Roinel, & De Rouffignac, 1993; Shareghi & Agus, 1982). The flux–voltage relationship indicates that, in the TAL, Mg^{2+} is passively reabsorbed from the lumen to the peritubular space through the paracellular pathway, driven by a lumen-positive V_{te}. The generation of this lumen-positive V_{te} in the TAL can be attributed to two mechanisms: (1) the active transport V_{te} owing to apical K^+ recycling through the potassium channel ROMK and basolateral Cl^- exit through ClC-Kb chloride channels, coupled with NaCl reabsorption via the apical $Na^+2Cl^-K^+$ cotransporter (NKCC2) (Fig. 1A); and (2) the diffusion V_{te} generated by the transepithelial

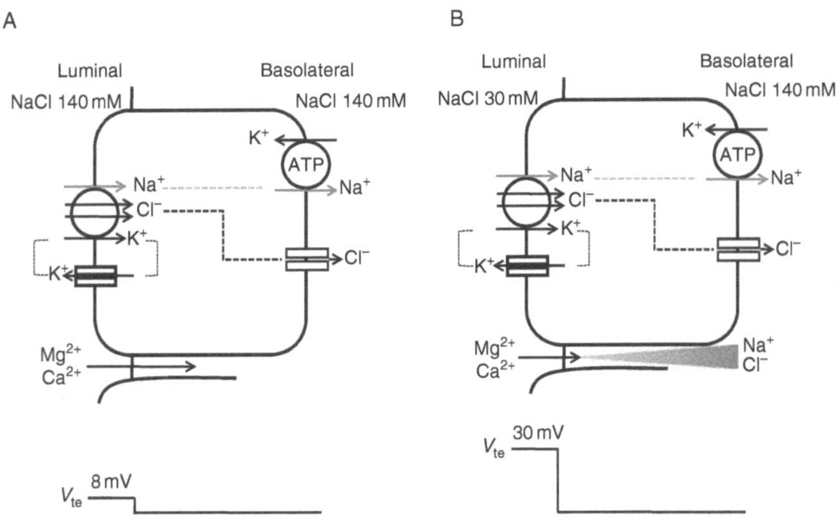

FIGURE 1 Transepithelial electrogenic ion transport and generation of lumen-positive potential. (A) When similar salt concentrations are present at the luminal and basolateral side, the luminal potential is generated by the concerted action of luminal K^+ channels, basolateral Cl^- channels, the $Na^+2Cl^-K^+$ cotransporter, and the Na^+,K^+-ATPase. (B) When a dilute luminal fluid is present after NaCl absorption along the water-tight TAL, the luminal potential is now generated as a diffusion voltage by the "backleak" of salt. The diffusion voltage depends on the permselectivity of the tight junction.

NaCl concentration gradient and the cation selectivity of the paracellular pathway of TAL (Fig. 1B). Both components are present in parallel along the TAL; however, their contributions vary at different parts of TAL and depend on tubular flow. At the beginning of the TAL segment or under high-flow conditions, it is the first mechanism that mainly provides a voltage around +8 mV. There is minimal contribution of diffusion potential at this stage since the concentration gradient has not yet been fully established or rapidly washed out. With continuous NaCl reabsorption along the axis of the TAL segment, the luminal fluid is gradually diluted and a large NaCl gradient is generated at the end of the TAL. Because the paracellular permeability of the TAL is cation-selective (with a P_{Na}/P_{Cl} value between 2 and 4) (Greger, 1981), the diffusion V_{te} is superimposed onto the active transport V_{te} and becomes the major source of the lumen-positive V_{te}, which now increases substantially up to +30 mV.

III. CLAUDINS AND HUMAN RENAL DISEASES OF MAGNESIUM WASTING

A. Mutations in CLDN16 Cause FHHNC

FHHNC (OMIM #248250) is a rare autosomal recessive tubular disorder. It was first described by Michelis, Drash, Linarelli, De Rubertis, and Davis (1972). Since then, numerous affected kindreds have been reported, which allowed a comprehensive characterization of the clinical spectrum of this disorder and discrimination from other Mg^{2+}-wasting tubular diseases (Benigno et al., 2000; Manz, Schärer, Janka, & Lombeck, 1978; Praga et al., 1995; Rodriguez-Soriano & Vallo, 1994). As a consequence of excessive renal Mg^{2+} and Ca^{2+} wasting, patients develop the characteristic triad of hypomagnesemia, hypercalciuria, and nephrocalcinosis. In early childhood, FHHNC patients may also have recurrent urinary tract infections, polyuria/polydipsia, nephrolithiasis, and/or failure to thrive. Clinical signs of severe hypomagnesemia such as tetanic episodes or seizures are less common. Extrarenal manifestations, especially ocular involvements (including severe myopia, nystagmus, or chorioretinitis), have been reported (Benigno et al., 2000; Praga et al., 1995; Rodriguez-Soriano & Vallo, 1994). Additional laboratory findings include elevated serum PTH (parathyroid hormone) levels before the onset of chronic renal failure, incomplete distal tubular acidosis, hypocitraturia, and hyperuricemia (Weber, Schneider, et al., 2001). Many FHHNC patients develop progressive chronic renal failure during the first two decades of life. On average, one-third of these patients reach end-stage renal failure during adolescence.

In addition to continuous Mg^{2+} supplementation, therapeutic treatments also aim to reduce the Ca^{2+} excretion by using thiazide diuretics to prevent the progression of nephrocalcinosis and stone formation. The degree of renal calcification has been correlated with the progression of chronic renal failure (Praga et al., 1995). In a short-term study, thiazide diuretics have been demonstrated to effectively reduce urinary calcium excretion in FHHNC patients (Zimmermann et al., 2006). However, whether these therapeutic strategies have the potential to slow down or prevent the progression of renal failure remains to be determined. Supportive therapy is important for protecting kidney functions and should include the provision of sufficient fluids. Prevention and/or effective treatment of urinary tract infections are also important. Renal transplantation does not result in the recurrence of this disease because the primary defect resides in the kidney.

Based on clinical observations and clearance studies, it had been postulated that the primary defect in FHHNC was related to disturbed Mg^{2+} and Ca^{2+} reabsorption in the TAL of the loop of Henle (Rodriguez-Soriano & Vallo, 1994). Simon et al. (1999) used the positional cloning strategy and identified a gene on 3q27 (CLDN16, formerly known as PCLN1), which is mutated in patients with FHHNC. Most mutations reported to date in FHHNC are simple missense mutations affecting the transmembrane domains and the ECL with a particular clustering in the first ECL (ECL1) composing the putative ion selectivity filter (for summary, see Günzel et al., 2009). Within this domain, patients from Germany and Eastern European countries exhibit a common mutation (L151F) due to a founder effect (Weber, Schneider, et al., 2001). As this mutation is present in approximately 50% of the mutant alleles, molecular diagnosis is greatly facilitated in patients from these countries.

Carriers of heterozygous CLDN16 mutations may also present with clinical symptoms. Two independent studies described a high prevalence of hypercalciuria, nephrolithiasis, and/or nephrocalcinosis in the first-degree relatives of FHHNC patients (Praga et al., 1995; Weber, Schneider, et al., 2001). Another study reported that there is a tendency toward mild hypomagnesemia in family members with heterozygous CLDN16 mutations (Blanchard et al., 2001). Thus, it is possible that CLDN16 mutations are involved in idiopathic hypercalciuric stone formation. Very recently, a genome-wide association study using the SNP-Chip technology has found that common sequence variations in CLDN14 conferred an increased risk for the development of kidney stones to individuals carrying these variants (Thorleifsson et al., 2009).

A peculiar homozygous CLDN16 mutation (T303R) affecting the C-terminal PDZ-binding domain has been identified in two families with disease phenotypes very different from FHHNC. In contrast to the other CLDN16 mutations, patients with the T303R mutation showed hypercalciuria and nephrocalcinosis but without any disturbance in renal Mg^{2+} handling

(Müller et al., 2003). Interestingly, the hypercalciuria disappeared after puberty. Transient transfection of the Madine Darby canine kidney (MDCK) cells with the CLDN16 (T303R) mutant revealed protein trafficking defects (Müller et al., 2003).

As mentioned earlier, many FHHNC patients develop chronic renal failure associated with progressive tubulointerstitial nephritis. The pathophysiology of these phenomena, which are far less common in other electrolyte wasting disorders, is still unclear. In the past, renal failure in FHHNC has been attributed to the concomitant hypercalciuria and nephrocalcinosis, but no correlation has been established. It was speculated that CLDN16 was not only involved in paracellular electrolyte reabsorption but also in tubular cell proliferation and differentiation (Lee, Huang, & Ward, 2006). This hypothesis is supported by the naturally occurring bovine CLDN16 knockout (KO) phenotypes, which exhibited early onset renal failure owing to interstitial nephritis with diffuse zonal fibrosis (Hirano et al., 2000; Ohba et al., 2000). Tubular epithelial cells were reported as "immature" with depolarization and detachment from the basement membrane.

B. Mutations in CLDN19 Cause FHHNC

FHHNC is a genetically heterogeneous disorder. Mutations in another TJ gene encoding CLDN19 have also been linked to this disease (Konrad et al., 2006). The renal tubular phenotypes are indistinguishable of patients with mutations in CLDN16 from those with CLDN19. CLDN19 mutations are invariably associated with severe ocular abnormalities (including severe myopia, nystagmus, or macular coloboma) (Benigno et al., 2000; Praga et al., 1995; Rodriguez-Soriano & Vallo, 1994). This association has been named FHHNC with severe ocular involvement (OMIM #248190). In contrast, only a small subset of FHHNC patients with CLDN16 mutations displayed severe myopia but neither nystagmus nor colobomata has ever been observed (Weber, Schneider, et al., 2001).

IV. CLAUDIN BIOLOGY AND PHYSIOLOGY

A. Biosynthesis, Trafficking, and Interaction of CLDN16 and CLDN19 Molecules

1. Translational Start Site of CLDN16

The human CLDN16 gene encodes a 305 amino acid protein that possesses two in-frame start codons (ATG: encoding methionine M1 and M71, respectively) (Fig. 2) at the 5′-end in a suitable Kozak consensus sequence.

```
Mouse   ------------------------------------------------------------
Rat     ------------------------------------------------------------
Human   MTSRTPLLVTACLYYSYCNSRHLQQGVRKSKRPVFSHCQVPETQKTDTRHLSGARAGVCP
        1

Mouse   ----------MKDLLQYAACFLAIFSTGFLIVATWTDCWMVNADDSLEVSTKCRGLWWEC
Rat     ----------MKDLLQYAACFLAIFSTGFLIVATRTDCWMVNADDSLEVSTKCRGLWWEC
Human   CCHPDGLLATMRDLLQYIACFFAFFSAGFLIVATWTDCWMVNADDSLEVSTKCRGLWWEC
                  *:***** ***:*:**:******* *************************
                  71
```

FIGURE 2 Comparison of amino acid sequence of CLDN16 across the species of mouse, rat, and human. Note that the human sequence possesses two in-frame methionines, with the second methionine highly conserved throughout the species.

The second ATG corresponds to the start codon of mouse and rat CLDN16 (Weber, Schlingmann, et al., 2001). The similarity of the sequence downstream of amino acid M71 is high among all three species. Genetic analysis of human CLDN16 revealed an insertion/deletion polymorphism at amino acid position 55 that would result in a frame shift and premature translation stop at position 90 (minor allele frequency 0.17 in Caucasians), indicating that the translation of human CLDN16 is initiated from the second ATG at M71 (Weber, Schneider, et al., 2001). To confirm the translational initiation start site of human CLDN16, Hou, Paul, and Goodenough (2005) have expressed the full-length human CLDN16 (FL) in a number of mammalian cell lines (including MDCK and HEK293), and have found that CLDN16 migrated as two separate bands (33 and 27 kDa) with different electrophoretic mobility. The 33 kDa band matched the predicted molecular weight of full-length CLDN16 (FL, amino acids 1–305) and the 27 kDa band matched its truncated form (δ70: amino acids 71–305). Using immunofluorescence imaging, Hou et al. (2005) revealed that δ70-CLDN16 concentrated at TJs, whereas the FL-CLDN16 was targeted to endosomes or lysosomes. Genetic analysis of FHHNC patients has identified a point mutation (M71T) linked to FHHNC that caused CLDN16 targeting to the lysosomes (Konrad et al., 2008), thus highlighting the functional significance of the second start codon. As FL-CLDN16 fails to translocate to TJs, it is possible that in humans, the native cellular environment of the TAL of the nephron contains regulatory factors to allow bypassing the first methionine (M1) in CLDN16 and ensure appropriate translation from the second methionine (M71).

2. Molecular Regulation of CLDN16 Localization in the Tight Junction

The cytoplasmic tail of CLDN16 contains a PDZ domain-binding motif (TRV, highlighted in bold; Fig. 3) that interacts with the PDZ domain of scaffolding proteins such as the zonula occludens (ZO) proteins. The

```
claudin-10   IGGVIFCFSIS-----DNNKTPRYTYNGATSVMSSRTKYHGGEDF--------KTTNPSKQFDKNAYV
claudin-15   LGGLCLCSACCCGSDEDPAASARRPYQAPVSVMPVATSDQEGD------------SSFGKYGRNAYV
claudin-11   VG-----------------GCVILCCAGDAQAFGENRFYYTAG------------SSSPTHAKSAHV
claudin-1    LGGALLCCSCP--RKTTS-YPTPRPYPKPAPSSGKDY---------------V-------------
claudin-7    LGGALLSCSCPGNESKAG-YRAPRSYPK--SNSSKEY---------------V-------------
claudin-19   LGGSFLCCTCP--EPERP-NSSPQPYRPGPSAAAREY---------------V-------------
claudin-2    IAGIILCFSCSSQRNRSNYYDAYQAQPLATRSSPRPG---QP---------PKVKSEFNS-YSLTGYV
claudin-14   IGGTLLCLSCQDEAP----YRPYQAPPRATTTTANTAPAYQPPAAYKDNRAPSVTSATHSGYRLNDYV
claudin-3    LGGALLCCSCPPR----EKKYTATKVVYSAPRSTGPGASLGT-----------------GYDRKDYV
claudin-4    LGGGLLCCNCPPRT---DKPYSAK---YSAARSAAASN-------------------------YV
claudin-6    LGGGLLCCTCPSGGSQGPSHYMARYSTSAPAISRGPSEYP----------------------TKNYV
claudin-9    LGGGLLCCTCPPPQVERPRG--PRLGYSIPSRS-GASGLD----------------------KRDYV
claudin-5    VGGCLLCCGAWVCTGRPDLSFPVK--YSAPRRPTATGDYD----------------------KKNYV
claudin-8    VGGALFCCVFCCNEKSSSYRYSIP-SHRTTQKSYHTGKKSPS-----------------VYSRSQYV
claudin-17   IGGGLLCGFCCC NRKKQGYRYPVP-GYRVPHTDKRRNTTMLS-----------------KTSTS-YV
claudin-16   LAGAVLTCCLYLFKDVGPERNYPYSLRKAYSAAGVSMAKSYSAPR----------TETAKMYAVDTRV
                        |→
                   Cytoplasmic tail
```

FIGURE 3 Alignment of the amino acid sequences of the cytoplasmic tails (arrow) among the members of the claudin superfamily. The PDZ domain-binding motif (TRV) in CLDN16 is highlighted as bold. The S217 site in CLDN16 is highlighted as underlined.

cytoplasmic tail of claudin plays a critical role in its trafficking and assembly into the TJ. Ikari et al. (2004) have shown that CLDN16 interacted with ZO-1 in MDCK cells and that deletion of the PDZ-binding motif (TRV) in CLDN16 abolished its interaction with ZO-1 and its localization in the TJ. Müller et al. (2003) identified a novel mutation in the PDZ-binding motif (TRV mutated to RRV) of CLDN16 in patients affected from childhood hypercalciuria (this phenotype is very different from FHHNC), which disrupted CLDN16 binding with ZO-1 and caused CLDN16 mislocalization in the lysosome. Ikari et al. (2006) reported CLDN16 phosphorylation at Ser217 (labeled as underlined; Fig. 3) by protein kinase A (PKA) under physiological conditions in MDCK cells. Dephosphorylation of CLDN16 with PKA inhibitors dissociated CLDN16 from ZO-1 and mistargeted CLDN16 to the lysosome.

3. CLDN16 is Internalized via a Clathrin-Dependent Pathway

To monitor cell-surface expression and internalization of Cldn16, Kausalya et al. (2006) generated an antibody against the ECL1 of CLDN16 (amino acids 52–66), which selectively labeled the cell-surface CLDN16 molecules. In MDCK cells transfected with CLDN16, Kausalya et al. (2006) found CLDN16 localization at the regions of cell–cell contact as well as in the intracellular vesicles of cells, suggesting that CLDN16 was endocytosed. To determine whether CLDN16 internalizes via a clathrin- or caveolae-dependent pathway, Kausalya et al. (2006) treated CLDN16-expressing cells with inhibitors that selectively block clathrin- or caveolae-mediated internalization. Hypertonicity and cytosol acidification, which selectively block clathrin-mediated endocytosis (Heuser & Anderson, 1989;

Sandvig, Olsnes, Petersen, & van Deurs, 1987), resulted in the accumulation of CLDN16 on the plasma membrane, whereas cholesterol oxidase, a selective inhibitor of caveolae-mediated endocytosis (Coconnier, Lorrot, Barbat, Laboisse, & Servin, 2000; Smart, Ying, Conrad, & Anderson, 1994), had no effects.

4. Human FHHNC Mutations Causing Trafficking Defects in CLDN16

Several studies have analyzed the subcellular localization of the different mutant CLDN16 proteins identified in patients with FHHNC (Hou et al., 2005; Kausalya et al., 2006; Konrad et al., 2008). As summarized in Table I, the CLDN16 mutants that failed to reach the cell surface localized either to the ER, the Golgi complex, or the lysosomes. The CLDN16 mutants that were retained in the ER were targeted for proteasomal degradation by the ER quality-control machinery (Kausalya et al., 2006). Cell-permeable chemical compounds that facilitate the folding of membrane proteins (pharmacological chaperones) have been shown to rescue the trafficking of some misfolded ER-retained proteins (Ulloa-Aguirre, Janovick, Brothers, & Conn, 2004). Using pharmacological chaperones, such as thapsigargin and 4-phenylbutyrate, Kausalya et al. (2006) have rescued the TJ localization of some misfolded CLDN16 mutants, in particular, R79L, G121R, and G169R.

5. The Molecular Interactions Between CLDN16 and CLDN19

Claudins interact with one another both intracellularly and intercellularly: they copolymerize linearly within the plasma membrane of the cell, together with the integral protein occludin, to form the classical intramembrane fibrils or strands visible in freeze-fracture replicas. These intramembrane interactions (side-to-side) can involve one claudin protein (homomeric or homopolymeric) or different claudins (heteromeric or heteropolymeric). In the formation of the intercellular junction, claudins may interact head-to-head with claudins in an adjacent cell, generating both homotypic and heterotypic claudin–claudin interactions (Furuse et al., 1999). Peripheral membrane proteins, such as ZO-1 (Stevenson, Siliciano, Mooseker, & Goodenough, 1986) and ZO-2 (Gumbiner, Lowenkopf, & Apatira, 1991), play key roles in recruiting claudins to TJ strands in polarized epithelial cells (Umeda et al., 2006), but are not required in fibroblasts (Furuse et al., 1999).

Using a novel split-ubiquitin yeast two-hybrid (Y2H) assay, Hou et al. (2008) have found a strong CLDN19 homomeric interaction and an equally strong CLDN19–16 heteromeric interaction. On the other hand, CLDN16 failed to show any homomeric interaction. The point mutations in CLDN19 (L90P and G123R) and in CLDN16 (L145P, L151F, G191R, A209T, and F232C) that are known to cause the inherited human disease FHHNC, disrupted the CLDN19–16 heteromeric interaction. In mammalian cells,

TABLE I
Mutations Affecting the Function of CLDN16

Construct	Position of mutation	Localization	P_{Na} (10^{-6} cm/s)	P_{Cl} (10^{-6} cm/s)	Function
Vector	–	–	6.381 ± 0.107	21.857 ± 0.107	–
WT	–	TJ	25.750 ± 0.092	21.310 ± 0.092	+
D97S	1st ECL	ER	9.065 ± 0.530	34.633 ± 0.530	–
D104S	1st ECL	TJ	19.260 ± 0.206	31.723 ± 0.208	Partial –
D105S	1st ECL	TJ	22.683 ± 0.223	26.917 ± 0.223	Partial –
E108T	1st ECL	TJ	22.937 ± 0.203	21.830 ± 0.200	+
E119T	1st ECL	TJ	22.917 ± 0.101	28.070 ± 0.098	Partial –
D126S	1st ECL	TJ	18.563 ± 0.047	24.120 ± 0.050	Partial –
D132S	1st ECL	TJ	28.983 ± 0.497	22.003 ± 0.500	+
E133T	1st ECL	TJ	23.307 ± 0.103	22.583 ± 0.103	+
D135S	1st ECL	TJ	25.003 ± 0.053	20.880 ± 0.050	+
E140T	1st ECL	TJ	17.063 ± 0.165	24.650 ± 0.162	Partial –
L145P	1st ECL	TJ	13.977 ± 0.174	29.723 ± 0.174	Partial –
R149L	1st ECL	ER	7.830 ± 0.186	26.157 ± 0.187	–
L151F	1st ECL	TJ	17.107 ± 0.093	24.603 ± 0.093	Partial –
L167P	2nd TMD	–	7.183 ± 0.038	27.450 ± 0.040	–
G191R	3rd TMD	TJ	17.070 ± 0.192	21.980 ± 0.192	Partial –
G198D	3rd TMD	–	6.656 ± 0.090	24.453 ± 0.091	–
A209T	2nd ECL	TJ	20.513 ± 0.156	31.927 ± 0.156	Partial –
R216T	2nd ECL	–	10.023 ± 0.159	29.873 ± 0.156	–
F232C	2nd ECL	TJ	31.570 ± 0.070	31.713 ± 0.073	+
G233D	2nd ECL	–	8.015 ± 0.106	28.690 ± 0.106	–
S235P	2nd ECL	ER	7.273 ± 0.366	26.717 ± 0.364	–
G239R	4th TMD	Golgi	8.906 ± 0.094	23.297 ± 0/094	–
K112S	1st ECL	TJ	27.587 ± 0.277	22.020 ± 0.280	+
R114T	1st ECL	TJ	26.203 ± 0.264	27.777 ± 0.264	+
R129T	1st ECL	TJ	30.387 ± 0.337	23.593 ± 0.337	+
K144S	1st ECL	TJ	25.570 ± 0.179	20.313 ± 0.176	+
R149T	1st ECL	ER	8.868 ± 0.140	26.430 ± 0.139	–

ECL, extracellular loop; ER, endoplasmic reticulum; Golgi, Golgi apparatus; TJ, tight junction; TMD, transmembrane domain; +, showing the function of CLDN16; –, abolishing the function of CLDN16.

such as the HEK293 cells, CLDN16 can be coimmunoprecipitated with CLDN19 (Hou et al., 2008). Freeze-fracture replicas revealed the assembly of TJ strands (homomeric interactions) in CLDN19-expressing L fibroblast cells, but not in CLDN16-expressing L cells (lack of homomeric interaction). In L cells coexpressing CLDN16 and CLDN19, CLDN16 was recruited to well-developed networks of TJ strands (Hou et al., 2008), supporting the heteromeric interaction between CLDN16 and CLDN19.

As L cells express no endogenous claudin, the forming of TJ strands in CLDN19-expressing cells indicated that CLDN19 was also capable of homotypic interaction. Coculture of L cells expressing CLDN16 and CLDN19 genes revealed that CLDN16 did not heterotypically interact with CLDN19 (Hou et al., 2008). Examination of TJ formation in cultured cells involving other claudins suggests a three-stage hypothesis of TJ assembly (Blasig et al., 2006; Piontek et al., 2008). First, CLDN16 *cis* associates with CLDN19 within the plane of the membrane into dimers, or higher oligomeric state. Second, *trans* interactions between CLDN19 in adjacent cells take place. Third, additional *cis* interactions between CLDN16 and CLDN19 occur, elaborating the TJ strands.

B. Biophysical Properties of CLDN16 and CLDN19 Paracellular Channels

1. *In Vitro* Studies of CLDN16 and CLDN19 Functions

In vitro studies of CLDN16 and CLDN19 function comprise the transfection of epithelial cell lines (such as MDCK and LLC-PK1 cells) that express no endogenous CLDN16 or CLDN19. Cells are grown to confluence on permeable filter supports allowing the measurement of transepithelial ionic permeabilities. To faithfully interpret the permeability values using these surrogate cell models, the following parameters need to be controlled: (1) the endogenous claudin expression is not affected by the overexpression of CLDN16 or CLDN19; (2) overexpressed CLDN16 or CLDN19 correctly translocates to the TJ; (3) any permeability change reflects the combined function of endogenous and exogenous claudins; and (4) the channel-dead mutations in CLDN16 and CLDN19 need to be identified and distinguished from the ones causing trafficking defects.

a. CLDN16 is a cation channel. The hypothesis that CLDN16 forms a selective paracellular Mg^{2+}/Ca^{2+} pore was tested in a number of *in vitro* studies. Ikari et al. (2004) transfected low-resistance MDCK cells with CLDN16 and reported that the Ca^{2+} flux in these cells was increased in the apical to basolateral direction, while the Ca^{2+} flux in the opposite direction remained unchanged. The Mg^{2+} flux was without any noticeable change. Kausalya et al. (2006) transfected the high-resistance MDCK-C7 cell line and found that CLDN16 only moderately increased Mg^{2+} permeability without any directional preference. The effects of CLDN16 on Mg^{2+}/Ca^{2+} permeation appeared so small (< 50%) that the Mg^{2+}/Ca^{2+} pore theory can hardly explain the dramatic effect of mutations in CLDN16 on Mg^{2+} and Ca^{2+} homoeostasis in FHNNC patients.

In contrast to these studies, Hou et al. (2005) transfected the anion-selective LLC-PK1 cells with CLDN16 and found a large increase in Na^+ permeability (P_{Na}) accompanied by an only moderately enhanced Mg^{2+} permeability (P_{Mg}). The increase in P_{Na} was not affected by a Na^+/K^+-ATPase inhibitor (1 mM ouabain) but greatly reduced or completely disappeared in all FHHNC relevant CLDN16 mutants (Table I). The permeability of CLDN16 to other alkali metal cations was found to be: $K^+ > Rb^+ > Na^+ \approx Li^+$. This sequence is quite different from the sequence of their free-solution mobilities and resembles Eisenman selectivity sequence V–VIII. This suggests that permeating cations have to dehydrate to enter the CLDN16 pore where negatively charged interaction sites within the pore stabilize the permeating cation (Diamond & Wright, 1969; Eisenman & Horn, 1983).

Functional studies of the CLDN2 channel show that there have to be at least two types of paracellular pores, distinguishing either between charged and uncharged solutes or according to molecular size (Van Itallie et al., 2008; Yu et al., 2009). The paracellular permeation is carried out through two distinct pathways: a high-capacity pathway with an estimated pore radius of about 4 Å and a low-capacity pathway with a pore radius of at least 7 Å. The high-capacity paracellular pores use intrapore electrostatic binding sites to achieve a high conductance with a high degree of charge selectivity. Charge selectivity is mediated by the electrostatic interaction of partially dehydrated permeating ions with charged sites within the pore. It is generally accepted that the charges on the ECL of claudins line the channel pore and electrostatically influence the passage of dehydrated ions (Colegio et al., 2003; Van Itallie et al., 2003). The first ECL of CLDN16 is enriched with 10 negatively charged amino acids (Fig. 4: Asp and Glu labeled as bold). Hou et al. (2005) have systematically mutated each of the negatively charged amino acids to serine or threonine to study the effects of charge upon the function of CLDN16 in LLC-PK1 cells. A summary of the physiological changes resulting from the mutations is shown in Table I. Mutational analysis identified a

```
CLDN16  WTDCWMVNADDSLEVS-TKCRGLWWECVTNAFDGIRTCDEYDSILAEHPLKLVVTRAL
        97--------------104-105--108-------112-114---------119------------ 126---129--132-133-135-----140------144-------149

CLDN19  WKQSSYAGDAIIT--AVGPYEGLWMSCASQST-GQVQCKLYDSLLALD-GHIQSARAL
        31--------------38---------------------------48------------------------------------------65-----68-----------74-----------------81
```

FIGURE 4 Alignment of the amino acid sequences of the first extracellular loop (ECL1) of CLDN16 with CLDN19.

locus of acidic amino acids (D104S, D105S, E119T, D126S, and E140T) that affected its cation selectivity and were interspersed with other acidic residues that had no effect. Mutation of each of the five functionally important residues had a modest effect (11–33% reduction in P_{Na}), and combining the mutations appeared to be additive (Hou et al., 2005).

b. CLDN19 is a Cl^- blocker. Using the LLC-PK1 cells, Hou et al. (2008) found that CLDN19 profoundly decreased absolute Cl^- permeability (P_{Cl}) without affecting absolute Na^+ permeability (P_{Na}). This effect was not inhibited by the Na^+/K^+-ATPase blocker, indicating a paracellular pathway for ion flux. The FHHNC mutations from human patients (Table II: G20D, Q57E, L90P, and G123R) either partially or completely abolished the CLDN19 effect on P_{Cl}. No significant effect of CLDN19 was found on P_{Mg}. A different study by Angelow, El-Husseini, Kanzawa, and Yu (2007) analyzed CLDN19-transfected MDCK cells and found that CLDN19 significantly decreased paracellular permeability to mono- and divalent cations (including Mg^{2+}). Collectively, both studies argue against the hypothesis that CLDN19 is part of a paracellular Mg^{2+}/Ca^{2+} pore.

c. Heteromeric CLDN16–CLDN19 interaction generates cation selectivity of the tight junction. When expressed separately in LLC-PK1 cells, CLDN16 increased P_{Na} while CLDN19 decreased P_{Cl} (Hou et al., 2005, 2008). A run of biochemical and genetic analyses had proven the physical interaction between CLDN16 and CLDN19 proteins (Hou et al., 2008). Coexpression of CLDN16 and CLDN19 in LLC-PK1 cells resulted in a dramatic upregulation of P_{Na} and downregulation of P_{Cl}, generating a highly

TABLE II

Mutations Affecting the Function of CLDN19

Construct	Position of mutation	Localization	P_{Na} (10^{-6} cm/s)	P_{Cl} (10^{-6} cm/s)	Function
Vector	–	–	6.866 ± 0.184	23.220 ± 0.182	–
CLDN19-WT	–	TJ	5.999 ± 0.014	6.836 ± 0.014	+
CLDN19-G20D	1st TMD	ER	5.866 ± 0.077	20.733 ± 0.076	–
CLDN19-Q57E	1st ECL	Apical membrane	7.996 ± 0.219	20.680 ± 0.219	–
CLDN19-L90P	2nd TMD	TJ	6.497 ± 0.077	12.623 ± 0.075	Partial –
CLDN19-G123R	3rd TMD	TJ	5.856 ± 0.035	9.831 ± 0.035	Partial –

ECL, extracellular loop; ER, endoplasmic reticulum; TJ, tight junction; TMD, transmembrane domain; +, showing the function of CLDN19; –, abolishing the function of CLDN19.

cation-selective paracellular pathway (Hou et al., 2008). The TJs in LLC-PK1 cells were transformed from having anion selectivity (at the LLC-PK1 cell baseline level) to having cation selectivity. Certain FHHNC mutations in CLDN16 (Table III: L145P, L151F, G191R, A209T, and F232C) or CLDN19 (Table III: L90P and G123R) that disrupted CLDN16–CLDN19 interaction, also abolished this physiological change (Hou et al., 2008). As CLDN16 colocalizes with CLDN19 in the TAL of the kidney (Angelow et al., 2007; Konrad et al., 2006), CLDN16 and CLDN19 association through heteromeric interactions generates cation selectivity of the TJ in the TAL. Human FHHNC mutations in CLDN16 or CLDN19 that abolish the cation selectivity diminish the lumen-positive diffusional potential (V_{te}) as the driving force for Mg^{2+} and Ca^{2+} reabsorption, readily explaining the devastating phenotypes in FHHNC patients.

According to Eisenman's theory (Diamond & Wright, 1969), the negative charge strength in the channel pore and its electrostatic interaction (the Coulomb forces) with permeating cations are what primarily control the cation selectivity. The ECL1 of both CLDN16 and CLDN19 are enriched

TABLE III

Coexpression of CLDN16 and CLDN19 Generates Cation Selectivity of the Tight Junction

Construct	P_{Na} (10^{-6} cm/s)	P_{Cl} (10^{-6} cm/s)	Function
Vector	6.866 ± 0.184	23.220 ± 0.182	Anion selectivity
CLDN19-WT	5.999 ± 0.014	6.836 ± 0.014	CLDN19 wild-type function
CLDN16-WT	33.363 ± 0.346	22.257 ± 0.346	CLDN16 wild-type function
CLDN19-WT + CLDN16-WT	28.563 ± 0.117	7.425 ± 0.118	Cation selectivity
CLDN19-WT + CLDN16-L145P	4.742 ± 0.011	5.627 ± 0.012	CLDN19 wild-type function
CLDN19-WT + CLDN16-L151F	2.964 ± 0.011	2.771 ± 0.011	CLDN19 wild-type function
CLDN19-WT + CLDN16-G191R	7.463 ± 0.030	8.092 ± 0.030	CLDN19 wild-type function
CLDN19-WT + CLDN16-A209T	8.267 ± 0.019	8.418 ± 0.019	CLDN19 wild-type function
CLDN19-WT + CLDN16-F232C	7.183 ± 0.018	8.639 ± 0.018	CLDN19 wild-type function
CLDN19-L90P + CLDN16-WT	15.463 ± 0.102	10.030 ± 0.103	CLDN16 wild-type function
CLDN19-G123R + CLDN16-WT	17.230 ± 0.030	10.167 ± 0.033	CLDN16 wild-type function

with negatively charged residues (Fig. 4; Asp and Glu labeled as bold: 10 in CLDN16, 4 in CLDN19), some of which may selectively bind cations while acting like a sieve to block anion passage. In an earlier study, Hou et al. (2005) identified a locus of acidic residues in the ECL1 of CLDN16 composing its cation selectivity filter. The acidic residues in the ECL1 of CLDN19 may block anion permeation and reduce anion selectivity, though the locus of its selectivity filter has not been identified. LLC-PK1 cells express endogenous claudins (e.g., CLDN1, -3, -4, and -7) and show anion-selective background (Fig. 5A). Addition of CLDN19 likely reduces the positive charge strength and the anion selectivity of the endogenous channel pore with its protruding negatively charged residues (Fig. 5B). Coexpression of CLDN16 and CLDN19 in LLC-PK1 cells generated a paracellular pathway that appeared to be the sum of the functional effects of CLDN16 and CLDN19

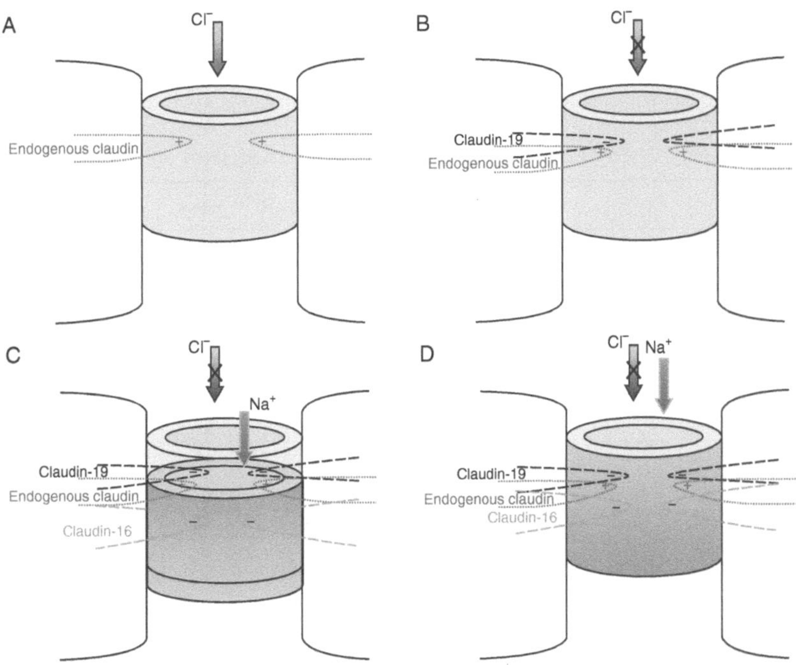

FIGURE 5 Structural models of claudin channels. (A) Claudin channels are depicted as cylinders joining two neighboring cell membranes, with charged residues protruding into the lumen of the cylinder. Endogenous LLC-PK1 cell background is anion-selective, permitting through Cl^-. (B) Addition of CLDN19 reduces the anion selectivity in LLC-PK1 cells and blocks Cl^- permeation. (C) CLDN16 and CLDN19 form two parallel homomeric channels each with its own physiologic signature. (D) CLDN16 and CLDN19 form a heteromeric channel with novel properties that require their synergy.

(i.e., increase in P_{Na} and decrease in P_{Cl}). There are two models of CLDN16 and CLDN19 channel structure and function: (1) CLDN16 and CLDN19 form two parallel homomeric channels each with its own physiologic signature (Fig. 5C); (2) CLDN16 and CLDN19 form a heteromeric channel with novel properties that require their interaction (Fig. 5D). The fact that CLDN16 lacked homomeric interaction on a Y2H assay and depended upon CLDN19 for assembly into TJ in a strand-forming assay argues against Model 1. The newly formed CLDN16–CLDN19 heteromeric channel retains the physiological characteristics of each homomeric channel. This is not directly in favor of Model 2. Nevertheless, any meaningful conclusion about the molecular basis of CLDN16–CLDN19 channel charge selectivity will require further studies on cells expressing both claudins.

2. *In Vivo* Analysis of CLDN16 Function

a. CLDN16 is required for TAL cation selectivity. CLDN16-deficient mice were recently generated using RNA interference (RNAi) technology by Hou et al. (2007). These CLDN16 knockdown (KD) mice showed significantly reduced plasma Mg^{2+} levels and excessive urinary excretions (approximately fourfold) of Mg^{2+} and Ca^{2+}. Furthermore, calcium deposits were observed in the basement membranes of the medullary tubules and the interstitium in the kidney of adult CLDN16 KD mice. These phenotypes of CLDN16 KD mice are very similar to the symptoms in human FHHNC patients.

As discussed in Section II, the paracellular reabsorption of Mg^{2+} and Ca^{2+} is driven by a lumen-positive V_{te} through two mechanisms: (1) the active transport V_{te} owing to apical K^+ recycling through ROMK and basolateral Cl^- exit through ClC-Kb channels, coupled with NaCl reabsorption via apical $Na^+2Cl^-K^+$ cotransporter (NKCC2) (blockable by furosemide); (2) the diffusion V_{te} generated by the transepithelial NaCl concentration gradient and the cation selectivity of the paracellular pathway of TAL. When isolated TAL segments were perfused *ex vivo* with symmetrical NaCl solutions, there was no difference in V_{te} between CLDN16 KD and wild-type (WT) mice (Hou et al., 2007). Thus, the mechanism "1" was normal in CLDN16 KD. However, in the presence of furosemide, the cation selectivity (P_{Na}/P_{Cl}) was significantly reduced from 3.1 ± 0.3 in WT to 1.5 ± 0.1 in CLDN16 KD (Hou et al., 2007). Loss of CLDN16 would dissipate the lumen-positive V_{te} through the mechanism "2." For example, with a NaCl gradient of 145 versus 30 mM, the resulting diffusion potential was 18 mV in WT, but only 6.6 mV in CLDN16 KD. The reduction in V_{te} well explained the Mg^{2+} and Ca^{2+} wasting in CLDN16 KD animals. P_{Na}/P_{Mg}, as measured by biionic potentials, was unchanged, suggesting that CLDN16 works as a nonselective cation channel in the TAL (Hou et al., 2007). These conclusions for the TAL are consistent with those obtained from the *in vitro* analyses of CLDN16-overexpressing LLC-PK1 epithelial cells by the same research group (Hou et al., 2005).

b. Defective paracellular cation selectivity in the TAL leads to salt loss. Renal handling of salt in CLDN16 KD mice is more complex. The mechanism "2" of V_{te} is an equilibrium potential. With the loss of CLDN16 and thereby the loss of cation selectivity without gain in TER, the diffusion potential of V_{te} was markedly reduced (Hou et al., 2007), that is, the voltage was below normal equilibrium. In consequence, substantial amounts of salt will backleak into the lumen. These changes will trigger complex compensation mechanisms. CLDN16 KD mice had increased the fractional excretion of Na^+ (FE_{Na}) and Cl^- (FE_{Cl}), indicating a net transport defect for Na^+ along the nephron (Himmerkus et al., 2008). These animals were also hypotensive and had elevated plasma aldosterone levels, suggesting that the total body Na^+ content was reduced. Aldosterone compensates Na^+ loss by the upregulation of Na^+ reabsorption in the collecting duct. In perfused CLDN16 KD mouse collecting ducts, amiloride-sensitive electrogenic Na^+ current was increased by fivefold (Himmerkus et al., 2008). Increases in luminal salt will be sensed by the macula densa located at the end of the TAL, leading to an activation of the tubuloglomerular feedback (TGF) (Thomson & Blantz, 2008). TGF allows changes in luminal salt to effect a reciprocal change in the glomerular filtration rate (GFR). The GFR was 27% lower in CLDN16 KD animals (Hou et al., 2007).

In the human FHHNC phenotype, there is also some evidence for significant salt and water loss. For example, polyuria and polydipsia are the most frequently reported symptoms from FHHNC patients (Weber, Schneider et al., 2001). Many patients receive their initial medical attention because of nocturnal enuresis. However, Blanchard et al. (2001) demonstrated normal salt excretion and normal response to loop diuretics in two siblings with proven CLDN16 mutations. It has to be noted that polyuria may also be caused by medullary nephrocalcinosis which affects the urinary diluting capacity of the loop of Henle, which in turn impairs the urinary concentrating ability because of the reduced medullary tonicity.

c. Homeostatic coupling of Na^+ handling with extracellular fluid volume and Ca^{2+}. The initial analysis showed that CLDN16 KD mice were able to cope with the renal loss of calcium. The plasma PTH concentrations were normal, whereas $1,25(OH)_2D_3$ levels were increased threefold in CLDN16 KD mice (Hou et al., 2007). Recent studies (Titze et al., 2004, 2005) show that Na^+ homeostasis is tightly coupled to Ca^{2+} homeostasis and that Ca^{2+} is retained by the kidney when the total body Na^+ content is reduced. This suggests that the extracellular fluid volume and the blood pressure are sacrificed by CLDN16 KD animals to maintain Ca^{2+} balance (Bukoski, 2004) when these animals are coping with Ca^{2+} and Na^{+} loss at the same time. In addition, the altered concentrations of luminal and interstitial Ca^{2+} along the distal nephron may increase urine volume production via Ca^{2+}-sensing receptor signaling (Hebert, Brown, & Harris, 1997; Sands et al., 1997). In FHHNC patients, this mechanism may contribute significantly to polyuria and increase their vulnerability toward restrictions in salt and water intake.

V. PERSPECTIVE

A. The Biochemical Nature of CLDN16 and CLDN19 Oligomerization

Though the molecular mechanism underlying claudin oligomerization is largely unknown, it has been hypothesized that claudin oligomerization occurs before strand assembly on the basis of CLDN4 expression studies in insect cells (Mitic et al., 2003) and exhibit ~10 nm-sized oligomers. Following trafficking to the cell surface, it is believed that oligomerized claudins then assemble into the TJ strands where they interact with cognate claudins in the adjacent cell (Piontek et al., 2008). Assembly of claudins into TJ strands requires the TJ scaffold proteins ZO-1 and ZO-2, which interact with both claudin PDZ-binding domains (Itoh et al., 1999) and the TJ peripheral proteins such as cingulin, Par-3, and Par-6 (Cordenonsi et al., 1999; Joberty, Petersen, Gao, & Macara, 2000). An elegant study by Mitic et al. (2003) has used the perfluoro-octanoic acid (PFO) as a detergent to solubilize the CLDN4 oligomer and found that CLDN4 has a hexameric quaternary configuration. The structural configuration of assembled CLDN16–CLDN19 oligomer is not known. A similar study using PFO solubilization, followed by sucrose gradient sedimentation, will allow the determination of the molecular size of the CLDN16–CLDN19 oligomer. To reveal the stoichiometry of CLDN16–CLDN19 oligomer, efforts will be made to purify the CLDN16–CLDN19 complex assembled in insect cell models. The structural insights of CLDN16–CLDN19 oligomerization will help elucidate the transport functions of the CLDN16–CLDN19 channel.

B. The Molecular Basis of CLDN16-CLDN19 Channel Charge Selectivity

The most elementary question about selectivity is how a channel preferentially binds its chosen ion and rejects other ions. It is generally accepted that the charges on the ECL1 of claudins line the channel pores and electrostatically influence the passage of soluble ions. In particular, an elegant study by Yu et al. (2009) found CLDN2 channel pore to be narrow and cation selective. The charge selectivity was mediated by the electrostatic interaction of permeating cations with a negatively charged site (D65) within ECL1 of CLDN2. The ECL1 of both CLDN16 and CLDN19 are enriched with negatively charged residues. In a preliminary study, Hou et al. (2005) found CLDN16 to be a cation-selective channel and that charge removal in D104, D105, E119, D126, and E140 in ECL1 of CLDN16 (Fig. 4) reduced its cation selectivity. Nevertheless, this study using mutagenesis to replace charged residues with neutral residues (Ser or Thr) did not exclude the steric effects of mutations on protein folding and interaction that could also influence ion permeation. Changes caused by charge removal in the electrostatic field in CLDN16 channel pore have not been recorded. Hou et al. (2008) found that CLDN19 functioned as a Cl^- blocker. However, the selectivity filter of its channel pore has not been identified. The negatively charged residues in ECL1 of CLDN19 (e.g., D38, E48, D68, and D74; Fig. 4) are likely important for its charge selectivity. The residues composing charge selectivity must satisfy two criteria: (1) charge removal leads to a loss of inhibition of Cl^- permeability; and (2) the residue directly interacts with permeating anions through its electrostatic field. The role of CLDN16–CLDN19 interaction in their function has not been fully elucidated. It is still debated whether CLDN16 and CLDN19 form two parallel homomeric channels each with its own physiologic signature (Model 1; Fig. 5C); or CLDN16 and CLDN19 form a heteromeric channel with novel properties that require their interaction (Model 2; Fig. 5D). No dominant negative mutation in CLDN16 or CLDN19 has ever been identified. These mutations must satisfy two criteria: (A) they will not affect CLDN16 and CLDN19 interaction; and (B) expressing mutant CLDN16 with WT CLDN19 or mutant CLDN19 with WT CLDN16 will antagonize WT CLDN19 and CLDN16 functions, respectively. Mutations in CLDN16 or CLDN19 selectivity filter are good candidates.

C. Mouse Models Deficient in Both CLDN16 and CLDN19

Both CLDN16 and CLDN19 are associated with human FHHNC. Miyamoto et al. (2005) have generated a CLDN19 KO mouse model and found behavioral abnormalities in these animals that could be attributed to peripheral nervous system deficits. The renal phenotypes in CLDN19 KO have not been analyzed. In cultured renal epithelial cells, CLDN16 and CLDN19 show similar effects of increasing the cation selectivity of the TJ (Hou et al., 2005, 2008). The cation-selective TJ is vital in the process of building up the lumen-positive diffusion potential in the TAL, which is the driving force for Mg^{2+} reabsorption. Given the fact that the FHHNC patients with CLDN19 mutations show renal abnormalities similar to those patients with CLDN16 mutations, CLDN19 KO mice will likely recapitulate the FHHNC phenotypes in mice similar to those of CLDN16 KD, described by Hou et al. (2007).

Generation of the mouse models deficient in both claudins will be critical for testing the two structural models of CLDN16 and CLDN19 functions as hypothesized in the section earlier on CLDN16 and CLDN19 interactions. Model 1 (*CLDN16 and CLDN19 form two parallel homomeric channels*) predicts that the double KD (KO) mice will have a severe phenotype resembling the additive effects of the two individual KDs (KO). The renal wasting of Mg^{2+} and Ca^{2+} will be more severe in the double KD (KO), with a more rapid progression of nephrocalcinosis and toward renal failure. If the interaction between the two claudins is required for normal function *in vivo*, as predicted by Model 2 (*CLDN16 and CLDN19 form a heteromeric channel*), the measured parameters such as the renal excretion of Mg^{2+} and Ca^{2+} in the double KD (KO) will be similar to those seen in individual KDs (KO).

Acknowledgments

This work was supported by American Heart Association (grant 0930050N to J. H.) and the Peter Foundation (to M. K.).

References

Amasheh, S., Meiri, N., Gitter, A. H., Schoneberg, T., Mankertz, J., Schulzke, J. D., et al. (2002). Claudin-2 expression induces cation-selective channels in tight junctions of epithelial cells. *Journal of Cell Science, 115*, 4969–4976.

Angelow, S., El-Husseini, R., Kanzawa, S. A., & Yu, A. S. (2007). Renal localization and function of the tight junction protein, claudin-19. *American Journal of Physiology. Renal Physiology, 293*, F166–F177.

Benigno, V., Canonica, C. S., Bettinelli, A., von Vigier, R. O., Truttmann, A. C., & Bianchetti, M. G. (2000). Hypomagnesaemia-hypercalciurianephrocalcinosis: A report of nine cases and a review. *Nephrology, Dialysis, Transplantation, 15*, 605–610.

Ben-Yosef, T., Belyantseva, I. A., Saunders, T. L., Hughes, E. D., Kawamoto, K., Van Itallie, C. M., et al. (2003). Claudin 14 knockout mice, a model for autosomal recessive deafness DFNB29, are deaf due to cochlear hair cell degeneration. *Human Molecular Genetics, 12*, 2049–2061.

Blanchard, A., Jeunemaitre, X., Coudol, P., Dechaux, M., Froissart, M., May, A., et al. (2001). Paracellin-1 is critical for magnesium and calcium reabsorption in the human thick ascending limb of Henle. *Kidney International, 59*, 2206–2215.

Blasig, I. E., Winkler, L., Lassowski, B., Mueller, S. L., Zuleger, N., Krause, E., et al. (2006). On the self-association potential of transmembrane tight junction proteins. *Cellular and Molecular Life Sciences, 63*, 505–514.

Bukoski, R. D. (2004). Linkage of Na^{+} and Ca^{2+} balance: Evidence that Na^{+} retention preserves Ca^{2+} balance and limits bone wasting. *Journal of Hypertension, 22*, 683–685.

Coconnier, M. H., Lorrot, M., Barbat, A., Laboisse, C., & Servin, A. L. (2000). Listeriolysin O-induced stimulation of mucin exocytosis in polarized intestinal mucin-secreting cells: Evidence for toxin recognition of membrane-associated lipids and subsequent toxin internalization through caveolae. *Cellular Microbiology, 2*, 487–504.

Colegio, O. R., Van Itallie, C. M., McCrea, H. J., Rahner, C., & Anderson, J. M. (2002). Claudins create charge-selective channels in the paracellular pathway between epithelial cells. *American Journal of Physiology. Cell Physiology, 283*, C142–C147.

Colegio, O. R., Van Itallie, C., Rahner, C., & Anderson, J. M. (2003). Claudin extracellular domains determine paracellular charge selectivity and resistance but not tight junction fibril architecture. *American Journal of Physiology. Cell Physiology, 284*, C1246–C1254.

Cordenonsi, M., D'Atri, F., Hammar, E., Parry, D. A., Kendrick-Jones, J., Shore, D., et al. (1999). Cingulin contains globular and coiled-coil domains and interacts with ZO-1, ZO-2, ZO-3, and myosin. *Journal of Cell Biology, 147*, 1569–1582.

Daugherty, B. L., Ward, C., Smith, T., Ritzenthaler, J. D., & Koval, M. (2007). Regulation of heterotypic claudin compatibility. *Journal of Biological Chemistry, 282*, 30005–30013.

Diamond, J. M., & Wright, E. M. (1969). Biological membranes: The physical basis of ion and nonelectrolyte selectivity. *Annual Review of Physiology, 31*, 581–646.

Di Stefano, A., Roinel, N., De Rouffignac, C., & Wittner, M. (1993). Transepithelial Ca^{2+} and Mg^{2+} transport in the cortical thick ascending limb of Henle's loop of the mouse is a voltage-dependent process. *Renal Physiology and Biochemistry, 16*, 157–166.

Ebnet, K., Suzuki, A., Ohno, S., & Vestweber, D. (2004). Junctional adhesion molecules (JAMs): More molecules with dual functions? *Journal of Cell Science, 117*, 19–29.

Eisenman, G., & Horn, R. (1983). Ionic selectivity revisited: The role of kinetic and equilibrium processes in ion permeation through channels. *Journal of Membrane Biology, 76*, 197–225.

Farquhar, M. G., & Palade, G. E. (1963). Junctional complexes in various epithelia. *Journal of Cell Biology, 17*, 375–412.

Furuse, M., Fujita, K., Hiiragi, T., Fujimoto, K., & Tsukita, S. (1998). Claudin-1 and -2: Novel integral membrane proteins localizing at tight junctions with no sequence similarity to occludin. *Journal of Cell Biology, 141*, 1539–1550.

Furuse, M., Hirase, T., Itoh, M., Nagafuchi, A., Yonemura, S., Tsukita, Sa., & Tsukita, Sh. (1993). Occludin—A novel integral membrane protein localizing at tight junctions. *Journal of Cell Biology, 123*, 1777–1788.

Furuse, M., Sasaki, H., & Tsukita, S. (1999). Manner of interaction of heterogeneous claudin species within and between tight junction strands. *Journal of Cell Biology, 147*, 891–903.

Ghassemifar, M. R., Sheth, B., Papenbrock, T., Leese, H. J., Houghton, F. D., & Fleming, T. P. (2002). Occludin TM4(-): An isoform of the tight junction protein present in primates lacking the fourth transmembrane domain. *Journal of Cell Science, 115*, 3171–3180.

Greger, R. (1981). Cation selectivity of the isolated perfused cortical thick ascending limb of Henle's loop of rabbit kidney. *Pflugers Archiv: European Journal of Physiology, 390*, 30–37.

Gumbiner, B., Lowenkopf, T., & Apatira, D. (1991). Identification of a 160 kDa polypeptide that binds to the tight junction protein ZO-1. *Proceedings of the National Academy of Sciences of the United States of America, 88*, 3460–3464.

Günzel, D., Haisch, L., Pfaffenbach, S., Krug, S. M., Milatz, S., Amasheh, S., et al. (2009). Claudin function in the thick ascending limb of Henle's loop. *Annals of the New York Academy of Sciences, 1165*, 152–162.

Hamazaki, Y., Itoh, M., Sasaki, H., Furuse, M., & Tsukita, S. (2001). Multi-PDZ-containing protein 1 (MUPP1) is concentrated at tight junctions through its possible interaction with claudin-1 and junctional adhesion molecule (JAM). *Journal of Biological Chemistry, 277*, 455–461.

Hebert, J. C., & Andreoli, T. E. (1986). Ionic conductance pathways in the mouse medullary thick ascending limb of Henle. *Journal of General Physiology, 87*, 567–590.

Hebert, S. C., Brown, E. M., & Harris, H. W. (1997). Role of the Ca^{2+}-sensing receptor in divalent mineral ion homeostasis. *Journal of Experimental Biology, 200*, 295–302.

Heuser, J. E., & Anderson, R. G. (1989). Hypertonic media inhibit receptor-mediated endocytosis by blocking clathrin-coated pit formation. *Journal of Cell Biology, 108*, 389–400.

Himmerkus, N., Shan, Q., Goerke, B., Hou, J., Goodenough, D. A., & Bleich, M. (2008). Salt and acid-base metabolism in claudin-16 knockdown mice: Impact for the pathophysiology of FHHNC patients. *American Journal of Physiology. Renal Physiology, 295*, F1641–F1647.

Hirano, T., Kobayashi, N., Itoh, T., Takasuga, A., Nakamaru, T., Hirotsune, S., et al. (2000). Null mutation of PCLN-1/Claudin-16 results in bovine chronic interstitial nephritis. *Genome Research, 10*, 659–663.

Hou, J., Paul, D. L., & Goodenough, D. A. (2005). Paracellin-1 and the modulation of ion selectivity of tight junctions. *Journal of Cell Science, 118*, 5109–5118.

Hou, J., Renigunta, A., Konrad, M., Gomes, A. S., Schneeberger, E. E., Paul, D. L., et al. (2008). Claudin-16 and claudin-19 interact and form a cation-selective tight junction complex. *Journal of Clinical Investigation, 118*, 619–628.

Hou, J., Shan, Q., Wang, T., Gomes, A. S., Yan, Q., Paul, D. L., et al. (2007). Transgenic RNAi depletion of claudin-16 and the renal handling of magnesium. *Journal of Biological Chemistry, 282*, 17114–17122.

Ikari, A., Hirai, N., Shiroma, M., Harada, H., Sakai, H., Hayashi, H., et al. (2004). Association of paracellin-1 with ZO-1 augments the reabsorption of divalent cations in renal epithelial cells. *Journal of Biological Chemistry, 279*, 54826–54832.

Ikari, A., Matsumoto, S., Harada, H., Takagi, K., Hayashi, H., Suzuki, Y., et al. (2006). Phosphorylation of paracellin-1 at Ser217 by protein kinase A is essential for localization in tight junctions. *Journal of Cell Science, 119*, 1781–1789.

Itoh, M., Furuse, M., Morita, K., Kubota, K., Saitou, M., & Tsukita, S. (1999). Direct binding of three tight junction-associated MAGUKs, ZO-1, ZO-2, and ZO-3, with the COOH termini of claudins. *Journal of Cell Biology, 147*, 1351–1363.

Joberty, G., Petersen, C., Gao, L., & Macara, I. G. (2000). The cell-polarity protein Par6 links Par3 and atypical protein kinase C to Cdc42. *Nature Cell Biology, 2*, 531–539.

Kausalya, P. J., Amasheh, S., Günzel, D., Wurps, H., Müller, D., Fromm, M., et al. (2006). Disease-associated mutations affect intracellular traffic and paracellular Mg^{2+} transport function of Claudin-16. *Journal of Clinical Investigation, 116*, 878–891.

Kerstan, D., & Quamme, G. (2002). Physiology and pathophysiology of intestinal absorption of magnesium. In S. G. Massry, H. Morii & Y. Nishizawa (Eds.), *Calcium in internal medicine* (pp. 171–183). London: Springer-Verlag.

Konrad, M., Hou, J., Weber, S., Dötsch, J., Kari, J. A., Seeman, T., et al. (2008). The *CLDN16* genotype predicts the progression of renal failure in familial hypomagnesemia with hypercalciuria and nephrocalcinosis. *Journal of the American Society of Nephrology, 19*, 171–181.

Konrad, M., Schaller, A., Seelow, D., Pandey, A. V., Waldegger, S., Lesslauer, A., et al. (2006). Mutations in the tight-junction gene claudin 19 (CLDN19) are associated with renal magnesium wasting, renal failure, and severe ocular involvement. *American Journal of Human Genetics, 79*, 949–957.

Lee, D. B., Huang, E., & Ward, H. J. (2006). Tight junction biology and kidney dysfunction. *American Journal of Physiology. Renal Physiology, 290*, F20–F34.

Mandon, B., Siga, E., Roinel, N., & De Rouffignac, C. (1993). Ca^{2+}, Mg^{2+} and K^{+} transport in the cortical and medullary thick ascending limb of the rat nephron: Influence of transepithelial voltage. *Pflugers Archiv: European Journal of Physiology, 424*, 558–560.

Manz, F., Schärer, K., Janka, P., & Lombeck, J. (1978). Renal magnesium wasting, incomplete tubular acidosis, hypercalciuria and nephrocalcinosis in siblings. *European Journal of Pediatrics, 128*(2), 67–79.

Michelis, M. F., Drash, A. L., Linarelli, L. G., De Rubertis, F. R., & Davis, B. B. (1972). Decreased bicarbonate threshold and renal magnesium wasting in a sibship with distal renal tubular acidosis: Evaluation of the pathophysiological role of parathyroid hormone. *Metabolism, 21*, 905–920.

Mitic, L. L., Unger, V. M., & Anderson, J. M. (2003). Expression, solubilization, and biochemical characterization of the tight junction transmembrane protein claudin-4. *Protein Science, 12*, 218–227.

Miyamoto, T., Morita, K., Takemoto, D., Takeuchi, K., Kitano, Y., Miyakawa, T., et al. (2005). Tight junctions in Schwann cells of peripheral myelinated axons: A lesson from claudin-19-deficient mice. *Journal of Cell Biology, 169*, 527–538.

Morita, K., Furuse, M., Fujimoto, K., & Tsukita, S. (1999). Claudin multigene family encoding four-transmembrane domain protein components of tight junction strands. *Proceedings of the National Academy of Sciences of the United States of America, 96*, 511–516.

Müller, D., Kausalya, P. J., Claverie-Martin, F., Meij, I. C., Eggert, P., Garcia-Nieto, V., et al. (2003). A novel claudin 16 mutation associated with childhood hypercalciuria abolishes binding to ZO-1 and results in lysosomal mistargeting. *American Journal of Human Genetics, 73*, 1293–1301.

Muresan, Z., Paul, D. L., & Goodenough, D. A. (2000). Occludin 1B, a variant of the tight junction protein occludin. *Molecular Biology of the Cell, 11*, 627–634.

Ohba, Y., Kitagawa, H., Kitoh, K., Sasaki, Y., Takami, M., Shinkai, Y., et al. (2000). A deletion of the paracellin-1 gene is responsible for renal tubular dysplasia in cattle. *Genomics, 68*, 229–236.

Piontek, J., Winkler, L., Wolburg, H., Müller, S. L., Zuleger, N., Piehl, C., et al. (2008). Formation of tight junction: Determinants of hemophilic interaction between classic claudins. *FASEB Journal, 22*, 146–158.

Praga, M., Vara, J., Gonzalez-Parra, E., Andres, A., Alamo, C., Araque, A., et al. (1995). Familial hypomagnesemia with hypercalciuria and nephrocalcinosis. *Kidney International, 47*, 1419–1425.

Quamme, G. A. (1997). Renal magnesium handling: New insights in understanding old problems. *Kidney International, 52*, 1180–1195.

Quamme, G. A., & Dirks, J. H. (1980). Magnesium transport in the nephron. *American Journal of Physiology, 239*, F393–F401.

Rodriguez-Soriano, J., & Vallo, A. (1994). Pathophysiology of the renal acidification defect present in the syndrome of familial hypomagnesaemia-hypercalciuria. *Pediatric Nephrology, 8*, 431–435.

Sands, J. M., Naruse, M., Baum, M., Jo, I., Hebert, S. C., Brown, E. M., et al. (1997). Apical extracellular calcium/polyvalent cation-sensing receptor regulates vasopressin-elicited water permeability in rat kidney inner medullary collecting duct. *Journal of Clinical Investigation, 99*, 1399–1405.

Sandvig, K., Olsnes, S., Petersen, O. W., & van Deurs, B. (1987). Acidification of the cytosol inhibits endocytosis from coated pits. *Journal of Cell Biology, 105*, 679–689.

Sasaki, H., Matsui, C., Furuse, K., Mimori-Kiyosue, Y., Furuse, M., & Tsukita, S. (2003). Dynamic behavior of paired claudin strands within apposing plasma membranes. *Proceedings of the National Academy of Sciences of the United States of America, 100*, 3971–3976.

Shareghi, G. R., & Agus, Z. S. (1982). Magnesium transport in the cortical thick ascending limb of Henle's loop of the rabbit. *Journal of Clinical Investigation, 69*, 759–769.

Simon, D. B., Lu, Y., Choate, K. A., Velazquez, H., Al Sabban, E., Praga, M., et al. (1999). Paracellin-1, a renal tight junction protein required for paracellular Mg^{2+} resorption. *Science, 285*, 103–106.

Smart, E. J., Ying, Y. S., Conrad, P. A., & Anderson, R. G. (1994). Caveolin moves from caveolae to the Golgi apparatus in response to cholesterol oxidation. *Journal of Cell Biology, 127*, 1185–1197.

Stevenson, B. R., Siliciano, J. D., Mooseker, M. S., & Goodenough, D. A. (1986). Identification of ZO-1: A high molecular weight polypeptide associated with the tight junction (zonula occludens) in a variety of epithelia. *Journal of Cell Biology, 103*, 755–766.

Tang, V. W., & Goodenough, D. A. (2003). Paracellular ion channel at the tight junction. *Biophysical Journal, 84*, 1660–1673.

Thomson, S. C., & Blantz, R. C. (2008). Glomerulotubular balance, tubuloglomerular feedback, and salt homeostasis. *Journal of the American Society of Nephrology, 19*, 2272–2275.

Thorleifsson, G., Holm, H., Edvardsson, V., Walters, G. B., Styrkarsdottir, U., Gudbjartsson, D. F., et al. (2009). Sequence variants in the CLDN14 gene associate with kidney stones and bone mineral density. *Nature Genetics, 41*, 926–930.

Titze, J., Bauer, K., Schafflhuber, M., Dietsch, P., Lang, R., Schwind, K. H., et al. (2005). Internal sodium balance in DOCA-salt rats: A body composition study. *American Journal of Physiology. Renal Physiology, 289*, F793–F802.

Titze, J., Rittweger, J., Dietsch, P., Krause, H., Schwind, K. H., Engelke, K., et al. (2004). Hypertension, sodium retention, calcium excretion and osteopenia in Dahl rats. *Journal of Hypertension, 22*, 803–810.

Tsukita, S., & Furuse, M. (2000). Pores in the wall. Claudins constitute tight junction strands containing aqueous pores. *Journal of Cell Biology, 149*, 13–16.

Tsukita, S., Furuse, M., & Itoh, M. (2001). Multifuctional strands in tight junctions. *Nature Reviews. Molecular Cell Biology, 2*, 285–293.

Ulloa-Aguirre, A., Janovick, J. A., Brothers, S. P., & Conn, P. M. (2004). Pharmacologic rescue of conformationally-defective proteins: Implications for the treatment of human disease. *Traffic, 5*, 821–837.

Umeda, K., Ikenouchi, J., Katahira-Tayama, S., Furuse, K., Sasaki, H., Nakayama, M., et al. (2006). ZO-1 and ZO-2 independently determine where claudins are polymerized in tight-junction strand formation. *Cell, 126*, 741–754.

Van Itallie, C. M., Fanning, A. S., & Anderson, J. M. (2003). Reversal of charge selectivity in cation or anion selective epithelial lines by expression of different claudins. *American Journal of Physiology. Renal Physiology, 285*, F1078–F1084.

Van Itallie, C. M., Holmes, J., Bridges, A., Gookin, J. L., Coccaro, M. R., Proctor, W., et al. (2008). The density of small tight junction pores varies among cell types and is increased by expression of claudin-2. *Journal of Cell Science, 121*, 298–305.

Van Itallie, C., Rahner, C., & Anderson, J. M. (2001). Regulated expression of claudin-4 decreases paracellular conductance through a selective decrease in sodium permeability. *Journal of Clinical Investigation, 107*, 1319–1327.

Watson, C. J., Rowland, M., & Warhurst, G. (2001). Functional modeling of tight junctions in intestinal cell monolayers using polyethylene glycol oligomers. *American Journal of Physiology. Cell Physiology, 281*, C388–C397.

Weber, S., Schlingmann, K. P., Peters, M., Nejsum, L. N., Nielsen, S., Engel, H., et al. (2001). Primary gene structure and expression studies of rodent paracellin-1. *Journal of the American Society of Nephrology, 12*, 2664–2672.

Weber, S., Schneider, L., Peters, M., Misselwitz, J., Ronnefarth, G., Boswald, M., et al. (2001). Novel paracellin-1 mutations in 25 families with familial hypomagnesemia with hypercalciuria and nephrocalcinosis. *Journal of the American Society of Nephrology, 12*, 1872–1881.

Wen, H., Watry, D. D., Marcondes, M. C., & Fox, H. S. (2004). Selective decrease in paracellular conductance of tight junctions: Role of the first extracellular domain of claudin-5. *Molecular and Cellular Biology, 24*, 8408–8417.

Yu, A. S., Cheng, M. H., Angelow, S., Günzel, D., Kanzawa, S. A., Schneeberger, E. E., et al. (2009). Molecular basis for cation selectivity in claudin-2-based paracellular pores: Identification of an electrostatic interaction site. *Journal of General Physiology, 133*, 111–127.

Yu, A. S., Enck, A. H., Lencer, W. I., & Schneeberger, E. E. (2003). Claudin-8 expression in MDCK cells augments the paracellular barrier to cation permeation. *Journal of Biological Chemistry, 278*, 17350–17359.

Zimmermann, B., Plank, C., Konrad, M., Stöhr, W., Gravou-Apostolatou, C., Rascher, W., et al. (2006). Hydrochlorothiazide in CLDN16 mutation. *Nephrology, Dialysis, Transplantation, 21*, 2127–2132.

CHAPTER 8

Claudins and Barrier Function of the Lung

Per Flodby,* Zea Borok,*,† Edward D. Crandall,*,‡,# and Kwang-Jin Kim*,§,¶,**

*Department of Medicine, Will Rogers Institute Pulmonary Research Center, Keck School of Medicine, University of Southern California, Los Angeles, California, USA
†Department of Biochemistry and Molecular Biology, Keck School of Medicine, University of Southern California, Los Angeles, California, USA
‡Department of Pathology, Keck School of Medicine, University of Southern California, Los Angeles, California, USA
§Department of Physiology and Biophysics, Keck School of Medicine, University of Southern California, Los Angeles, California, USA
¶Department of Pharmacology and Pharmaceutical Sciences, School of Pharmacy, University of Southern California, Los Angeles, California, USA
#Department of Chemical Engineering and Materials Science, Viterbi School of Engineering, University of Southern California, Los Angeles, California, USA
**Department of Biomedical Engineering, Viterbi School of Engineering, University of Southern California, Los Angeles, California, USA

Current Topics in Membranes, Volume 65

1063-5823/10 $35.00
DOI: 10.1016/S1063-5823(10)65008-9

I. OVERVIEW

Claudins are constituents of tight junctional protein complexes and are thought to manifest the unique permselectivity properties of a given epithelial or endothelial barrier. Although several claudin isoforms have been found in the lung (claudin-1, -3, -4, -5, -7, -10, and -18), not much lung-specific information on claudin function has been obtained so far. High expression levels in pneumocytes of claudin-3, -4, and -18 are likely to contribute to the maintenance of the tight alveolar epithelial barrier, which helps prevent leakage of excess liquid (and hydrophilic solutes including serum proteins) from the interstitial and vascular spaces into the alveolar airspaces. The underlying mechanisms for the regulation of claudin expression and localization in the lung are not known in great detail, although TNFα, IFN-γ, TGFβ, IL-13, and matrix metalloproteinases are known to affect the expression of various tight junctional proteins in the lung and/or other tissues. Negative regulation of claudins (e.g., by the transcription factor Snail in mammary epithelial cells) may also be in operation in the lung, especially for lung cells undergoing inflammation and epithelial–mesenchymal transition. Posttranslational modification of claudins by phosphorylation or palmitoylation reported in other tissues for the regulation of the efficiency of insertion and stable integration of claudins into tight junctional complexes may play important role(s) in epithelial cell transdifferentiation and inflammation/injury/recovery in lung epithelial/endothelial barriers. More studies are needed to define the distribution/expression profile of various claudin isoforms in different cell types in the lung and to elucidate claudin-dictated permselectivity properties of pulmonary epithelial and endothelial barriers in normal and diseased lungs.

II. INTRODUCTION

Conducting airways (trachea, bronchi, and bronchioles) and respiratory air saccules (alveoli) of the lung are lined by a number of different types of epithelial cells (e.g., surface epithelial cells, basal cells, goblet cells for trachea and bronchi/bronchioles; nonciliated Clara cells in the terminal region of bronchioles; and alveolar epithelial type I (AT1) and type II (AT2) cells in alveoli). These lung epithelial cells are connected via several types of junctional protein complexes. The apical-most aspect of these complexes constitutes the tight junction (TJ). A slew of tight junctional proteins have been reported to date, including occludin, claudins, tricellulins, and junction adhesion molecules (JAM), which connect epithelial cells, whereas several so-called junction-associated intracellular structural proteins (e.g., zonula occludens (ZO) proteins) are also congregated at TJ. The first description of the claudin family of proteins appeared in 1998 from the laboratory of the late S. Tsukita (Furuse, Fujita, Hiiragi, Fujimoto, & Tsukita, 1998). We now know that the claudin family includes at least 24 members. It can be noted, however, that some of the claudin family members were identified (in retrospect) as far back as 1991 as rat ventral prostate 1, RVP1 (the same as claudin-3; Briehl & Miesfeld, 1991). In this regard, others include *Clostridium perfringens* enterotoxin receptor, CPE-R (claudin-4; Katahira, Inoue, Horiguchi, Matsuda, & Sugimoto, 1997; Katahira et al., 1997), and oligodendrocyte-specific protein, OSP (claudin-11; Bronstein et al., 1996), although at the time of identification, their role(s) in tight junctional structure/function were unclear. Following the seminal report by Tsukita's group that transfection into fibroblasts of claudin-1 or claudin-2 can establish connections between these cells that lack TJ (Furuse, Sasaki, Fujimoto, & Tsukita, 1998), structure–function relations of claudins have been extensively studied, as described in excellent recent reviews (Angelow, Ahlstrom, & Yu, 2008; Angelow & Yu, 2007; Findley & Koval, 2009; Terry, Nie, Matter, & Balda; Van Itallie & Anderson, 2006).

The major function of claudins as the constituents of tight junctional proteins appears to be to provide selective permeability to small inorganic ions (and perhaps to hinder the unrestricted passage of small hydrophilic nonelectrolytes), which is unique among the various TJ-constituting proteins including occludin and JAM (Angelow, Schneeberger, & Yu, 2007). Studies of claudin expression profiles in various epithelia indicate that claudins are expressed in an epithelial cell type-specific manner. Widely different expression profiles for claudins in various epithelia determine either leaky or tight (low or high transepithelial electrical resistance) epithelial phenotypes, where the permeability of TJ dictates the phenotype. In this regard, detailed studies in the kidney have revealed that the expression levels of individual claudins can dramatically change the paracellular bioelectrical and transport properties of different segments of the renal epithelium. For example, claudin-2 is highly expressed in the proximal tubule and thin descending limb, generating a leaky epithelium capable of high Na^+ conductance. In contrast, more distal segments of the renal tubular epithelium express high levels of claudin-4 but low levels of claudin-2, resulting in an epithelium with low cation conductance (i.e., tight epithelium) (reviewed in Angelow & Yu, 2007; Van Itallie & Anderson, 2006).

Epithelial TJ constitute the rate-limiting barrier to the paracellular transport of hydrophilic solutes, including inorganic ions. The ionic permselectivity properties of TJ are thought to be almost exclusively defined by the claudins expressed in the TJ complex, although other constituents (e.g., occludin and tricellulins) may play indirect (or direct) role(s). In TJ strands of the same cell, claudin molecules interact with one another in a so-called side-by-side fashion. On the other hand, interactions between claudins located on different cells take place in a head-to-head fashion, which can be either homotypic (between same isoforms) or heterotypic (between different isoforms) (Angelow & Yu, 2007; Van Itallie & Anderson, 2006). Thus, the permselectivity properties of small pores formed by claudins are determined by both the specific claudin interactions that predominate and the particular isoforms that are involved.

The amino acid sequences of the first extracellular domain of claudins vary widely for different members and are thought to confer the unique permselectivity properties when pores are formed by adjoining claudins from two cells, lending credence to the thesis that the function of a TJ barrier between two epithelial cells is determined by the specifically expressed claudin members (and by their levels of expression) (Angelow & Yu, 2007). As discussed earlier, the formation of tight junctional pores involves the interaction between claudins emanating from two neighboring cells in a homotypic and/or heterotypic fashion. However, several studies have provided evidence that there are only a limited number of combinations of heterotypic interactions between claudin family members that actually occur. In other words, only specific combinations (e.g., between claudin-1 and claudin-3, but not between claudin-1 and claudin-2) of claudins take place (Coyne, Ribeiro, Boucher, & Johnson, 2003; Daugherty, Ward, Smith, Ritzenthaler, & Koval, 2007; Furuse, Sasaki, & Tsukita, 1999). Since the expression levels of claudin family members in a given (epithelial) cell type are widely variable, a very complex picture of TJ pore formation by claudins (and thus resultant permselectivity) arises, adding another level of complexity. In order to understand the properties and functions of TJ in different parts of the lung, we begin with a review of the expression profile of claudins, including the identification of claudin family members expressed, levels of claudin expression, and localization of various claudins at TJ versus intracellular locale(s) versus elsewhere (e.g., basal–lateral aspects of an epithelium) in the lung.

III. PROFILE OF CLAUDIN EXPRESSION IN THE LUNG

A. Expression of Claudins in Proximal Airways

Coyne, Gambling, Boucher, Carson, and Johnson (2003) evaluated the expression of 12 different claudins in human proximal airways by immunostaining and found that both bronchi and bronchioles express claudin-1, -3, -4, -5, and -7, while claudin-2, -6, -9, -10, -11, -15, and -16 were not detected. Interestingly, this study revealed the localization of some claudin isoforms outside of the TJ in the airway epithelium. While claudin-3 and claudin-5 were exclusively localized to the apical TJs, claudin-1 and claudin-4 were also detected at lateral intercellular junctions located below the TJ of airway epithelia. Claudin-7 could barely, if at all, be detected at (apical) TJs and was found almost entirely at lateral intercellular junctions formed below TJ. Expression of claudin-7 outside TJ was also reported in mouse airway epithelium (Blackman, Russell, Nordeen, Medina, & Neville, 2005), where expression was detected in punctate cytoplasmic structures close to (or associated with) the basolateral surfaces. Moldvay et al. (2007) observed the expression of claudin-1, -2, -3, -4, and -7 proteins in human bronchial epithelium, although claudin-2 was only found in cytoplasmic granules and not at TJ. Similar claudin expression in human bronchiolar epithelium was presented in a recent study by Kaarteenaho-Wiik and Soini (2009), who demonstrated the expression of claudin-1, -2, -3, -4, and -7, while claudin-5 expression was weak. Claudin-10 expression has been described in Clara cells (Zemke et al., 2009), while no expression was detected in any other cell types (e.g., in the airways or alveoli). The claudin-10 gene is subject to alternative splicing, resulting in proteins with different properties (Gunzel et al., 2009; Van Itallie et al., 2006) due to sequence differences in the first extracellular loop. The roles of the different claudin-10 splice forms in Clara cells are unclear.

B. Expression of Claudins in Distal Lung

Wang et al. (2003) reported that very low expression levels of claudin-1 and claudin-2 proteins were detectable in rat lung sections. Consistent with this, Chen et al. (2005) reported that low expression levels of claudin-1 were found in whole rat lung, while claudin-2 mRNA was undetectable. Chen et al. (2005) also showed detectable levels of claudin-3 and claudin-4 proteins specifically in alveolar epithelial type II (AT2) cells. Morita, Sasaki, Furuse, and Tsukita (1999) detected claudin-5 protein at the cell–cell borders of endothelial cells in blood vessels in the mouse lung, while no claudin-5 expression was detected in the alveolar epithelium, although subsequent studies have detected claudin-5 protein in both alveolar epithelium and in purified AT2 cells from rat lungs (Chen et al., 2005; Wang et al., 2003). The claudin-18 gene was identified in 2001 and shown to be expressed in the epithelia of lung and stomach in the mouse (Niimi et al., 2001). Another study in the mouse lung demonstrated a widespread expression of claudin-3 and claudin-18 in the alveolar epithelium, while claudin-4 was more heterogeneous in distribution but was found in both alveolar epithelial type I (AT1) and AT2 cells (Wray et al., 2009). A recent study in human alveolar epithelium (Kaarteenaho-Wiik & Soini, 2009) revealed that claudin-3, -4, and -7 were expressed mainly in AT2 cells, while claudin-1, -2, and -5 were not detected. However, claudin-5 was found in the endothelial cells of capillaries, arteries, veins, and lymphatic vessels of these human lung specimens. Morita, Furuse, Fujimoto, and Tsukita (1999) reported the presence of mRNA for claudin-8 in the lung, although no further confirmatory studies on the expression of this claudin family member in the lung have been published to date. Other claudin family members have not been studied extensively to date, although claudin-19 mRNA expression was investigated in mouse and rat lung and found to be below the detection limit (Lee et al., 2006). As can be seen from these reports, several claudin family members have been identified at both mRNA and protein levels in proximal and distal lung epithelium, although their relative distribution in specific epithelial cell types (e.g., AT1 vs. AT2 cells) has not been well characterized.

IV. REGULATION OF CLAUDIN EXPRESSION IN THE LUNG

Most available information describing the regulation of claudin gene expression comes from other tissues or cell culture models, but likely applies to the regulation of claudin expression in lungs as well. Claudins appear to be regulated at both transcriptional and posttranslational levels. Overexpression of the transcription factor Snail in the mouse mammary epithelial cell line EpH4 resulted in drastic reductions in the mRNA levels of claudin-3, -4, and -7 (Ikenouchi, Matsuda, Furuse, & Tsukita, 2003). Similarly, claudin-1 was repressed by Snail (Martinez-Estrada et al., 2006). Posttranslational modification mechanisms such as phosphorylation and palmitoylation are also important in regulating claudins (Angelow et al., 2008; Tanaka, Kamata, & Sakai, 2005a, 2005b; Van Itallie, Gambling, Carson, & Anderson, 2005). Several kinases, such as protein kinases A (PKA) and C (PKC), mitogen-activated protein kinase (MAPK), and lysine-deficient protein kinase (WNK) (Angelow et al., 2008), have been shown to phosphorylate claudins. One important function of claudin phosphorylation appears to be in regulating the insertion and stable integration of claudin proteins into TJ. For example, the EphA4-mediated phosphorylation of claudin-4 leads to a weaker interaction between claudin-4 and ZO-1, resulting in decreased amounts of claudin-4 protein in the TJ (Tanaka et al., 2005a), while the palmitoylation of claudin-14 increases the incorporation of this isoform into TJ complexes in MDCK cells (Van Itallie et al., 2005). Palmitoylation of claudin-2 and claudin-4 was demonstrated in the same study, although no structure/function data were provided. Other members of the claudin family also contain membrane-proximal cysteines that could potentially be modified by *S*-acylation with palmitic acid, making it attractive to posit that this type of posttranslational modification is of wide importance for the correct localization of claudins into TJ. Trafficking of claudin proteins, probably interconnected to posttranslational modifications of claudins, presents another level of regulation. The importance of trafficking as a regulatory mechanism of claudin expression has been demonstrated in differentiating fetal alveolar cells *in vitro*, where an increase in the localization of claudin proteins to TJ was associated with a decreased rate of clathrin-mediated endocytosis of claudins from TJ into the cell interior (Daugherty et al., 2004).

Claudin-18 was identified in a suppressive subtractive hybridization screen designed to identify genes affected by thyroid transcription factor-1 (TTF-1) deficiency in knockout lungs at E16.5 (Niimi et al., 2001). The presence of lung- and stomach-specific forms of claudin-18 that are the products of alternative promoter usage was found in the same study. The lung-specific promoter of claudin-18 includes two binding sites for TTF-1 (also known as Nkx2.1) which is expressed in AT2 cells, but not in AT1 cells, in the adult lung. Since claudin-18 was downregulated in TTF-1 knockout lung, it would be reasonable to assume that claudin-18 expression in the adult alveolar epithelium is limited to AT2 cells. However, it can be noted that when isolated rat AT2 cells were cultured for 5 days on different matrices allowing them to transdifferentiate to become AT1-like cells (known to have very low TTF-1 levels), all cultured cells expressed the claudin-18 protein (Koval et al., 2010). Whether or not a remaining low level of TTF-1 is enough to confer the necessary regulation of claudin-18 gene in primary cultured rat AT1-like cell monolayers is unknown. However, the recent observation by Wray et al. (2009) that claudin-18 protein is found in the entire mouse alveolar epithelium, actually suggests that the transcription of the claudin-18 gene, at least in AT1 cells, may be independent of TTF-1, which would be in line with high claudin-18 expression in AT1-like cells transdifferentiated from AT2 cells *in vitro*. It is also possible that claudin-18 gene expression is TTF-1-dependent in embryonic lung cells but not in adult cells. An alternative explanation for the lack of claudin-18 expression in the TTF-1 knockout could be that TTF-1-deficient E16.5 lung cells were arrested in development and never reached the point where the expression of claudin-18 begins.

V. CLAUDINS AND BARRIER FUNCTIONS OF THE LUNG

Claudin expression profiles in the different airway and alveolar epithelial cell types have not been studied systematically to date. It is highly likely that airway TJ properties are dictated by the differential expression of various claudin isoforms in specific airway cell types. Similarly, in the alveolar epithelium, differences could also be anticipated in TJ properties formed between AT1–AT1 versus AT2–AT2 versus AT1–AT2 cells. Several lines of evidence suggest that the properties of claudin-3 are similar to those of claudin-4 with regard to paracellular transport. No change in transepithelial resistance (TER) was observed when claudin-3 was overexpressed in MDCK I cells, which normally express claudin-1 and claudin-4 (Furuse, Furuse, Sasaki, & Tsukita, 2001). This may be due to the fact that both claudin-1 and claudin-4 expression result in high TER and addition of excess claudin-3, which exhibits properties similar to claudin-4, would not alter barrier function. In fact, claudin-3 and claudin-4 have a very high degree of sequence homology in the first extracellular domain that dictates the permselective properties of the pore formed by claudins (Angelow et al., 2008, 301; Daugherty et al., 2007, 374). Due to this likely functional redundancy, it would be reasonable to expect that claudin-4 knockout mice would show a very mild, if any, phenotype at baseline.

When claudin-5 was knocked out in mice, it was found that the blood–brain barrier function was compromised to allow the passage of small solutes that are normally excluded from penetrating into the brain (Nitta et al., 2003). In keeping with this latter finding, the overexpression of claudin-5 in MDCK-II cells resulted in increased TER and decreased cation permeability (Wen, Watry, Marcondes, & Fox, 2004). Disruption of the barrier function in neural blood vessels and in retinal microvasculature caused by hypoxia was correlated with a decrease in claudin-5 expression in these endothelial cells (Koto et al., 2007). Thus, it would be reasonable to expect that hypoxia caused by pulmonary edema might augment the leaky phenotype of lung capillaries through the repression of claudin-5 expression. Interestingly, the coexpression of claudin-1 and claudin-3 in NIH/3T3 cells, which do not express any endogenous claudins, resulted in a greater TER than that in 3T3 cells coexpressing claudin-1, -3, and -5 together (Coyne, Gambling, et al., 2003). Furthermore, direct heterophilic interactions between claudin-1 and claudin-3 and between claudin-3 and claudin-5 were confirmed in coimmunoprecipitation experiments in 3T3 cells, and these interactions were correlated with low- and high paracellular permeability, respectively (Coyne, Gambling, et al., 2003). Thus, claudin-5 appears to be involved in lowering the TER in these cells by replacing claudin-3 in heterophilic claudin-1/claudin-3 interactions. In line with this role of claudin-5, rat AT2 cells cultured on permeable filters and treated with methanandamide exhibited low TER correlating with a specific increase in claudin-5 expression (Wang et al., 2003). In addition, the treatment of cultured primary cell monolayers of rat AT2 cells with epidermal growth factor (EGF) resulted in increased TER, which was paralleled by elevated expression of claudin-4 and claudin-7 (but decreased expression of claudin-3 and claudin-5) (Chen et al., 2005). It can be pointed out that claudin-5 in endothelial cells appears to contribute to tighter TJ phenotype, while claudin-5 in epithelial cells may be important for the manifestation of a leakier TJ function. The reason for such differences is not clear but could be based on differences in claudin expression profiles between the specific cell types, resulting in the formation of TJ pores with different properties due to different heterophilic claudin interactions. Other possible reasons could be differences in the post-translational modifications of claudins (e.g., phosphorylation) leading to differences in heterophilic interactions and the formation of pores with different properties.

It is well established that Na^+,K^+-ATPase activity is important for both TJ formation and maintenance of barrier function (Rajasekaran & Rajasekaran, 2003; Rajasekaran, Palmer, Moon, et al., 2001; Rajasekaran, Palmer, Quan, et al., 2001). During TJ formation the polymerization of actin is a prerequisite for the assembly of TJ complexes. Inhibition of Na^+,K^+-ATPase activity with either ouabain or by potassium depletion in bathing fluids prevented TJ formation (Rajasekaran, Palmer, Moon, et al., 2001). Moreover, both RhoA activity and actin polymerization were decreased with sodium pump inhibition, suggesting that TJ formation involves the RhoA-signaling pathway leading to the polymerization of actin (Rajasekaran, Beyenbach, & Rajasekaran, 2008). Na^+,K^+-ATPase activity is also needed to maintain the barrier properties of established TJ complexes (Rajasekaran et al., 2003). The TJ-associated subpopulation of sodium pumps is conjectured to provide an optimal local ion concentration necessary for TJ function (Krupinski & Beitel, 2009). Rajasekaran et al. (2007) demonstrated that the Na^+,K^+-ATPase could be found at TJ in a human pancreatic cell line and that it was associated with the phosphatase PP2A. Inhibition of sodium pump activity resulted in reduced PP2A activity, leading to the hyperphosphorylation of occludin and increased tight junctional permeability (Rajasekaran et al., 2007). Inhibition of the sodium pump in the human bronchial cell line Calu-3 with ouabain resulted in decreased TER and increased paracellular flux of mannitol and inulin. The observed Calu-3 barrier dysfunction was correlated with the downregulation of occludin, JAM-1 (now categorized as JAM-A), claudin-2, and claudin-4, while ZO-1 and claudin-14 were upregulated (Go et al., 2006). Hypoxia also causes downregulation of the sodium pump. The underlying mechanism appears to involve the generation of reactive oxygen species and PKC-zeta-mediated phosphorylation of Na^+,K^+-ATPase $\alpha 1$, resulting in endocytosis and degradation of this subunit (Dada et al., 2003). It is well established that decreased sodium pump activity results in decreased alveolar fluid clearance (AFC), leading to edema. In addition, it is likely that impaired TJ function as a result of decreased Na^+,K^+-ATPase activity will also contribute to hypoxia-induced edema due to increased paracellular leakage across the alveolar epithelium from interstitial and vascular spaces.

As discussed earlier, different intercellular and subcellular localization patterns for claudins have been evident in several studies (Blackman et al., 2005; Coyne, Gambling, et al., 2003; Moldvay et al., 2007). Whereas some claudins are localized at the TJ only (claudin-3 and claudin-5), others (e.g., claudin-1 and claudin-4) are also found at lateral and basolateral locations. It can be noted that claudin-2 and claudin-7 are expressed almost exclusively at lateral/basolateral or cytoplasmic locations in human and mouse airway epithelial cells (Blackman et al., 2005; Coyne, Gambling, et al., 2003; Moldvay et al., 2007). The role of the claudins expressed at basal/lateral and cytoplasmic locales is currently unclear, but the localization of claudins outside the TJ suggests that as yet unknown roles in addition to the barrier functions at the apical TJ for such claudins may exist. Possible additional functions include interactions with receptors, involvement in vesicle trafficking to the basolateral membrane, cell–matrix interactions, and proliferation. In this regard, claudin-11 has been shown to induce proliferation when overexpressed in an oligodendrocyte cell line, while cell migration was negatively affected in primary oligodendrocytes deficient in claudin-11 (Tiwari-Woodruff et al., 2001). These effects on proliferation and migration were linked to claudin-11 interactions with $\alpha 1$ integrin and OSP/claudin-11-associated protein (OAP)1 (Tiwari-Woodruff et al., 2001). It is interesting that tight junctional proteins (e.g., claudin-1 and claudin-7) are expressed in dendritic cells (DC) of the lung (Sung et al., 2006). The presence of TJ proteins in DC is believed to allow these cells to traverse epithelial cell layers via paracellular pathways and to efficiently elicit a response to antigenic challenges to the respiratory system without breaching the functional fence of the alveolar epithelium (Sung et al., 2006).

VI. CLAUDINS IN LUNG PATHOLOGY

Cleavage sites for a cysteine protease from dust mites in the protein sequences of both occludin and claudin-1 have been reported (Wan et al., 1999). Degradation of claudin-1 could result in a leakier phenotype if heterophilic claudin-1/claudin-3 interactions were replaced by claudin-3/claudin-5 interactions. Such mechanisms may be responsible for TJ disruption and allowing allergens (e.g., from dust mites) to gain access to sentinel dendritic antigen-presenting cells. Inflammation in conditions like asthma leads to increased levels of matrix metalloprotein proteases (MMP). In a study using primary human proximal airway epithelial cells, treatment with MMP-9 resulted in increased epithelial permeability associated with decreased levels of claudin-1 and occludin localized to TJ complexes (Vermeer et al., 2009).

Lung fibrosis involves the progressive accumulation of fibroblasts in the pulmonary interstitium and epithelial–mesenchymal transition (EMT) has been suggested to contribute to this process (Kim et al., 2006; Willis, duBois, & Borok, 2006; Willis et al., 2005). TJ disassembly is a crucial step in EMT and the actions of transforming growth factor-β (TGFβ) and its receptor are known to regulate this process at several levels. Since Snail expression is induced by TGFβ during EMT, the previously described transcriptional downregulation of claudins by Snail likely contributes to TJ disassembly. Another important mechanism during EMT directly involves the TGFβ receptor in the downregulation of RhoA and its associated signaling pathway, crucial for both TJ formation and maintenance. The TGFβ receptor 1 (TGFβR1) can localize to TJ complexes by interacting directly with occludin (Barrios-Rodiles et al., 2005). After TGFβ binding and recruitment of TGFβR2, the receptor-ligand complex phosphorylates PAR6 which in turn recruits the ubiquitin ligase SMURF1, leading to the ubiquitination and degradation of RhoA and subsequent disassembly of the TJ complex (recently reviewed in Thiery & Sleeman, 2006; Zavadil & Bottinger, 2005). EGF and hepatocyte growth factor (HGF) both have the capability to increase TER in MDCK cells (Lipschutz, Li, Arisco, & Balkovetz, 2005; Singh & Harris, 2004). As mentioned, rat AT2 cells treated with EGF feature higher TER, paralleled by increases in claudin-4 and claudin-7 and decreases in claudin-3 and claudin-5 (Chen et al., 2005). The fact that both EGF and HGF promote a tighter epithelium and thus oppose the effects of TGFβ suggests the possibility that these growth factors may be protective against EMT and fibrosis.

Several other soluble factors can affect the expression or localization of claudins. For example, IFN-γ induces the removal of occludin, JAM-A, and claudin-1 from the TJ in intestinal epithelial cells (Bruewer et al., 2005). When intestinal epithelial T84 cells were treated with TNFα/IFN-γ together, redistribution of claudin-4 and an accompanying decrease in TER took place, while IL-13 induced an increase in claudin-2 expression and lower resistance (Prasad et al., 2005). It is reasonable to believe that these cytokines would have the same effects on claudin expression and cause similar changes in bioelectrical properties in the lung. Indeed, the importance of TNFα in mediating carrageenan-induced alterations of immunohistochemical localization signals for ZO-1, claudin-2, claudin-4, claudin-5, and β-catenin in the lung was proven in TNF-α receptor 1 (TNF-αR1) knockout mice (Mazzon & Cuzzocrea, 2007). However, the effect of TNF-α on TJ protein expression might be different in human airway epithelial cells since no effect on the expression of claudin-1 and claudin-4 was observed when these cells were treated with TNF-α (Coyne et al., 2002). Increased presence of claudin-2 in TJ complexes has been shown to result in a leaky phenotype by the formation of cation-selective channels (Amasheh et al., 2002; Angelow et al., 2007; Hou, Gomes, Paul, & Goodenough, 2006; Van Itallie, Fanning, & Anderson, 2003). Thus, IL-13-mediated induction of claudin-2 under pathological conditions in the alveolar epithelium would be anticipated to lead to increased paracellular leakiness.

High alcohol consumption increases lung epithelial permeability, which is paralleled by changes in claudin expression (Fernandez, Koval, Fan, & Guidot, 2007). Western analysis of claudin expression revealed decreases in claudin-1, -3, and -7 in whole lung lysates, while claudin-5 expression was increased. However, when freshly isolated AT2 cells from alcohol-fed rats were analyzed, only the protein level of claudin-5 was altered (increased), while the other claudins remained unchanged. Interestingly, AT1-like cells derived from these isolated AT2 cells featured the same pattern of decreases in claudin-1 and claudin-7 and an increase in claudin-5 as in whole lung. Decreased levels of claudin-1 and claudin-3 and an increase in claudin-5 in association with increased epithelial permeability are in line with the observations made in airway epithelial cells referred to earlier (Coyne, Gambling, et al., 2003), suggesting that claudin-1 and claudin-3 contribute to tighter TJ, while claudin-5 induces a leaky phenotype.

The response to lung injury involves changes in the composition of the extracellular matrix (ECM), where fibronectin and type I collagen appear to be the two dominant ECM proteins that are increased after lung injury. When freshly isolated rat AT2 cells were cultured on fibronectin, the development of TJ complexes, including high levels of claudin-3 and claudin-18 and accompanying high TER, took place in a much shorter time (2 days) compared to cells grown on laminin or type I collagen (Koval et al., 2010). Although AT2 cells grown on laminin were slower in forming high TER monolayers, these cells had considerably higher TER and lower levels of claudin-5 after 5 days in culture compared to cells grown on fibronectin. On day 5, the expression of claudin-7 had increased considerably compared to day 2 in cells grown on fibronectin, which was paralleled by decreased chloride conductance. [Of note is that claudin-7 was reported to act as a chloride ion barrier while increasing permeability to Na^+ (Alexandre, Jeansonne, Renegar, Tatum, & Chen, 2007; Alexandre, Lu, & Chen, 2005; Hou et al., 2006).] Thus, a fibronectin-rich ECM environment, which can be found soon after an injury to the alveolar epithelial barrier, may allow a fast recovery of epithelial TJ function to quickly form an epithelium with high TER. At the same time, initial high paracellular chloride permeability would allow an increased absorption of chloride ions (i.e., counterions for actively transported sodium ions) to accommodate increased active Na absorption and increased AFC after injury. In the healing alveolar epithelium, laminin expression would increase with time and replace the fibronectin-rich ECM (prevalent in the early phase of injury) with an accompanying suppression of claudin-5 expression to generate an epithelium with a high TER.

Mechanical ventilation of the lung at high tidal volumes causes ventilator-induced lung injury (VILI) characterized by damage to the alveolar epithelial barrier and ensuing edema. In a recent report, claudin-4 has been suggested to provide a protective function during VILI (Wray et al., 2009), in that claudin-4 expression in injured mice was induced at both mRNA and protein levels, while both claudin-3 and claudin-18 levels were unaffected. Interestingly, pretreatment of mice before VILI with a *C. perfringens* enterotoxin-derived peptide CPE_{BD} (which binds and decreases claudin-4 expression) caused lung injury even at moderate tidal volumes that normally do not induce injury. Both basal and β-adrenergic agonist-stimulated AFC rates were significantly lower in mice treated with CPE_{BD}. These findings suggest that, consistent with claudin-4 being a selective sodium barrier (Colegio, Van Itallie, McCrea, Rahner, & Anderson, 2002; Van Itallie, Rahner, & Anderson, 2001), the decreased presence of claudin-4 in TJ complexes increases the paracellular leakage of Na^+ in the presence of CPE_{BD}. However, it is also possible that the CPE_{BD}-induced TJ changes could lead to decreased Na^+,K^+-ATPase activity, resulting in lower fluid clearance. As discussed, direct interactions between TJ and the sodium pump have been demonstrated (Krupinski & Beitel, 2009; Rajasekaran et al., 2008), making it attractive to postulate that a change in the TJ composition could affect the activity of associated sodium pumps. Another possibility could be that the sodium pump directly interacts with claudin-4, both of which are localized at sites below TJ.

Evaluation of the expression of claudin-1, -2, -3, -4, -5, and -7 in tissue samples from patients with pathologic usual interstitial pneumonia (UIP) and sarcoidosis (Kaarteenaho-Wiik & Soini, 2009) showed increased immunoreactivity for all these claudins in metaplastic alveolar epithelium (especially in UIP samples), while expression levels in normal parts of the alveolar and bronchial epithelial tissues (described in Section IV) remained unchanged. The significance of a broad increase in claudin isoform expression in these pathological conditions remains unclear but merits further investigation. Claudins show aberrant expression in a number of cancers and is the topic of a separate chapter in this issue.

VII. CONCLUDING REMARKS

In order to delineate the *in vivo* barrier functions of claudins in the lung, cell-specific knockout or transgenic overexpression strategies in the mouse will be very useful. The knockdown technology of *in vivo* transfection of the rat lung with plasmids expressing siRNA (Li & Folkesson, 2006; Li, Koshy, & Folkesson, 2007, 2008) and the recent development of transgenic mouse models that allow cell-specific deletion in the alveolar epithelium (Flodby et al., 2009; Perl, Tichelaar, & Whitsett, 2002) can be utilized to address the specific roles of claudin isoforms of interest. Permselectivity properties contributed by claudins (and likely contributed as well by occludin, tricellulins, and JAM) of airway or alveolar epithelial TJ complexes in health and disease remain largely undetermined.

Acknowledgments

This work was supported in part by Hastings Foundation, Whittier Foundation, and research grants from the National Institutes of Health (ES017034, ES018782, HL038578, HL038621, HL062569, HL064365, and HL089445). Zea Borok is Ralph Edgington Chair in Medicine and Edward D. Crandall is Hastings Professor and Kenneth T. Norris Jr. Chair of Medicine.

References

Alexandre, M. D., Jeansonne, B. G., Renegar, R. H., Tatum, R., & Chen, Y. H. (2007). The first extracellular domain of claudin-7 affects paracellular Cl-permeability. *Biochemical and Biophysical Research Communications, 357*(1), 87–91.

Alexandre, M. D., Lu, Q., & Chen, Y.-H. (2005). Overexpression of claudin-7 decreases the paracellular Cl-conductance and increases the paracellular Na + conductance in LLC-PK1 cells. *Journal of Cell Science, 118*(12), 2683–2693.

Amasheh, S., Meiri, N., Gitter, A. H., Schoneberg, T., Mankertz, J., Schulzke, J. D., et al. (2002). Claudin-2 expression induces cation-selective channels in tight junctions of epithelial cells. *Journal of Cell Science, 115*(Pt 24), 4969–4976.

Angelow, S., Ahlstrom, R., & Yu, A. S. (2008). Biology of claudins. *American Journal of Physiology. Renal Physiology, 295*(4), F867–F876.

Angelow, S., Schneeberger, E. E., & Yu, A. S. (2007). Claudin-8 expression in renal epithelial cells augments the paracellular barrier by replacing endogenous claudin-2. *The Journal of Membrane Biology, 215*(2–3), 147–159.

Angelow, S., & Yu, A. S. (2007). Claudins and paracellular transport: An update. *Current Opinion in Nephrology and Hypertension, 16*(5), 459–464.

Barrios-Rodiles, M., Brown, K. R., Ozdamar, B., Bose, R., Liu, Z., Donovan, R. S., et al. (2005). High-throughput mapping of a dynamic signaling network in mammalian cells. *Science, 307* (5715), 1621–1625.

Blackman, B., Russell, T., Nordeen, S. K., Medina, D., & Neville, M. C. (2005). Claudin 7 expression and localization in the normal murine mammary gland and murine mammary tumors. *Breast Cancer Research, 7*(2), R248–R255.

Briehl, M. M., & Miesfeld, R. L. (1991). Isolation and characterization of transcripts induced by androgen withdrawal and apoptotic cell death in the rat ventral prostate. *Molecular Endocrinology, 5*(10), 1381–1388.

Bronstein, J. M., Kozak, C. A., Chen, X. N., Wu, S., Danciger, M., Korenberg, J. R., et al. (1996). Chromosomal localization of murine and human oligodendrocyte-specific protein genes. *Genomics, 34*(2), 255–257.

Bruewer, M., Utech, M., Ivanov, A. I., Hopkins, A. M., Parkos, C. A., & Nusrat, A. (2005). Interferon-gamma induces internalization of epithelial tight junction proteins via a macropinocytosis-like process. *FASEB Journal, 19*(8), 923–933.

Chen, S. P., Zhou, B., Willis, B. C., Sandoval, A. J., Liebler, J. M., Kim, K.-J., et al. (2005). Effects of transdifferentiation and EGF on claudin isoform expression in alveolar epithelial cells. *Journal of Applied Physiology, 98*(1), 322–328.

Colegio, O. R., Van Itallie, C. M., McCrea, H. J., Rahner, C., & Anderson, J. M. (2002). Claudins create charge-selective channels in the paracellular pathway between epithelial cells. *American Journal of Physiology. Cell Physiology, 283*(1), C142–C147.

Coyne, C. B., Gambling, T. M., Boucher, R. C., Carson, J. L., & Johnson, L. G. (2003). Role of claudin interactions in airway tight junctional permeability. *American Journal of Physiology. Lung Cellular and Molecular Physiology, 285*(5), L1166–L1178.

Coyne, C. B., Ribeiro, C. M., Boucher, R. C., & Johnson, L. G. (2003). Acute mechanism of medium chain fatty acid-induced enhancement of airway epithelial permeability. *The Journal of Pharmacology and Experimental Therapeutics, 305*(2), 440–450.

Coyne, C. B., Vanhook, M. K., Gambling, T. M., Carson, J. L., Boucher, R. C., & Johnson, L. G. (2002). Regulation of airway tight junctions by proinflammatory cytokines. *Molecular Biology of the Cell, 13*(9), 3218–3234.

Dada, L. A., Chandel, N. S., Ridge, K. M., Pedemonte, C., Bertorello, A. M., & Sznajder, J. I. (2003). Hypoxia-induced endocytosis of Na,K-ATPase in alveolar epithelial cells is mediated by mitochondrial reactive oxygen species and PKC-zeta. *The Journal of Clinical Investigation, 111*(7), 1057–1064.

Daugherty, B. L., Mateescu, M., Patel, A. S., Wade, K., Kimura, S., Gonzales, L. W., et al. (2004). Developmental regulation of claudin localization by fetal alveolar epithelial cells. *American Journal of Physiology. Lung Cellular and Molecular Physiology, 287*(6), L1266–L1273.

Daugherty, B. L., Ward, C., Smith, T., Ritzenthaler, J. D., & Koval, M. (2007). Regulation of heterotypic claudin compatibility. *The Journal of Biological Chemistry, 282*(41), 30005–30013.

Fernandez, A. L., Koval, M., Fan, X., & Guidot, D. M. (2007). Chronic alcohol ingestion alters claudin expression in the alveolar epithelium of rats. *Alcohol, 41*(5), 371–379.

Findley, M. K., & Koval, M. (2009). Regulation and roles for claudin-family tight junction proteins. *IUBMB Life, 61*(4), 431–437.

Flodby, P., Borok, Z., Banfalvi, A., Zhou, B., Gao, D., Minoo, P., et al. (2010). Directed expression of Cre in alveolar epithelial type 1 cells. *American Journal of Respiratory Cell and Molecular Biology,* in press, Sep 18 (Epub ahead of print).

Furuse, M., Fujita, K., Hiiragi, T., Fujimoto, K., & Tsukita, S. (1998). Claudin-1 and -2: Novel integral membrane proteins localizing at tight junctions with no sequence similarity to occludin. *The Journal of Cell Biology, 141*(7), 1539–1550.

Furuse, M., Furuse, K., Sasaki, H., & Tsukita, S. (2001). Conversion of zonulae occludentes from tight to leaky strand type by introducing claudin-2 into Madin-Darby canine kidney I cells. *The Journal of Cell Biology, 153*(2), 263–272.

Furuse, M., Sasaki, H., Fujimoto, K., & Tsukita, S. (1998). A single gene product, claudin-1 or -2, reconstitutes tight junction strands and recruits occludin in fibroblasts. *The Journal of Cell Biology, 143*(2), 391–401.

Furuse, M., Sasaki, H., & Tsukita, S. (1999). Manner of interaction of heterogeneous claudin species within and between tight junction strands. *The Journal of Cell Biology, 147*(4), 891–903.

Go, M., Kojima, T., Takano, K.-i., Murata, M., Koizumi, J., Kurose, M., et al. (2006). Connexin 26 expression prevents down-regulation of barrier and fence functions of tight junctions by Na+/K + -ATPase inhibitor ouabain in human airway epithelial cell line Calu-3. *Experimental Cell Research, 312*(19), 3847.

Gunzel, D., Stuiver, M., Kausalya, P. J., Haisch, L., Krug, S. M., Rosenthal, R., et al. (2009). Claudin-10 exists in six alternatively spliced isoforms that exhibit distinct localization and function. *Journal of Cell Science, 122*(Pt 10), 1507–1517.

Hou, J., Gomes, A. S., Paul, D. L., & Goodenough, D. A. (2006). Study of claudin function by RNA interference. *The Journal of Biological Chemistry, 281*(47), 36117–36123.

Ikenouchi, J., Matsuda, M., Furuse, M., & Tsukita, S. (2003). Regulation of tight junctions during the epithelium-mesenchyme transition: Direct repression of the gene expression of claudins/occludin by Snail. *Journal of Cell Science, 116*(Pt 10), 1959–1967.

Kaarteenaho-Wiik, R., & Soini, Y. (2009). Claudin-1, -2, -3, -4, -5, and -7 in usual interstitial pneumonia and sarcoidosis. *The Journal of Histochemistry and Cytochemistry, 57*(3), 187–195.

Katahira, J., Inoue, N., Horiguchi, Y., Matsuda, M., & Sugimoto, N. (1997). Molecular cloning and functional characterization of the receptor for *Clostridium perfringens* enterotoxin. *The Journal of Cell Biology, 136*(6), 1239–1247.

Katahira, J., Sugiyama, H., Inoue, N., Horiguchi, Y., Matsuda, M., & Sugimoto, N. (1997). *Clostridium perfringens* enterotoxin utilizes two structurally related membrane proteins as functional receptors in vivo. *The Journal of Biological Chemistry, 272*(42), 26652–26658.

Kim, K. K., Kugler, M. C., Wolters, P. J., Robillard, L., Galvez, M. G., Brumwell, A. N., et al. (2006). Alveolar epithelial cell mesenchymal transition develops in vivo during pulmonary fibrosis and is regulated by the extracellular matrix. *Proceedings of the National Academy of Sciences of the United States of America, 103*(35), 13180–13185.

Koto, T., Takubo, K., Ishida, S., Shinoda, H., Inoue, M., Tsubota, K., et al. (2007). Hypoxia disrupts the barrier function of neural blood vessels through changes in the expression of claudin-5 in endothelial cells. *American Journal of Pathology, 170*(4), 1389–1397.

Koval, M., Ward, C., Findley, M. K., Roser-Page, S., Helms, M. N., & Roman, J. (2010). Extracellular matrix influences alveolar epithelial claudin expression and barrier function. *American Journal of Respiratory Cell and Molecular Biology, 42*(2), 172–180.

Krupinski, T., & Beitel, G. J. (2009). Unexpected roles of the Na–K-ATPase and other ion transporters in cell junctions and tubulogenesis. *Physiology, 24*(3), 192–201.

Lee, N. P. Y., Tong, M. K., Leung, P. P., Chan, V. W., Leung, S., Tam, P.-C., et al. (2006). Kidney claudin-19: Localization in distal tubules and collecting ducts and dysregulation in polycystic renal disease. *FEBS Letters, 580*(3), 923.

Li, T., & Folkesson, H. G. (2006). RNA interference for alpha-ENaC inhibits rat lung fluid absorption in vivo. *American Journal of Physiology. Lung Cellular and Molecular Physiology, 290*(4), L649–L660.

Li, T., Koshy, S., & Folkesson, H. G. (2007). Involvement of {alpha}ENaC and Nedd4-2 in the conversion from lung fluid secretion to fluid absorption at birth in the rat as assayed by RNA interference analysis. *American Journal of Physiology. Lung Cellular and Molecular Physiology, 293*(4), L1069–L1078.

Li, T., Koshy, S., & Folkesson, H. G. (2008). RNA interference for CFTR attenuates lung fluid absorption at birth in rats. *Respiratory Research, 9*, 55.

Lipschutz, J. H., Li, S., Arisco, A., & Balkovetz, D. F. (2005). Extracellular signal-regulated kinases 1/2 control claudin-2 expression in Madin-Darby canine kidney strain I and II cells. *The Journal of Biological Chemistry, 280*(5), 3780–3788.

Martinez-Estrada, O. M., Culleres, A., Soriano, F. X., Peinado, H., Bolos, V., Martinez, F. O., et al. (2006). The transcription factors slug and snail act as repressors of claudin-1 expression in epithelial cells. *Biochemical Journal, 394*(Pt 2), 449–457.

Mazzon, E., & Cuzzocrea, S. (2007). Role of TNF-alpha in lung tight junction alteration in mouse model of acute lung inflammation. *Respiratory Research, 8*, 75.

Moldvay, J., Jackel, M., Paska, C., Soltesz, I., Schaff, Z., & Kiss, A. (2007). Distinct claudin expression profile in histologic subtypes of lung cancer. *Lung Cancer, 57*(2), 159–167.
Morita, K., Furuse, M., Fujimoto, K., & Tsukita, S. (1999). Claudin multigene family encoding four-transmembrane domain protein components of tight junction strands. *Proceedings of the National Academy of Sciences of the United States of America, 96*(2), 511–516.
Morita, K., Sasaki, H., Furuse, M., & Tsukita, S. (1999). Endothelial claudin: Claudin-5/TMVCF constitutes tight junction strands in endothelial cells. *The Journal of Cell Biology, 147*(1), 185–194.
Niimi, T., Nagashima, K., Ward, J. M., Minoo, P., Zimonjic, D. B., Popescu, N. C., et al. (2001). Claudin-18, a novel downstream target gene for the T/EBP/NKX2.1 homeodomain transcription factor, encodes lung- and stomach-specific isoforms through alternative splicing. *Molecular and Cellular Biology, 21*(21), 7380–7390.
Nitta, T., Hata, M., Gotoh, S., Seo, Y., Sasaki, H., Hashimoto, N., et al. (2003). Size-selective loosening of the blood-brain barrier in claudin-5-deficient mice. *The Journal of Cell Biology, 161*(3), 653–660.
Perl, A. K., Tichelaar, J. W., & Whitsett, J. A. (2002). Conditional gene expression in the respiratory epithelium of the mouse. *Transgenic Research, 11*(1), 21–29.
Prasad, S., Mingrino, R., Kaukinen, K., Hayes, K. L., Powell, R. M., MacDonald, T. T., et al. (2005). Inflammatory processes have differential effects on claudins 2, 3 and 4 in colonic epithelial cells. *Laboratory Investigation, 85*(9), 1139–1162.
Rajasekaran, A. K., & Rajasekaran, S. A. (2003). Role of Na–K-ATPase in the assembly of tight junctions. *American Journal of Physiology. Renal Physiology, 285*(3), F388–F396.
Rajasekaran, S. A., Barwe, S. P., Gopal, J., Ryazantsev, S., Schneeberger, E. E., & Rajasekaran, A. K. (2007). Na–K-ATPase regulates tight junction permeability through occludin phosphorylation in pancreatic epithelial cells. *American Journal of Physiology. Gastrointestinal Liver Physiology, 292*(1), G124–G133.
Rajasekaran, S. A., Beyenbach, K. W., & Rajasekaran, A. K. (2008). Interactions of tight junctions with membrane channels and transporters. *Biochimica et Biophysica Acta (BBA)—Biomembranes, 1778*(3), 757.
Rajasekaran, S. A., Hu, J., Gopal, J., Gallemore, R., Ryazantsev, S., Bok, D., et al. (2003). Na, K-ATPase inhibition alters tight junction structure and permeability in human retinal pigment epithelial cells. *American Journal of Physiology. Cell Physiology, 284*(6), C1497–C1507.
Rajasekaran, S. A., Palmer, L. G., Moon, S. Y., Peralta Soler, A., Apodaca, G. L., Harper, J. F., et al. (2001). Na,K-ATPase activity is required for formation of tight junctions, desmosomes, and induction of polarity in epithelial cells. *Molecular Biology of the Cell, 12*(12), 3717–3732.
Rajasekaran, S. A., Palmer, L. G., Quan, K., Harper, J. F., Ball , W. J., Jr., Bander, N. H., et al. (2001). Na,K-ATPase beta-subunit is required for epithelial polarization, suppression of invasion, and cell motility. *Molecular Biology of the Cell, 12*(2), 279–295.
Singh, A. B., & Harris, R. C. (2004). Epidermal growth factor receptor activation differentially regulates claudin expression and enhances transepithelial resistance in Madin-Darby canine kidney cells. *The Journal of Biological Chemistry, 279*(5), 3543–3552.
Sung, S. S., Fu, S. M., Rose , C. E., Jr., Gaskin, F., Ju, S. T., & Beaty, S. R. (2006). A major lung CD103 (alphaE)-beta7 integrin-positive epithelial dendritic cell population expressing Langerin and tight junction proteins. *Journal of Immunology, 176*(4), 2161–2172.
Tanaka, M., Kamata, R., & Sakai, R. (2005a). EphA2 phosphorylates the cytoplasmic tail of Claudin-4 and mediates paracellular permeability. *The Journal of Biological Chemistry, 280*(51), 42375–42382.
Tanaka, M., Kamata, R., & Sakai, R. (2005b). Phosphorylation of ephrin-B1 via the interaction with claudin following cell–cell contact formation. *The EMBO Journal, 24*(21), 3700–3711.

Terry, S., Nie, M., Matter, K., & Balda, M. S. (2010). Rho signaling and tight junction functions. *Physiology (Bethesda), 25*(1), 16–26.

Thiery, J. P., & Sleeman, J. P. (2006). Complex networks orchestrate epithelial-mesenchymal transitions. *Nature Reviews. Molecular Cell Biology, 7*(2), 131–142.

Tiwari-Woodruff, S. K., Buznikov, A. G., Vu, T. Q., Micevych, P. E., Chen, K., Kornblum, H. I., et al. (2001). OSP/claudin-11 forms a complex with a novel member of the tetraspanin super family and beta1 integrin and regulates proliferation and migration of oligodendrocytes. *The Journal of Cell Biology, 153*(2), 295–305.

Van Itallie, C. M., & Anderson, J. M. (2006). Claudins and epithelial paracellular transport. *Annual Review of Physiology, 68*(1), 403.

Van Itallie, C. M., Fanning, A. S., & Anderson, J. M. (2003). Reversal of charge selectivity in cation or anion-selective epithelial lines by expression of different claudins. *American Journal of Physiology. Renal Physiology, 285*(6), F1078–F1084.

Van Itallie, C. M., Gambling, T. M., Carson, J. L., & Anderson, J. M. (2005). Palmitoylation of claudins is required for efficient tight-junction localization. *Journal of Cell Science, 118*(Pt 7), 1427–1436.

Van Itallie, C., Rahner, C., & Anderson, J. M. (2001). Regulated expression of claudin-4 decreases paracellular conductance through a selective decrease in sodium permeability. *The Journal of Clinical Investigation, 107*(10), 1319–1327.

Van Itallie, C. M., Rogan, S., Yu, A., Vidal, L. S., Holmes, J., & Anderson, J. M. (2006). Two splice variants of claudin-10 in the kidney create paracellular pores with different ion selectivities. *American Journal of Physiology. Renal Physiology, 291*(6), F1288–F1299.

Vermeer, P. D., Denker, J., Estin, M., Moninger, T. O., Keshavjee, S., Karp, P., et al. (2009). MMP9 modulates tight junction integrity and cell viability in human airway epithelia. *American Journal of Physiology. Lung Cellular and Molecular Physiology, 296*(5), L751–L762.

Wan, H., Winton, H. L., Soeller, C., Tovey, E. R., Gruenert, D. C., Thompson, P. J., et al. (1999). Der p 1 facilitates transepithelial allergen delivery by disruption of tight junctions. *Journal of Clinical Investigation, 104*(1), 123.

Wang, F., Daugherty, B., Keise, L. L., Wei, Z., Foley, J. P., Savani, R. C., et al. (2003). Heterogeneity of claudin expression by alveolar epithelial cells. *American Journal of Respiratory Cell and Molecular Biology, 29*(1), 62–70.

Wen, H., Watry, D. D., Marcondes, M. C., & Fox, H. S. (2004). Selective decrease in paracellular conductance of tight junctions: Role of the first extracellular domain of claudin-5. *Molecular and Cellular Biology, 24*(19), 8408–8417.

Willis, B. C., duBois, R. M., & Borok, Z. (2006). Epithelial origin of myofibroblasts during fibrosis in the lung. *Proceedings of the American Thoracic Society, 3*(4), 377–382.

Willis, B. C., Liebler, J. M., Luby-Phelps, K., Nicholson, A. G., Crandall, E. D., du Bois, R. M., et al. (2005). Induction of epithelial-mesenchymal transition in alveolar epithelial cells by transforming growth factor-beta1: Potential role in idiopathic pulmonary fibrosis. *American Journal of Pathology, 166*(5), 1321–1332.

Wray, C., Mao, Y., Pan, J., Chandrasena, A., Piasta, F., & Frank, J. A. (2009). Claudin-4 augments alveolar epithelial barrier function and is induced in acute lung injury. *American Journal of Physiology. Lung Cellular and Molecular Physiology, 297*(2), L219–L227.

Zavadil, J., & Bottinger, E. P. (2005). TGF-beta and epithelial-to-mesenchymal transitions. *Oncogene, 24*(37), 5764–5774.

Zemke, A. C., Snyder, J. C., Brockway, B. L., Drake, J. A., Reynolds, S. D., Kaminski, N., et al. (2009). Molecular staging of epithelial maturation using secretory cell-specific genes as markers. *American Journal of Respiratory Cell and Molecular Biology, 40*(3), 340–348.

CHAPTER 9

Claudins in Intestinal Function and Disease

Roland Bücker,[*] Michael Schumann,[†] Salah Amasheh,[‡] and Jörg-Dieter Schulzke[*,†]

[*]Department of General Medicine, Campus Benjamin Franklin, Charité, Berlin, Germany
[†]Department of Gastroenterology, Campus Benjamin Franklin, Charité, Berlin, Germany
[‡]Institute of Clinical Physiology, Campus Benjamin Franklin, Charité, Berlin, Germany

Current Topics in Membranes, Volume 65

1063-5823/10 $35.00
DOI: 10.1016/S1063-5823(10)65009-0

I. OVERVIEW

Epithelial tight junction (TJ) strands consist of TJ proteins including occludin, JAM, claudins, and tricellulin. Especially the members of the claudin family are key determinants in the regulation of epithelial barrier function in the intestine. Multiple differences in claudin expression along the gastrointestinal (GI) tract define the epithelial barrier properties in each particular intestinal segment and can—when changed—contribute to the pathophysiology of different diseases. Such changes in TJ protein composition can be due to expression regulation from the gene, modified TJ protein degradation, regulation of TJ assembly/disassembly via the cytoskeleton or by effects on TJ protein endocytosis. Various regulatory mechanisms are responsible for alterations in claudin expression reported here in detail for inflammatory bowel disease (IBD), including ulcerative colitis and Crohn's disease (CD), for celiac disease as well as for intestinal infections like norovirus, enteropathogenic *Escherichia coli*, *Campylobacter*, or *Giardia lamblia*. These analyses revealed that specific modifications in claudin expression and distribution are found in diseases in combination with their distinct inflammatory appearance, as a result of which leak-flux diarrhea can be induced or the entry of luminal antigens into the mucosa is facilitated, which perpetuates or can even initiate inflammatory processes.

II. CLAUDINS IN INTESTINAL EPITHELIA

Epithelial TJ, and especially the strand-forming members of the claudin family, are among the key determinants in the regulation of epithelial barrier function in the GI tract. This chapter describes common GI disorders with respect to their role in the onset and maintenance of the disease.

A. Barrier and Channel Function of Intestinal Claudins

In intestine, a number of tetraspan proteins have been identified within TJ strands, which are regarded primary contributors to barrier function. The first landmark study, which at the same time proved a protein-based TJ strand composition, was published in 1993, introducing occludin (Furuse et al., 1993). Subsequently, the family of claudins was identified (Furuse, Fujita, Hiiragi, Fujimoto, & Tsukita, 1998; Morita, Furuse, Fujimoto, & Tsukita, 1999), and more recently, tricellulin (Ikenouchi et al., 2005).

The function of occludin is still a matter of debate, since no perturbation of intestinal barrier properties was observed in occludin-deficient mice (Saitou et al., 2000; Schulzke et al., 2005). In contrast, in many studies, expression of claudins has been identified to primarily determine the charge- and size-selective barrier against ions and molecules in intestine (Rahner, Mitic, & Anderson, 2001). Since then, numerous studies showed that claudins play a significant role in intestinal diseases. Apart from analyses of intestinal epithelia, studies involved knockout approaches and experiments employing a variety of epithelial cell models, including the human intestinal cell lines HT-29, Caco-2, and T84, and Madin Darby canine kidney (MDCK) cells.

Mammalian intestine reveals segment-specific expression and localization of a number of claudins: In a pioneer work, claudin-2, -3, -4, and -5 were identified in rat intestine (Rahner et al., 2001). Studies on human intestine showed expression of claudin-1, -2, -3, -4, -5, -7, and -8 in the colon (Bürgel et al., 2002; Zeissig et al., 2007). In addition, marked expression of further claudins, including claudin-12 and -15, was found in mouse intestine (Fujita et al., 2006; Holmes, Van Itallie, Rasmussen, & Anderson, 2006).

Whereas many claudins functionally contribute to a tightening of the paracellular pathway, some members of the claudin family have been shown to form paracellular channels, or contribute to distinct transport functions of TJs. Accordingly, it is helpful to functionally attribute claudins as (i) tightening, (ii) permeability-mediating, or (iii) claudins with ambiguous functions depending on the model systems studied. In intestine, for example, claudin-1, -3, -4, -5, and -8 belong to the first group, claudin-2 belongs to the second, and claudin-7, -12, and -15 belong to the third group.

Claudin-1 represents a classical example of the first group, as claudin-1-deficient mice die within hours after birth because of a lack of epidermal barrier properties (Furuse et al., 2002). This function was confirmed by stable transfection of cell lines (Weng, Beyenbach, & Quaroni, 2005).

Clostridium perfringens enterotoxin (CPE) binds to claudin-3 and -4, resulting in a disturbed barrier (Fujita et al., 2000; Sonoda et al., 1999). These studies, along with transfection experiments, demonstrated that both claudin-3 and -4 can also be regarded as major intestinal tightening TJ proteins (Van Itallie, Rahner, & Anderson, 2001).

Claudin-5 was initially detected in blood vessels (Morita, Sasaki, Furuse, & Tsukita, 1999), but was later identified as a tightening TJ protein in intestinal epithelial cells as well, as claudin-5 cDNA was cloned from the human colon epithelial cell line HT-29/B6 and increased barrier properties of Caco-2 monolayers after stable transfection (Amasheh et al., 2005).

Claudin-8 forms a barrier to cations, including protons, and also to ammonium and bicarbonate ions (Angelow, Kim, & Yu, 2006; Yu, Enck, Lencer, & Schneeberger, 2003). Moreover, claudin-8 was shown to be induced in parallel to the induction of Na^+ absorption via the epithelial sodium channel ENaC in the colon, thereby preventing increased Na^+ leak flux and demonstrating a synergistic physiological strategy of transport and barrier regulation (Amasheh, Milatz, et al., 2009).

In contrast to the intestinal tightening TJ proteins mentioned earlier, claudin-2 represents a paracellular channel that is selective for small cations but almost impermeable for anions or uncharged solutes of any size (Amasheh et al., 2002).

In addition, claudins with ambiguous function can be detected in the intestine, the function of which could be even more dependent on the overall expression background of other claudins, namely claudin-7, -12, and -15.

Claudin-7 was demonstrated to induce an increase in paracellular permeability for small cations and a decrease in anion permeability (Alexandre, Lu, & Chen, 2005). The overall effect in NaCl-based buffer solutions appears to be a tightening, which was confirmed in a knockout mouse model (Tatum et al., 2010). Claudin-12 has been proposed to increase paracellular permeability, as it is involved in the paracellular transport of Ca^{2+} between enterocytes (Fujita et al., 2008). In contrast, a tightening function was concluded from its expression in the blood vessels of the brain (Nitta et al., 2003) and in the urinary bladder (Acharya et al., 2004).

Claudin-15 has a specific function in the intestine apart from its barrier formation, as it is involved in morphogenesis: Claudin-15-deficient mice showed an enhanced crypt cell proliferation and a megaintestine phenotype (Tamura et al., 2008). Concerning barrier properties, transfection studies revealed no effect in MDCK II cells, but in the porcine kidney cell line LLC-PK1, a cation permeability was induced by overexpression (Van Itallie, Fanning, & Anderson, 2003).

B. Claudins Along the Gut Axis

Quite recently, claudin expression was directly compared with barrier properties in rat intestine: The colon showed highest epithelial resistance and strongest expression of tightening claudin-1, -3, -4, -5, and -8. Among small intestinal segments, the duodenum exhibited highest epithelial resistance and also a marked expression of tightening claudins, which corresponds to the role of the duodenum as the first and vanguard segment following the stomach and thus facing low pH, bile, and hypoosmolarity (Markov, Veshnyakova, Fromm, Amasheh, & Amasheh, 2010). A special role is attributed to claudin-2, which could be strongly detected in all intestinal segments except for the colon, where, if detectable at all, the protein is localized only in crypt bases. This localization is in accordance with the function of claudin-2 that has been identified as a paracellular channel permeable for small cations (Amasheh et al., 2002).

C. Two Effects of Claudin Dysfunction

Generally, reduced expression of claudins, and dysregulation or redistribution of the TJ can result in two unwanted effects: (i) paracellular secretion of small solutes into the gut lumen and (ii) paracellular uptake of molecules from the gut lumen.

The first effect, paracellular secretion of small solutes, including ions, causes water to follow for osmotic reasons and results, in clinical terms, in a so-called leak-flux diarrhea.

During inflammation, for example, in Crohn's patients or after stimulation with TNFα, an increase of claudin-2 contributes to increased permeability and diarrhea (Zeissig et al., 2007). However, claudin-2 may not necessarily only trigger the circulus vitiosus of mucosal inflammation and barrier dysfunction observed in IBD but may also play a protective role. In this context, upregulation of claudin-2 might even represent a mechanism to rinse the mucosa free of noxious agents and antigens, as Na^+ secretion results in a passive Cl^- movement followed by water, which also permeates the paracellular claudin-2 channel (Rosenthal et al., 2010).

As functional counterparts of permeability mediators such as claudin-2, a decrease of tightening TJ proteins, including claudin-1, -4, -5, and -8, has been demonstrated to correlate with barrier dysfunction in IBD (Amasheh, Dullat, et al., 2009; Bürgel et al., 2002; Heller et al., 2005, Zeissig et al., 2007).

For many gastrointestinal infectious pathogens, claudins are the molecular target, as for example, for *C. perfringens*, *Salmonella enterica*, or enteropathogenic *E. coli*. Furthermore, a diarrheal mechanism by disruption of claudins can be induced by protozoa or viruses, as for example, by *G. lamblia* or norovirus.

The second effect of claudin dysfunction resulting in an impaired epithelial barrier relates to larger molecules. Luminal antigens or other potentially noxious agents, which normally cannot cross the epithelial barrier, can enter the mucosal compartment in large amounts initiating immune response and subsequently cause inflammation.

Thus, intestinal barrier dysfunction can be regarded as both cause and consequence of many intestinal diseases in which claudins are involved.

III. CHRONIC INFLAMMATORY DISORDERS OF THE INTESTINE

IBD, which includes CD and ulcerative colitis, is an autoimmune disorder of the GI tract that presumably is caused by an immunological mismanagement of the commensal GI flora resulting in sometimes fatal conditions of severe colitis, enteritis, or gastritis where either the whole wall of the respective GI segment can be affected (as in CD) or the inflammation is restricted to the mucosa (as in ulcerative colitis). Treatment options are so far mainly based on the suppression of the immune system. Although a number of potent immunosuppressants are at hand, surgery is frequently needed to resect pharmacologically untreatable inflamed GI segments.

Epithelial TJs make up a significant part of the intestinal barrier and its functional integrity. Dysregulation or functional impairment of TJs is found in many intestinal diseases including infectious as well as noninfectious inflammatory entities. Intestinal barrier function can seriously be affected in chronic IBD, a change which has—as outlined above—two main consequences. First, small solutes and water can arrive at the lumen and cause leak-flux diarrhea in IBD. The second consequence regards larger molecules, which under normal conditions do not or only marginally cross the epithelial barrier, as uptake of high amounts of antigens through an impaired intestinal barrier initiates or aggravates intestinal inflammation in IBD. Besides changes in TJs, other structural features, such as epithelial cell apoptosis, epithelial lesions, or transcytosis, can contribute to this barrier defect. Furthermore, in some parts along the GI tract, the passage through M-cells which can be found in the follicle-associated epithelium (FAE) above Peyer's patches and via transepithelial extrusions of dendritic cells is assumed to contribute significantly to the overall antigen transport through the epithelial barrier.

A. Crohn's Disease

The intestinal inflammation in CD can affect almost any part of the GI tract. Also, this can affect more than one segment which can even be located far away from each other (skip lesions). Clinical symptoms include diarrhea and, in case of a significant involvement of small intestinal segments, also malabsorption and weight loss. The stool can be contaminated with blood when the large intestine is involved. Abdominal pain is frequent, especially when the transmural inflammation and fibrosis has led to stenosis. Indirect experimental evidence for disturbed epithelial barrier properties in CD has been available now for decades from *in vivo* permeability tests with drinking solutions containing usually two permeability markers of different sizes as mannitol and lactulose which are subsequently quantified in the urine. However, the underlying cellular mechanisms and the regulation of barrier properties involved in this barrier dysfunction have only recently been elucidated in more detail. Direct evidence for an impaired epithelial resistance in inflamed intestinal segments of Crohn's patients comes from impedance spectroscopy measurements and has been detected already in mild to moderate disease activity (Zeissig et al., 2004). As structural correlate to this barrier defect, alterations in the molecular composition of the epithelial TJ were identified, as well as an increase in apoptotic rate within the epithelium (Zeissig et al., 2007). TJ structure was shown to be downregulated with a lower TJ strand number, a reduced depth of the tight junctional meshwork, and the appearance of strand discontinuities. Strand discontinuities might be important for the passage of macromolecules like food antigens and bacterial lipopolysaccharides from the intestinal lumen into the mucosa and to the blood circulation (Anderson, Van Itallie, & Fanning, 2004).

In Western blot analysis, the paracellular channel claudin-2 was upregulated, and tightening claudin-5 and -8 were distributed off the TJ into a basolateral membrane compartment and into the cytoplasm, whereas claudin-1 and -4 were unchanged (Zeissig et al., 2007).

In addition, epithelial apoptotic rate was found to be upregulated in CD (Schulzke et al., 2006; Zeissig et al., 2007). Apoptoses in the inflamed mucosa represent conductive leaks (Gitter, Bendfeldt, Schulzke, & Fromm, 2000) and can change the properties of an epithelium from tight towards leaky (Bojarski et al., 2001). However, epithelial gross lesions like ulcers or erosions are not present in early or mild stages of inflammation in CD which is different from ulcerative colitis (Gitter, Wullstein, Fromm, & Schulzke, 2001). Luminal antigens can also be absorbed transcellularly via endocytotic uptake. This transcellular uptake seems also to be intensified in CD (Schürmann et al., 1999; Söderholm et al., 2004). However, the molecular mechanisms that regulate this transport are still far from being understood but seem to include the input from proinflammatory cytokines as TNFα and interferon-gamma (IFN-γ). Finally, bacterial translocation is increased in IBD, although the mechanisms and routes are unknown and could also differ from strain to strain. Recent data indicates, for example, that α-hemolysin-producing *E. coli* strains can penetrate the epithelium paracellularly after the induction of focal leaks (Troeger, Richter, et al., 2007), a phenomenon that could also allow other bacteria to be taken up (Fig. 1).

B. Ulcerative Colitis

In this type of IBD, the inflammation is limited to the large intestine. Profuse diarrhea is the main clinical feature, often accompanied by blood contamination of the stool. Epithelial TJ structure and function is disturbed not only in CD but also in ulcerative colitis. Again, TJ strand number and meshwork depth are reduced in ulcerative colitis as in CD and this concerns surface as well as crypts. However, strand discontinuities are not as frequent

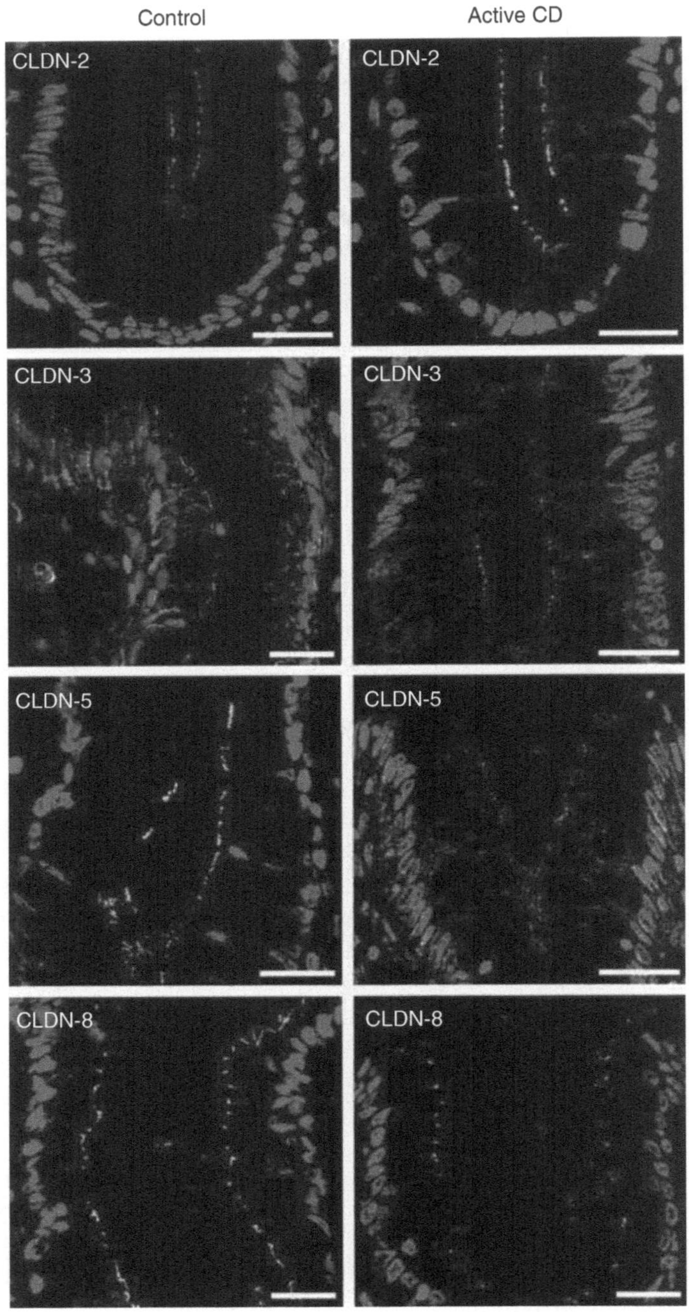

FIGURE 1 Merged pictures of ZO-1 (green), the respective claudin (red) and nuclei (blue) as obtained by immunofluorescence analysis of biopsy specimens from sigmoid colon of control and active Crohn's disease (CD). Pictures show the upper part of a crypt, except for claudin-2,

as in CD. One component in the TJ that can induce discontinuous strands is claudin-2 (Furuse, Sasaki, & Tsukita, 1999). However, it has to be assumed that other TJ proteins also contribute to this structural feature. In addition, apoptotic rate was found to be upregulated in ulcerative colitis, thereby contributing to the barrier defect (Heller et al., 2005). The key effector cytokine was identified to be IL-13 which is a potent inductor of apoptosis as well as of TJ changes (Schulzke et al., 2006). Furthermore, IL-13 can hamper epithelial restitution, which may explain that microerosions are very common already at early stages of inflammation in ulcerative colitis (Gitter et al., 2001).

C. Collagenous Colitis

Epithelial barrier function is also altered in collagenous colitis (Bürgel et al., 2002). Watery diarrhea is the predominant symptom of collagenous colitis. In contrast to CD and ulcerative colitis, mucosal surface area parameters as crypt depth or crypt number are not altered in collagenous colitis when compared to control colon. Also, epithelial apoptotic rate is not significantly changed. Both findings may reflect the rather low intensity of the microscopic inflammation. However, the epithelial TJ is affected in collagenous colitis, since occludin and claudin-4 expression has been detected to be reduced, while claudin-1 is unaltered and the channel-forming TJ protein claudin-2 is upregulated at least in some of the patients (Bürgel et al.). Taken together, TJ dysregulation but without a change in epithelial apoptosis or an induction of epithelial gross lesions is the structural correlate of the barrier dysfunction contributing to diarrhea and supporting the inflammatory process in collagenous colitis.

---

which is shown at the base of a crypt. Claudin-2 was restricted to a subset of crypt cells in active CD (tight junctional localization) and was not detectable in controls. Claudin-3 showed an intense and predominantly lateral membrane staining in controls, whereas claudin-3 signal was distinctly reduced in active CD with a diffuse cytoplasmic staining. In controls, claudin-5 showed tight junctional staining in both crypts and surface, whereas it was redistributed off the tight junction to the lateral plasma membrane in active CD. Claudin-8 was strictly localized tight junctionally in surface and crypts of controls, whereas it stained weakly in the apical cytoplasm in active CD mostly not colocalizing with ZO-1. Bar = 20 μm. Reproduced from Zeissig et al. (2007). Copyright © 2007 BMJ Publishing Group Ltd. (See Color Insert.)

D. Role of Cytokines in Altered Epithelial Barrier Function in IBD

There is a huge amount of evidence that epithelial barrier function is disturbed by proinflammatory cytokines like TNFα, IFN-γ, and IL-13 (Bruewer, Samarin, & Nusrat, 2006; Chiba et al., 2006; Clayburgh, Shen, & Turner, 2004; Gibson, 2004; Hollander, 2002; Laukoetter, Bruewer, & Nusrat, 2006) and this has to be assumed to play a central role in IBD. Cytokine levels are elevated in inflamed intestinal segments of patients with IBD (MacDonald, Hutchings, Choy, Murch, & Cooke, 1990; Raddatz, Bockemuhl, & Ramadori, 2005; Stallmach et al., 2004). Furthermore, freeze-fracture electron microscopy revealed that TJ strand architecture is altered in a similar pattern in both IBD and intestinal epithelial cell monolayers after exposure to TNFα and IFN-γ (Schmitz, Barmeyer, et al., 1999; Schmitz, Fromm, et al., 1999).

IFN-γ causes redistribution of occludin, JAM-A, claudin-1, and claudin-4 off the TJ domain in T84 (Bruewer et al., 2003). The mechanism of redistribution seems to depend on endocytosis of TJ proteins. This is indicated by the finding of subapical vesicle-like structures containing TJ proteins in biopsies from patients with active ulcerative colitis (Bruewer et al., 2005). Calcium depletion induces clathrin-mediated endocytosis of TJ proteins into a subapical cytoplasmic compartment (Ivanov, Nusrat, & Parkos, 2004). Furthermore, the internalization of TJ proteins in response to IFN-γ activates a macropinocytosis-like process resulting in the formation of vacuolar apical compartments (VACs) likely driven by myosin-II. VAC formation can be prevented by inhibition of Rho-associated protein kinase (ROCK) and not by myosin-light-chain kinase (MLCK), indicating Rho/ROCK signaling as the underlying mechanism in IFN-γ-induced VAC formation (Utech et al., 2005). This is corroborated by evidence from CD. Inhibition of Rho kinase can prevent inflammation via NF-κB inhibition (Segain et al., 2003). The TNFα-dependent increase in permeability was shown to be due to an NF-κB-dependent increase in both transcription and activation of MLCK in Caco-2 monolayers (Ma, Boivin, Ye, Pedram, & Said, 2005; Ye, Ma, & Ma, 2006).

Furthermore, there is an NF-κB-independent barrier effect of TNFα by transcriptional activation of MLCK via TNF receptor II leading to cytoskeletal TJ dysregulation. MLCK is thought be an important effector for TNF-induced TJ modulation, especially in IBD (Graham et al., 2006; Ma et al., 2005; Wang et al., 2005). MLCK inhibition can prevent TNF-induced barrier effects *in vivo* and *in vitro* as, for example, evident with the MLCK inhibitor PIK (amino acid sequence: -RKKYKYRRK-NH2) (Clayburgh et al., 2005; Zolotarevsky et al., 2002). MLCK expression and activity are increased in intestinal epithelia of experimental models of IBD (Wang et al., 2006) and in IBD patients (Blair, Kane, Clayburgh, & Turner, 2006). Another interesting correlation is that upregulation of MLCK in IBD patients parallels disease activity, which is consistent with the hypothesis that increased mucosal cytokine production contributes to MLCK-mediated barrier dysfunction (Blair et al., 2006). Also other proinflammatory cytokines can influence barrier function by modulation of MLCK activity. For example, LIGHT, another TNF family member that regulates T-cell activation and causes experimental IBD, induces TJ disruption via MLCK (Schwarz et al., 2007).

In addition to the redistribution of TJ proteins, their expression level is affected by proinflammatory cytokines. Decreased transepithelial resistance of intestinal epithelial HT-29/B6 cells was accompanied by a reduction of occludin-specific mRNA after TNFα treatment. Reporter gene analysis of the human occludin promoter revealed its downregulation by TNFα and IFN-γ, indicating a regulation of paracellular barrier-defining proteins at the transcriptional level (Mankertz et al., 2000). While occludin deficiency does not cause a significant barrier disturbance in a knockout mouse model (Saitou et al., 2000; Schulzke et al., 2005), it may become relevant when its substitution by other TJ proteins is prevented during intestinal inflammation. In parallel, the expression of the pore-forming TJ protein claudin-2 is upregulated by TNFα in HT-29/B6 monolayers, which is consistent with elevated claudin-2 levels in patients with activated CD (Zeissig et al., 2007). Additionally, IFN-γ leads to the generation of a 10 kDa claudin-2-specific fragment (Willemsen, Hoetjes, van Deventer, & van Tol, 2005).

E. Celiac Disease

Celiac disease is an autoimmune disease triggered by the dietary intake of gluten, a heterogeneous protein compound of several grains (e.g., wheat, barley, rye) resulting in a severe architectural change of the small intestinal mucosa that includes crypt hyperplasia, villous atrophy, and an intraepithelial lymphocytosis. Consecutively, patients suffer from a malabsorption syndrome that is causative for chronic diarrhea, body weight loss, microcytic anemia, and a number of vitamin deficiencies. So far, the only established therapeutic strategy for celiac patients is a strict, lifelong gluten-free diet.

Since celiac disease is accompanied by a severe alteration of small intestinal histology, several studies have addressed epithelial barrier function and have found significant alterations. Initially, intestinal permeability for small macromolecules like disaccharides (MW approximately 350 Da) as measured by the *in vivo* lactulose/mannitol permeability test as well as the permeability for macromolecules of an MW as high as 5.2 kDa were shown to be increased in celiac patients (Vogelsang, Schwarzenhofer, & Oberhuber, 1998). Moreover, impedance spectroscopy measurements of celiac specimens mounted in

Ussing chambers revealed a decrease in epithelial resistance (from 20 ± 2 in control to $9 \pm 1\ \Omega\ cm^2$ in active celiac disease), although the overall surface area is reduced in celiac disease which by itself should rather elevate epithelial resistance (Schulzke, Schulzke, Fromm, & Riecken, 1995). This points to a severe epithelial barrier defect in celiac disease that is probably important for both the paracellular passage of ions and of macromolecules like smaller gliadin fragments which normally do not cross the epithelial barrier. However, mucosal-to-serosal transcytosis of gliadin peptides was introduced in two recent studies as another mechanism to penetrate the epithelial barrier in celiac disease (Matysiak-Budnik et al., 2008; Schumann et al., 2008).

The impact of the aforementioned functional studies on epithelial barrier and associated TJ alterations in celiac disease was highlighted by a recent genetic association study that linked celiac disease to single nucleotide polymorphisms (SNPs) in two genes that are believed to be associated with the epithelial barrier. Firstly, PAR-3 (partitioning defective, synonymous PARD-3) was found to be associated with celiac disease (Wapenaar et al., 2008). PAR-3 is a PDZ-domain protein that is recruited to epithelial TJs via JAM and is a central part of a protein complex that, by interacting with the Rho-family GTPase Cdc42, regulates the assembly of TJs (Itoh et al., 2001; Lin et al., 2000). The same association study also identified MAGI-2 (membrane-associated guanylate kinase-2), a TJ adaptor, and scaffolding protein, to be associated with celiac disease (Wapenaar et al., 2008).

Studies of our own group have already pointed to alterations in TJ structure showing a reduced number of horizontal TJ strands and a higher number of TJ strand discontinuities in jejunal specimens from children (Schulzke, Bentzel, Schulzke, Riecken, & Fromm, 1998). Data collected in gliadin-exposed Caco-2 cells revealed a reduced ZO-1, occludin, claudin-3, and claudin-4 protein level that paralleled the reduction in transepithelial resistance and the increase in dextran-4000 flux in response to gliadin exposure (Sander, Cummins, Henshall, & Powell, 2005). Ciccocioppo et al. (2006) examined the duodenal biopsy specimen of celiac disease patients by Western blotting of immunoprecipitates and described a reduction in tyrosine-phosphorylated ZO-1- and ZO-1-bound occludin. Two other studies evaluated the ZO-1 level in celiac disease biopsies from the duodenum and found it to be reduced on the protein as well as the RNA level (Montalto et al., 2002; Pizzuti et al., 2004). However, data on claudin expression or localization to the TJ in celiac disease mucosae are currently missing.

When celiac disease patients follow a strict gluten-free diet, TJ parameters return to control values at the top of the villi, while strand counts are still slightly diminished in crypts (Schulzke et al., 1998). Because the crypt base is the most conductive position along the surface-crypt-axis, a disturbance in TJ structure is functionally most important at this site. However, symptom alleviation in celiac disease under gluten-free diet indicates the relevance of the restoration of strand discontinuities and of villus TJ properties in spite of the still altered crypt properties. In impedance spectroscopy, the hampered TJ structure in celiac disease was accompanied by a decreased epithelial resistance (−56%) in patients with acute celiac disease (Marsh IIIc), while patients on a gluten-free diet had a reduced epithelial resistance (−23%) (Schulzke et al., 1995). Thus, in patients under a gluten-free diet, recovery was not complete, neither morphologically nor functionally (Fig. 2).

IV. INTESTINAL INFECTIONS

Some enteropathogens can invade the intestinal mucosa but can also cause disruption of the intestinal barrier. TJ proteins of the claudin family have been shown to be affected during this process. The more we know about claudins the more we recognize their relevance for GI diseases. There are many studies on permeability alterations of intestinal epithelial cell monolayers for diarrhea-inducing bacteria, but information on structural changes of epithelial TJ proteins is still scarce. For example, *Bacteroides fragilis* toxin has been shown to induce cell swelling, stimulated Cl^--secretion, and altered F-actin structure in HT-29/C1 and T84 cells (Chambers et al., 1997; Koshy, Montrose, & Sears, 1996). Another *B. fragilis* toxin (fragilysin, a zinc-metalloproteinase) was reported to increase epithelial permeability in HT-29 cells, and it was proposed that disruption of TJ was mediated by degradation of ZO-1 (Obiso, Azghani, & Wilkins, 1997). However, no information on putative claudin changes is available in this context.

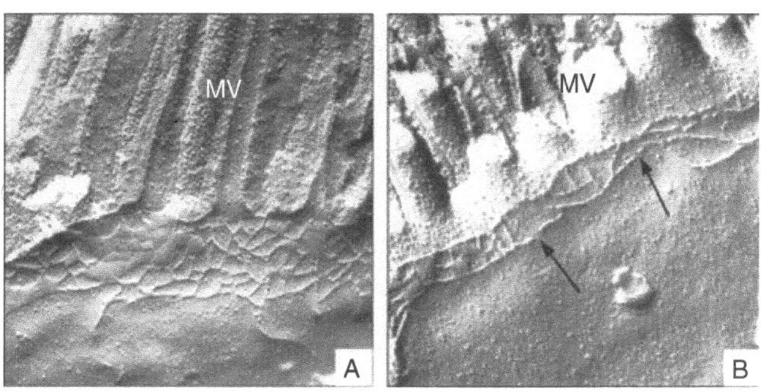

FIGURE 2 Freeze-fracture electron microscopy of tight junctions from surface epithelium in control (A) and acute celiac disease (B) duodenum. Magnification, 70,000×; MV, microvillus; arrows indicate tight junction strand discontinuities as found in celiac disease. Reproduced from Schulzke et al. (1998). © International Pediatrics Research Foundation, Inc., 1998.

An essential step in TJ research was the discovery of zonula occludens toxin (ZOT) from *Vibrio cholerae*. In Caco-2 cells, ZOT led to a redistribution of occludin and ZO-1 by alteration of the actin filament regulatory protein PKCα (Fasano et al., 1995). The cellular homologue of ZOT, zonulin, was identified in the small intestine, which was increased following *Salmonella* infection (El Asmar et al., 2002; Wang, Uzzau, Goldblum, & Fasano, 2000). A recent study showed the physiological importance of zonulin as modulator of TJs and intestinal permeability and furthermore identified human zonulin as the precursor for haptoglobin-2. Tripathi et al. (2009) found that the single-chain zonulin contains an EGF-like motif that leads to transactivation of EGF receptor via proteinase-activated receptor-2 activation, which increases intestinal permeability in mice. Whether or not this has implications beyond celiac disease-related barrier impairment—as for example, for infectious disease of the GI tract—remains to be elucidated.

A. *Salmonella enterica*

S. enterica serovar Typhimurium (*Salmonella typhimurium*) is a common cause of food poisoning and is one major zoonotic cause of diarrhea in man. *S. typhimurium* affects barrier function as indicated by a lowered transepithelial electrical resistance and increased paracellular permeability in different epithelial cell models (T84, Caco-2, MDCK) and in mice. Actions on TJs were identified in T84 cells as a pronounced decrease in ZO-1 expression due to the bacterial virulence factors SopB, SopE, SopE2, and SipA, all of which are encoded by the pathogenicity island SPI-1 (Bertelsen et al., 2004; Köhler et al., 2007). Bertelsen and coworkers also showed induction of chloride secretion by SopB via antagonization of inhibitory pathways through its function as inositol phosphatase. Furthermore, the decrease in epithelial resistance was shown to be PKC-dependent (Bertelsen et al.). ZO-1 degradation by these bacterial factors could not be blocked by the PKC inhibitor chelerythrine (Köhler et al., 2007). However, occludin was phosphorylated and claudin-1 was increased in Western blots, although claudin-1 appeared dysregulated in immunostainings (Köhler et al., 2007). Moreover, increased cell death was reported earlier by *S. typhimurium* in infected swine (Schauser, Olsen, & Larsson, 2005), both in a caspase-3-dependent and in a caspase-independent manner. An increased apoptotic rate could indeed contribute to the barrier impairment (Bojarski et al., 2004). However, another study showed a decrease of claudin-1, occludin, and ZO-1 in T84 cells and in mice by the *S. typhimurium* virulence factor AvrA, an effector also secreted by the T3SS (Liao et al., 2008). In immunostainings of AvrA treated cells, ZO-1 appeared thinner and claudin-1 less organized.

B. *Clostridium perfringens, C. difficile, C. botulinum*

The common food poisoning pathogen *C. perfringens* produces the enterotoxin CPE, which directly interacts with claudins. CPE is a β-pore forming toxin with a lipid raft-independent assembly into the host cell membrane, leading to rapid cell death (Caserta, Hale, Popoff, Stiles, & McClane, 2008). The discovery of claudins as receptors for pathogenic antigens was an important milestone in TJ research (Sonoda et al., 1999). Claudins of adjacent cells interact through two heterogeneous extracellular loops, EL1 and EL2. Similar to hepatitis C virus which uses EL1 for the entry into eukaryotic cells (see Chapter 12), CPE binds to EL2 of claudin-3 and -4. Molecular analysis of the toxin revealed a noncytotoxic fragment of CPE that contains the claudin binding site. By now, CPE is a well-characterized modulator of barrier integrity and has been suggested as an enhancer for vaccination. Moreover, CPE binds to EL2 of other claudins, which may have implications for therapeutic use, for example, as modulators of barrier function or as drug delivery enhancer (Ebihara et al., 2006; Winkler et al., 2009).

Another important pathogen, *C. difficile*, which is the causative agent in antibiotic-associated colitis (pseudomembranous colitis), disrupts the epithelial barrier by modifying the distribution of occludin, ZO-1, and ZO-2 via F-actin disorganization (Borriello, 1998; Nusrat et al., 2001). Barrier-forming claudins were so far not investigated in this context, but it is likely that they can also be influenced by the effects of *C. difficile* on F-actin. The *C. difficile* toxins A and B are glucosyltransferases and act via inactivation of Rho GTPases which leads to a variety of effects, including cellular rounding, disruption of cell adhesion, induction of apoptosis, and cytoskeletal depolymerization, which affects TJ integrity and leads to increased permeability and diarrhea (Narumiya, Ishizaki, & Watanabe, 1997; Pothoulakis, 1996). Another food poisoning species, *C. botulinum*, produces the C2 and C3 toxins that also act on actin filaments by ADP-ribosylation of G-actin (C2) or Rho proteins (C3) (Wilde & Aktories, 2001). The subsequent disassembly of actin leads to the dissociation of ZO-1 from the perijunctional complex, similar to the effects described for *C. difficile* with the same consequence of an increase in epithelial permeability (Nusrat et al., 1995).

C. *Escherichia coli*

E. coli strains can help to maintain or disturb barrier function. We do not know much about the probiotic enhancement of the TJ by commensal *E. coli* or, for example, by *E. coli* Nissle 1917, but we know a lot about the action of pathogenic *E. coli*. There are detailed studies showing how different *E. coli* strains can induce defined alterations of claudins in the host cells. Attaching and effacing bacteria (E/A) like, for example, enterohemorrhagic *E. coli* O157:H7 (EHEC) or enteropathogenic *E. coli* O127:H6 (EPEC), induce diarrhea by delivering effector proteins into the host cells via their T3SS secretion system to alter functions of organelles and the cytoskeleton as well as of membrane channels and the TJ. Studies on EPEC infection revealed a decrease in epithelial resistance and an increased permeability in Caco-2, T84, and MDCK cells by the bacterial effectors EspF, EspG, and MAP (Dean & Kenny, 2004; Matsuzawa, Kuwae, & Abe, 2005; McNamara et al., 2001). Thereby, occludin, ZO-1, and claudins were dissociated from the TJ. Thus, claudin-1 was redistributed in T84 monolayers during the EPEC infection (Muza-Moons, Schneeberger, & Hecht, 2004). Furthermore, freeze-fracture electron microscopy revealed morphological alterations in TJ distribution (Muza-Moons et al., 2004).

A recent study showed similar effects on epithelial function in T84 and MDCK cells after infection with the adherent-invasive *E. coli* (AIEC) strain LF82 (O83:H1) which has been discussed to interfere with IBD (Wine, Ossa, Gray-Owen, & Sherman, 2009). Thereby, epithelial permeability to 10 kDa dextran was increased and ZO-1 was redistributed from the TJ. These findings provide data for links between microbes and IBD, the intestinal epithelial cell barrier, and disease pathogenesis. Besides the known E/A effects, another "nonpathogenic" *E. coli* strain C25 can also alter barrier function. It was shown that claudin-1 was redistributed from the TJ in *E. coli* C25-infected T84 cells, but ZO-1 and claudin-4 were unchanged (Zareie et al., 2005). The authors proposed other mechanisms responsible for the barrier impairment than those observed with the aforementioned pathogenic *E. coli* strains. *E. coli* C25 activates NF-κB, induces IL-8 secretion and mitochondrial swelling as well as changes the distribution of cytoskeletal α-actinin (Zareie et al., 2005).

D. *Shigella flexneri*

As described for *E. coli*, also *S. flexneri* delivers effector proteins via its T3SS into the host cells, as a result of which severe inflammation and diarrhea are observed. Sakaguchi and coworkers showed that ZO-1, claudin-1, and the phosphorylation status of occludin were changed after infection in T84 cells. Interestingly, in the first minutes of infection, claudin-1 protein expression was initially severely depressed, while claudin-1 levels associated with the membrane returned to 50% of the control levels after 90 min of exposure to *S. flexneri* wild-type (Sakaguchi, Köhler, Gu, McCormick, & Reinecker, 2002). At this time point, claudin-1 appears redistributed off the TJ. Unfortunately, there is neither information on TJ effects of the more common strains *Shigella sonnei* or *Shigella dysenteriae* nor on specific enterotoxins of these germs.

E. Campylobacter jejuni

Campylobacter enteritis is the most common cause of bacterial diarrhea in humans worldwide. The predominant pathogenic species found in human gastroenteritis are *C. jejuni* and *Campylobacter coli*, but recently there is information about new emerging species like *Campylobacter concisus*, which may also be involved in the development of inflammatory diseases (Lastovica, 2009). The main source for human infection is contaminated meat from poultry. In most poultry farms, almost 100% of the animals are naturally colonized with *Campylobacter*. Although beyond humans many other species can be colonized by these bacteria, diarrhea is observed only in primates and pigs. Initially, TJ effects of *C. jejuni* were described in Caco-2 monolayers with a decrease in epithelial resistance and an impaired occludin integrity (MacCallum, Hardy, & Everest, 2005). Chen, Ge, Fox, and Schauer (2006) additionally reported phosphorylation of occludin and accumulation of claudin-1 in lipid rafts in an attempt to explain the barrier impairment in T84 cells. TJ remodeling was found to be mediated through PKC and tyrosine kinases. The *Campylobacter* infection is characterized by invasion and translocation through the epithelium with a concomitant induction of inflammatory processes which might be a major pathological mechanism in this infection (Hu, Tall, Curtis, & Kopecko, 2008). Unfortunately, less is known about the mechanisms of barrier dysfunction and the role of other claudins in this relevant human infection. Only a study in a chicken model has provided information about a claudin-4 decrease after *C. jejuni* colonization (Lamb-Rosteski, Kalischuk, Inglis, & Buret, 2008). A more detailed analysis of the role of claudins in this infection might be worthwhile, especially with regard to *Campylobacter* as a putative etiological factor in the development of chronic inflammation in the intestine (Gradel et al., 2009). *Campylobacter*-related complications like reactive arthritis and Guillian-Barré syndrome underline the invasive properties of these germs.

F. Arcobacter butzleri

The emerging human pathogen *A. butzleri* causes mainly watery diarrhea. The bacteria are relatives of the *Campylobacter* genus and share the same hosts, but are genetically quite different. In HT-29/B6 cells, *Arcobacter* infection was shown to affect the TJ and to induce epithelial apoptosis. As a result of this barrier disturbance, leak-flux diarrhea is observed (Bücker, Troeger, Kleer, Fromm, & Schulzke, 2009). In this study, no evidence for active ion secretion could be obtained and therefore the effect on the TJ seems to be the predominant pathomechanism in this infection. *A. butzleri* induced sealing claudins, like claudin-1, -5, and -8, to be redistributed off the TJ, while claudin-2, -3, and -4 remained unaltered. Claudin-8 and in particular claudin-1 changed their localization and appeared in intracellular aggregations. A similar distribution pattern was, for example, observed by Muza-Moons et al. (2004) for *E. coli*-infected T84 cells. Furthermore, claudin-5 was decreased in Western blots, in confocal laser-scanning micrographs as well as on mRNA level, suggesting expression regulation from the gene. Upregulated epithelial apoptosis is also a significant part of the *A. butzleri*-dependent barrier dysfunction which is induced in a caspase-3-dependent manner (Fig. 3; Bücker et al., 2009).

G. *Helicobacter pylori*

H. pylori chronically colonizes the stomach and is the major cause of gastroduodenal ulcers. More than half the world's population is infected. In 1994, the WHO declared *H. pylori* as class I carcinogen for gastric cancer. During chronic colonization of the mucus, the bacteria can adhere to the epithelium and induce pathological changes by their virulence factors via, for example, the T4SS. The bacterial factors CagA and VacA are also known to influence the TJ (Amieva et al., 2003; Bagnoli, Buti, Tompkins, Covacci, & Amieva, 2005; Papini et al., 1998). Increased paracellular permeability and TJ effects on occludin were reported from a mouse model with *H. pylori* gastritis (Suzuki et al., 2002). Fedwick, Lapointe, Meddings, Sherman, and Buret (2005) found a redistribution of claudin-4, claudin-5, and occludin in a mouse infection model. Also, claudin-4 was redistributed when human gastric AGS cells were infected with the same *H. pylori* strain SS1 (Fedwick et al., 2005). Moreover, the authors reported other virulence factors than cagA and vacA to be effectors of disrupting the TJ by activation of MLCK. Additionally, in MKN28 gastric epithelial cells, it was shown that *H. pylori* induces barrier dysfunction by occludin internalization via activation of Rho-kinase or MLCK which requires functional urease activity (Wroblewski et al., 2009). It was frequently reported that only direct attachment of the bacteria to epithelial cells can induce several of the changes as, for example, the induction of cytokine/chemokine response (IL-1β, IL-6, IL-8, TNF-α, or GM-CSF) or the activation of NF-κB. A mouse model with *Helicobacter*-induced gastric carcinoma showed an altered

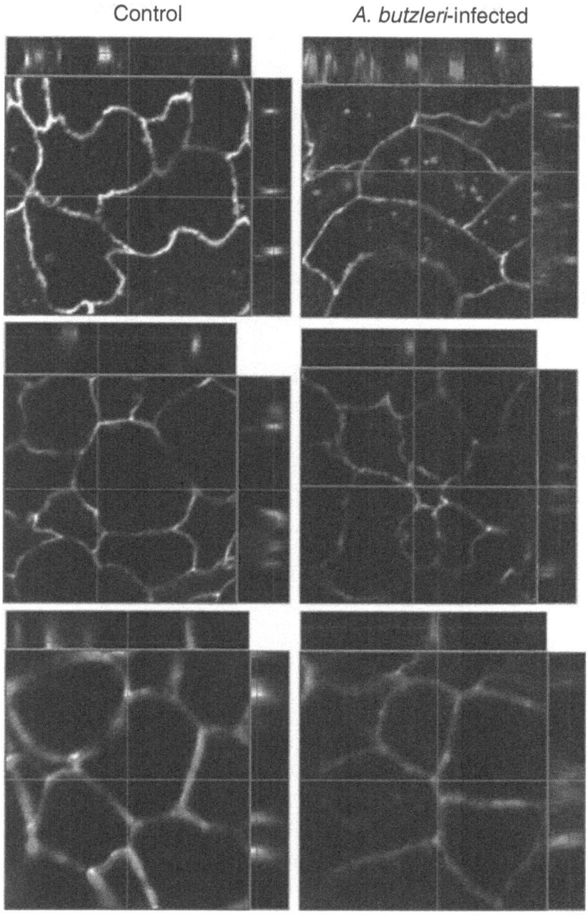

FIGURE 3 Confocal laser-scanning micrographs of *Arcobacter butzleri*-infected HT-29/B6 monolayers compared to controls. Zonula occludens protein-1 (ZO-1, green) is unaltered and has been used as TJ reference. Colocalization of a claudin (red) with ZO-1 appears yellow (merge). Claudin-1 shows a pronounced redistribution off the TJ to intracellular compartments (with the shape of cytosolic aggregates) in response to *A. butzleri*. Obvious is also the reduced claudin-5 signal intensity in the TJ and the redistribution of claudin-8 off the TJ. Reproduced from Bücker et al. (2009). Copyright © 2009 by the University of Chicago Press. (See Color Insert.)

claudin-7 expression (Takaishi, & Wang, 2007). A related emerging enterohepatic pathogen, *Helicobacter pullorum*, seems to be involved in inflammatory processes in the intestine and may also influence barrier function (Laharie et al., 2009; Varon et al., 2009).

H. Human Immunodeficiency Virus (HIV)

Diarrhea can be the result of opportunistic infections in patients with acquired immune deficiency syndrome (AIDS). In addition, the HI-virus can act directly or indirectly on the enteric mucosa which also leads to diarrhea, the so-called HIV-enteropathy due to HIV *per se*. Thereby, the mucosal architecture is changed to flattened and fused villi and a reduced villus/crypt ratio, accompanied by increased epithelial apoptosis (Epple et al., 2009). Furthermore, epithelial barrier function is hampered as indicated by an increased permeability as, for example, for mannitol and a reduced epithelial resistance in impedance spectroscopy. Thus, the diarrhea induced by HIV *per se* is mainly due to a leak-flux mechanism which could be partially induced via increased cytokine release from infected immune cells (Schmitz et al., 2002). Duodenal biopsy specimens from HIV patients showed an upregulation of claudin-2 and a downregulation of claudin-1, while claudin-4 and occludin were not different from controls (Epple et al., 2009). Moreover, HIV patients treated by highly active antiretroviral therapy (HAART) had less symptoms and were found to be normal in claudin expression as well as in villus architecture.

I. Norovirus

The norovirus (formely Norwalk virus) is the major cause of epidemic outbreaks of gastroenteritis worldwide. It spreads through fecal contamination of food and water and also through direct person-to-person contact. Within 48 h after exposure, patients develop characteristic symptoms of nausea, vomiting, diarrhea, and abdominal pain. The infection rises from the small intestine, where viruses multiply. In human duodenal biopsies from norovirus-infected patients, it was shown that a pronounced barrier dysfunction accompanies the infection (Troeger et al., 2009). Expression of the TJ proteins occludin, claudin-4, and claudin-5 was reduced, whereas claudin-1, -2, and -3 were unaffected. Furthermore, epithelial apoptosis was increased in the proximal small intestine which was induced in a caspase-3-dependent manner. These effects were accompanied and most likely partially mediated by an increased number of cytotoxic intraepithelial lymphocytes (IELs). Furthermore, a reduction in villus surface area was observed which paralleled the increase in the IELs which resulted from upregulated epithelial apoptosis despite a higher proliferation rate as shown by Ki-67/CD-103 stainings (Troeger et al., 2009).

J. Rotavirus

Rotavirus is the most common cause of gastroenteritis in children. Till the age of five, nearly every child was infected at least once (Parashar, Hummelman, Bresee, Miller, & Glass, 2003). The infection of the small intestine leads to severe diarrhea and in some cases to death by dehydration. The pathomechanisms of this disease and especially the effects of the enterotoxin NSP4 are induction of malabsorption with inhibition of Na^{+}-glucose symport activity due to SGLT1 disruption and destruction of enterocytes. It has been proposed that also active ion secretion is induced by NSP4 via phospholipase C-dependent calcium signaling, triggering the release of mediators from the enteric nervous system (ENS), contributing to diarrhea (for review, see Lorrot & Vasseur, 2007). However, also a leak-flux diarrhea component contributes to the diarrhea induced by rotavirus infection. A direct effect on TJs has been shown by Dickman and coworkers with a decreased transepithelial electrical resistance and an increased permeability to 4 kDa FITC-dextran in rotavirus-infected Caco-2 cells. The redistribution of occludin, ZO-1, and especially of claudin-1 was the pronounced feature of barrier dysfunction (Dickman et al., 2000). Another viral surface protein VP8 fragment increased permeability of MDCK cells to 70 kDa FITC-dextran and was responsible for the appearance of TJ strand breaks and loose end filaments in freeze-fracture EM micrographs. The authors could additionally show a pronounced redistribution of claudin-3 and also confirm the effects on occludin and ZO-1 (Nava, López, Arias, Islas, & González-Mariscal, 2004). Moreover, VP8 was able to prevent TJ formation in Ca^{2+}-switch assays.

K. Giardia lamblia (Giardiasis)

The protozoan parasite *G. lamblia* causes acute and chronic diarrhea accompanied by abdominal pain and nausea. In duodenal biopsies of patients with giardiasis, structural abnormalities of the mucosa were characterized. Villus surface area was reduced by 50% with no change in crypt morphology. Epithelial TJs were disrupted by downregulation of claudin-1, but with small effects on occludin, claudin-4, and claudin-7 (Troeger, Epple, et al., 2007). Disruption of F-actin and ZO-1 as well as the resulting increase in epithelial permeability seems to be modulated at least in part by MLCK and proapoptotic caspase-3 (Buret, Mitchell, Muench, & Scott, 2002; Chin et al., 2002; Scott, Meddings, Kirk, Lees-Miller, & Buret, 2002). A slightly increased epithelial apoptotic rate observed in TUNEL stainings seems to play a minor role in chronic giardiasis but can also contribute to diarrhea by a leak-flux mechanism. In addition, however, Na^+-dependent D-glucose absorption was impaired and active electrogenic anion secretion was activated. Therefore, the mechanisms of diarrhea in human giardiasis comprise *leak flux*, malabsorptive, and secretory components (Buret, 2007; Troeger, Epple, et al., 2007).

L. *Entamoeba histolytica* (Amebiasis)

Amebiasis is a common cause of intestinal dysfunction with severe dehydration. Like in giardiasis, the villus structure is disturbed in amebiasis and epithelial permeability is increased. Besides the known malabsorptive dysfunctions, the trophozoites of the protozoans cause disturbance of TJs in Caco-2 and T84 cells. Thereby, ZO-1 was released from ZO-2 and ZO-1 was degraded, while ZO-2 was phosphorylated (Lauwaet et al., 2004; Leroy, Lauwaet, De Bruyne, Cornelissen, & Mareel, 2000). The probable impact of *E. histolytica* on occludin and claudins, however, remains to be characterized so far.

M. Protection of Intestinal Barrier Function in Intestinal Diseases

Enhanced epithelial barrier function from nutritional factors was observed for the flavonoid *Quercetin* which considerably upregulates claudin-4 (Amasheh et al., 2008). The protective function of probiotics is due to different mechanisms which, in part, still need more investigations. First, nonpathogenic commensal microflora compete with pathogenic bacteria and can prevent its growth and colonization. Another effect of probiotics was shown, for example, for *Bifidobacterium infantis* that enhances epithelial barrier function directly by downregulating the pore-forming claudin-2 and upregulating ZO-1 and occludin (Ewaschuk et al., 2008). Furthermore, antiinflammatory effects were obtained in *Lactobacillus rhamnosus*-treated rats, which helps to maintain TJ integrity (Li et al., 2009). Moreover, transforming growth factor (TGF)-β which has been therapeutically added to nutrition formula was shown to protect epithelial barrier function by maintaining claudin-2, ZO-1, and occludin levels in EHEC-infected T84 cells (Howe, Reardon, Wang, Nazli, & McKay, 2005). Also, epidermal growth factor (EGF) can function as an antiinvasive factor as shown for *C. jejuni*, thereby preventing pathological changes like claudin-4 downregulation (Lamb-Rosteski et al., 2008).

Protective effects of *E. coli* Nissle 1917 on TJ function could also be shown for ZO-1 upregulation and maintenance in experimental colitis models (Ukena et al., 2007) or for ZO-2 redistribution resulting in TJ repair (Zyrek et al., 2007), but there is a lack of data on claudins. We have observed probiotic effects as, for example, a slightly improved epithelial barrier function after addition of *E. coli* Nissle 1917 in HT-29/B6 monolayers preexposed either with *E. coli* K12 or with inactivated pathogens like heat-killed *A. butzleri* (unpublished data). Moreover, the probiotic yeast *Saccharomyces boulardii* has protective effects on barrier function by inhibiting inflammatory processes and pathological changes in signal transduction in the host cells of, for example, EPEC- or *S. flexneri*-infected T84 cells. As one potential mechanism, the yeast preserves TJ integrity as shown by ZO-1 and ZO-2 analyses (Czerucka, Dahan, Mograbi, Rossi, & Rampal, 2000; Mumy, Chen, Kelly, & McCormick, 2008). *S. boulardii* enhances also the barrier function in Crohn's patients (Garcia Vilela et al., 2008). The influence of other yeasts and fungi on barrier function is still underestimated and needs further investigation, especially in respect to pathogenic *Candida* infections, microsporidiosis (Lima et al., 1997), and other mycoses.

V. CONCLUSION

GI disorders such as bacterial or parasite-induced enteritis, CD, and celiac disease share a common pathomechanism, namely a disturbed epithelial barrier with distinct changes in barrier-forming claudins, which has been shown to contribute to diarrhea. Moreover, barrier dysfunction like epithelial TJ alterations, apoptotic leaks, mucosal lesions, and epithelial restitution arrest are the result of immune dysregulation and are intensified by an increased luminal antigen uptake. Epithelial barrier function is an important first frontier of the immune defense and claudins are important features preserving this function, but they are also targets of pathogenic attacks and involved in the pathogenesis of several intestinal diseases. Pathogens are well adapted to their hosts and sometimes even to specific claudins which they misuse, for example, for the entry into the organism. On the other hand, intestinal TJs can become permeable to ions as a protective mechanism to rinse off noxious agents from the mucosa. Claudins with their special distribution pattern among different tissues along the GI-tract interact in a defined way, thereby modulating epithelial barrier function to their typical physiological requirements in each GI segment. Thus, they are indeed the first barrier feature in the immune defense. Therefore, the research on claudins and their contributions to barrier function is still in need of more attention in the future.

References

Acharya, P., Beckel, J., Ruiz, W. G., Wang, E., Rojas, R., Birder, L., et al. (2004). Distribution of the tight junction proteins ZO-1, occludin, and claudin-4, -8, and -12 in bladder epithelium. *American Journal of Physiology. Renal Physiology, 287*, F305–F318.

Alexandre, M. D., Lu, Q., & Chen, Y. H. (2005). Overexpression of claudin-7 decreases the paracellular Cl^- conductance and increases the paracellular Na^+ conductance in LLC-PK1 cells. *Journal of Cell Science, 118*, 2683–2693.

Amasheh, M., Schlichter, S., Amasheh, S., Mankertz, J., Zeitz, M., Fromm, M., et al. (2008). *Quercetin* enhances epithelial barrier function and increases claudin-4 expression in Caco-2 cells. *The Journal of Nutrition, 138*(6), 1067–1073.

Amasheh, S., Dullat, S., Fromm, M., Schulzke, J. D., Buhr, H. J., & Kroesen, A. J. (2009). Inflamed pouch mucosa possesses altered tight junctions indicating recurrence of inflammatory bowel disease. *International Journal of Colorectal Disease, 24*, 1149–1156.

Amasheh, S., Meiri, N., Gitter, A. H., Schöneberg, T., Mankertz, J., Schulzke, J. D., et al. (2002). Claudin-2 expression induces cation-selective channels in tight junctions of epithelial cells. *Journal of Cell Science, 115*(Pt 24), 4969–4976.

Amasheh, S., Milatz, S., Krug, S. M., Bergs, M., Amasheh, M., Schulzke, J. D., et al. (2009). Na^+ absorption defends from paracellular back-leakage by claudin-8 upregulation. *Biochemical and Biophysical Research Communication, 378*, 45–50.

Amasheh, S., Schmidt, T., Mahn, M., Florian, P., Mankertz, J., Tavalali, S., et al. (2005). Expression of claudin-5 contributes to barrier properties in tight junctions of epithelial cells. *Cell and Tissue Research, 321*, 89–96.

Amieva, M. R., Vogelmann, R., Covacci, A., Tompkins, L. S., Nelson, W. J., & Falkow, S. (2003). Disruption of the epithelial apical-junctional complex by *Helicobacter pylori* CagA. *Science, 300*(5624), 1430–1434.

Anderson, J. M., Van Itallie, C. M., & Fanning, A. S. (2004). Setting up a selective barrier at the apical junction complex. *Current Opinion in Cell Biology, 16*(2), 140–145.

Angelow, S., Kim, K. J., & Yu, A. S. (2006). Claudin-8 modulates paracellular permeabil-ity to acidic and basic ions in MDCK II cells. *Journal of Physiology, 571*, 15–26.

Bagnoli, F., Buti, L., Tompkins, L., Covacci, A., & Amieva, M. R. (2005). *Helicobacter pylori* CagA induces a transition from polarized to invasive phenotypes in MDCK cells. *Proceedings of the National Academy of Sciences of the United States of America, 102*(45), 16339–16344.

Bertelsen, L. S., Paesold, G., Marcus, S. L., Finlay, B. B., Eckmann, L., & Barrett, K. E. (2004). Modulation of chloride secretory responses and barrier function of intestinal epithelial cells by the *Salmonella* effector protein SigD. *American Journal of Physiology Cell Physiology, 287*(4), C939–C948.

Blair, S. A., Kane, S. V., Clayburgh, D. R., & Turner, J. R. (2006). Epithelial myosin light chain kinase expression and activity are upregulated in inflammatory bowel disease. *Laboratory Investigation, 86*, 191–201.

Bojarski, C., Gitter, A. H., Bendfeldt, K., Mankertz, J., Schmitz, H., Wagner, S., et al. (2001). Permeability of human HT-29/B6 colonic epithelium as a function of apoptosis. *Journal of Physiology, 535*(Pt 2), 541–552.

Bojarski, C., Weiske, J., Schöneberg, T., Schröder, W., Mankertz, J., Schulzke, J. D., et al. (2004). The specific fates of tight junction proteins in apoptotic epithelial cells. *Journal of Cell Science, 117*(Pt 10), 2097–2107.

Borriello, S. P. (1998). Pathogenesis of *Clostridium difficile* infection. *The Journal of Antimicrobial Chemotheraphy, 41*, 13–19.

Bruewer, M., Luegering, A., Kucharzik, T., Parkos, C. A., Madara, J. L., Hopkins, A. M., et al. (2003). Proinflammatory cytokines disrupt epithelial barrier function by apoptosis-independent mechanisms. *Journal of Immunology, 171*(11), 6164–6172.

Bruewer, M., Samarin, S., & Nusrat, A. (2006). Inflammatory bowel disease and the apical junctional complex. *Annals of the New York Academy of Sciences, 1072*, 242–252.

Bruewer, M., Utech, M., Ivanov, A. I., Hopkins, A. M., Parkos, C. A., & Nusrat, A. (2005). Interferon-gamma induces internalization of epithelial tight junction proteins via a macropinocytosis-like process. *FASEB Journal, 19*(8), 923–933.

Bücker, R., Troeger, H., Kleer, J., Fromm, M., & Schulzke, J. D. (2009). *Arcobacter butzleri* induces barrier dysfunction in intestinal HT-29/B6 cells. *The Journal of Infectious Diseases, 200*(5), 756–764.

Buret, A. G. (2007). Mechanisms of epithelial dysfunction in giardiasis. *Gut, 56*(3), 328–335.

Buret, A. G., Mitchell, K., Muench, D. G., & Scott, K. G. (2002). *Giardia lamblia* disrupts tight junctional ZO-1 and increases permeability in non-transformed human small intestinal epithelial monolayers: Effects of epidermal growth factor. *Parasitology, 125*(Pt 1), 11–19.

Bürgel, N., Bojarski, C., Mankertz, J., Zeitz, M., Fromm, M., & Schulzke, J. D. (2002). Mechanisms of diarrhea in collagenous colitis. *Gastroenterology, 123*(2), 433–443.

Caserta, J. A., Hale, M. L., Popoff, M. R., Stiles, B. G., & McClane, B. A. (2008). Evidence that membrane rafts are not required for the action of *Clostridium perfringens* enterotoxin. *Infection and Immunity, 76*(12), 5677–5685.

Chambers, F. G., Koshy, S. S., Saidi, R. F., Clark, D. P., Moore, R. D., & Sears, C. L. (1997). *Bacteroides fragilis* toxin exhibits polar activity on monolayers of human intestinal epithelial cells (T84 cells) in vitro. *Infection and Immunity, 65*(9), 3561–3570.

Chen, M. L., Ge, Z., Fox, J. G., & Schauer, D. B. (2006). Disruption of tight junctions and induction of proinflammatory cytokine responses in colonic epithelial cells by *Campylobacter jejuni*. *Infection and Immunity, 74*(12), 6581–6589.

Chiba, H., Kojima, T., Osanai, M., & Sawada, N. (2006). The significance of interferon-gamma-triggered internalization of tight-junction proteins in inflammatory bowel disease. In *Science STKE 2006(316)*. pe1.

Chin, A. C., Teoh, D. A., Scott, K. G., Meddings, J. B., Macnaughton, W. K., & Buret, A. G. (2002). Strain-dependent induction of enterocyte apoptosis by *G. lamblia* disrupts epithelial barrier function in a caspase-3-dependent manner. *Infection and Immunity, 70*, 3673–3680.

Ciccocioppo, R., Finamore, A., Ara, C., Di Sabatino, A., Mengheri, E., & Corazza, G. R. (2006). Altered expression, localization, and phosphorylation of epithelial junctional proteins in celiac disease. *American Journal of Clinical Pathology, 125*, 502–511.

Clayburgh, D. R., Barrett, T. A., Tang, Y., Meddings, J. B., Van Eldik, L. J., Watterson, D. M., et al. (2005). Epithelial myosin light chain kinase-dependent barrier dysfunction mediates T cell activation-induced diarrhea in vivo. *The Journal of Clinical Investigation, 115*, 2702–2715.

Clayburgh, D. R., Shen, L., & Turner, J. R. (2004). A porous defense: The leaky epithelial barrier in intestinal disease. *Laboratory Investigation, 84*(3), 282–291.

Czerucka, D., Dahan, S., Mograbi, B., Rossi, B., & Rampal, P. (2000). *Saccharomyces boulardii* preserves the barrier function and modulates the signal transduction pathway induced in enteropathogenic *Escherichia coli*-infected T84 cells. *Infection and Immunity, 68*(10), 5998–6004.

Dean, P., & Kenny, B. (2004). Intestinal barrier dysfunction by enteropathogenic *Escherichia coli* is mediated by two effector molecules and a bacterial surface protein. *Molecular Microbiology, 54*, 665–675.

Dickman, K. G., Hempson, S. J., Anderson, J., Lippe, S., Zhao, L., Burakoff, R., et al. (2000). Rotavirus alters paracellular permeability and energy metabolism in Caco-2 cells. *American Journal of Physiology. Gastrointestinal and Liver Physiology, 279*(4), G757–G766.

Ebihara, C., Kondoh, M., Hasuike, N., Harada, M., Mizuguchi, H., Horiguchi, Y., et al. (2006). Preparation of a claudin-targeting molecule using a C-terminal fragment of *Clostridium perfringens* enterotoxin. *The Journal of Pharmacology and Experimental Therapeutics, 316*(1), 255–260.

El Asmar, R., Panigrahi, P., Bamford, P., Berti, I., Not, T., Coppa, G. V., et al. (2002). Host-dependent zonulin secretion causes the impairment of the small intestine barrier function after bacterial exposure. *Gastroenterology, 123*, 1607–1615.

Epple, H. J., Schneider, T., Troeger, H., Kunkel, D., Allers, K., Moos, V., et al. (2009). Impairment of the intestinal barrier is evident in untreated but absent in suppressively treated HIV-infected patients. *Gut, 58*, 220–227.

Ewaschuk, J. B., Diaz, H., Meddings, L., Diederichs, B., Dmytrash, A., Backer, J., et al. (2008). Secreted bioactive factors from *Bifidobacterium infantis* enhance epithelial cell barrier function. *American Journal of Physiology. Gastrointestinal and Liver Physiology, 295*(5), G1025–G1034.

Fasano, A., Fiorentini, C., Donelli, G., Uzzau, S., Kaper, J. B., Margaretten, K., et al. (1995). Zonula occludens toxin modulates tight junctions through protein kinase C-dependent actin reorganization, in vitro. *The Journal of Clinical Investigation, 96*, 710–720.

Fedwick, J. P., Lapointe, T. K., Meddings, J. B., Sherman, P. M., & Buret, A. G. (2005). *Helicobacter pylori* activates myosin light-chain kinase to disrupt claudin-4 and claudin-5 and increase epithelial permeability. *Infection and Immunity, 73*(12), 7844–7852.

Fujita, H., Chiba, H., Yokozaki, H., Sakai, N., Sugimoto, K., Wada, T., et al. (2006). Differential expression and subcellular localization of claudin-7, -8, -12, -13, and -15 along the mouse intestine. *The Journal of Histochemistry and Cytochemistry, 54*, 933–944.

Fujita, H., Sugimoto, K., Inatomi, S., Maeda, T., Osanai, M., Uchiyama, Y., et al. (2008). Tight junction proteins claudin-2 and -12 are critical for vitamin D-dependent Ca^{2+} absorption between enterocytes. *Molecular Biology of the Cell, 19*, 1912–1921.

Fujita, K., Katahira, J., Horiguchi, Y., Sonoda, N., Furuse, M., & Tsukita, S. (2000). *Clostridium perfringens* enterotoxin binds to the second extracellular loop of claudin-3, a tight junction integral membrane protein. *FEBS Letters, 476*, 258–261.

Furuse, M., Fujita, K., Hiiragi, T., Fujimoto, K., & Tsukita, S. (1998). Claudin-1 and -2: Novel integral membrane proteins localizing at tight junctions with no sequence simi-larity to occludin. *The Journal of Cell Biology, 141*, 1539–1550.

Furuse, M., Hata, M., Furuse, K., Yoshida, Y., Haratake, A., Sugitani, Y., et al. (2002). Claudin-based tight junctions are crucial for the mammalian epidermal barrier: A lesson from claudin-1-deficient mice. *The Journal of Cell Biology, 156*(6), 1099–1111.

Furuse, M., Hirase, T., Itoh, M., Nagafuchi, A., Yonemura, S., Tsukita, S., et al. (1993). Occludin: A novel integral membrane protein localizing at tight junctions. *The Journal of Cell Biology, 123*, 1777–1788.

Furuse, M., Sasaki, H., & Tsukita, S. (1999). Manner of interaction of heterogeneous claudin species within and between tight junction strands. *The Journal of Cell Biology, 147*(4), 891–903.

Garcia Vilela, E., De Lourdes De Abreu Ferrari, M., Oswaldo Da Gama Torres, H., Guerra Pinto, A., Carolina Carneiro Aguirre, A., Paiva Martins, F., et al. (2008). Influence of *Saccharomyces boulardii* on the intestinal permeability of patients with Crohn's disease in remission. *Scandinavian Journal of Gastroenterology, 43*(7), 842–848.

Gibson, P. R. (2004). Increased gut permeability in Crohn's disease: Is TNF the link? *Gut, 53*(12), 1724–1725.

Gitter, A. H., Bendfeldt, K., Schulzke, J. D., & Fromm, M. (2000). Leaks in the epithelial barrier caused by spontaneous and TNF-alpha-induced single-cell apoptosis. *FASEB Journal, 14*(12), 1749–1753.

Gitter, A. H., Wullstein, F., Fromm, M., & Schulzke, J. D. (2001). Epithelial barrier defects in ulcerative colitis: Characterization and quantification by electrophysiological imaging. *Gastroenterology, 121*(6), 1320–1328.

Gradel, K. O., Nielsen, H. L., Schønheyder, H. C., Ejlertsen, T., Kristensen, B., & Nielsen, H. (2009). Increased short- and long-term risk of inflammatory bowel disease after *Salmonella* or *Campylobacter gastroenteritis*. *Gastroenterology, 137*(2), 495–501.

Graham, W. V., Wang, F., Clayburgh, D. R., Cheng, J. X., Yoon, B., Wang, Y., et al. (2006). Tumor necrosis factor-induced long myosin light chain kinase transcription is regulated by differentiation dependent signaling events. Characterization of the human long myosin light chain kinase promoter. *The Journal of Biological Chemistry, 281*, 26205–26215.

Heller, F., Florian, P., Bojarski, C., Richter, J. F., Christ, M., Hillenbrand, B., et al. (2005). Interleukin-13 is the key effector Th2 cytokine in ulcerative colitis that affects epithelial tight junctions, apoptosis, and cell restitution. *Gastroenterology, 129*(2), 550–564.

Hollander, D. (2002). Crohn's disease, TNF-alpha, and the leaky gut. The chicken or the egg? *American Journal of Gastroenterology, 97*(8), 1867–1868.

Holmes, J. L., Van Itallie, C. M., Rasmussen, J. E., & Anderson, J. M. (2006). Claudin profiling in the mouse during postnatal intestinal development and along the gastrointestinal tract reveals complex expression patterns. *Gene Expression Patterns, 6*, 581–588.

Howe, K. L., Reardon, C., Wang, A., Nazli, A., & McKay, D. M. (2005). Transforming growth factor-beta regulation of epithelial tight junction proteins enhances barrier function and blocks enterohemorrhagic *Escherichia coli* O157:H7-induced increased permeability. *American Journal of Pathology, 167*(6), 1587–1597.

Hu, L., Tall, B. D., Curtis, S. K., & Kopecko, D. J. (2008). Enhanced microscopic definition of *Campylobacter jejuni* 81-176 adherence to, invasion of, translocation across, and exocytosis from polarized human intestinal Caco-2 cells. *Infection and Immunity, 76*(11), 5294–5304.

Ikenouchi, J., Furuse, M., Furuse, K., Sasaki, H., Tsukita, S., & Tsukita, S. (2005). Tricellulin constitutes a novel barrier at tricellular contacts of epithelial cells. *The Journal of Cell Biology, 171*, 939–945.

Itoh, M., Sasaki, H., Furuse, M., Ozaki, H., Kita, T., & Tsukita, S. (2001). Junctional adhesion molecule (JAM) binds to PAR-3: A possible mechanism for the recruitment of PAR-3 to tight junctions. *The Journal of Cell Biology, 154*, 491–497.

Ivanov, A. I., Nusrat, A., & Parkos, C. A. (2004). Endocytosis of epithelial apical junctional proteins by a clathrin-mediated pathway into a unique storage compartment. *Molecular Biology of the Cell, 15*(1), 176–188.

Köhler, H., Sakaguchi, T., Hurley, B. P., Kase, B. A., Reinecker, H. C., & McCormick, B. A. (2007). *Salmonella enterica* serovar Typhimurium regulates intercellular junction proteins and facilitates transepithelial neutrophil and bacterial passage. *American Journal of Physiology. Gastrointestinal and Liver Physiology, 293*(1), G178–G187.

Koshy, S. S., Montrose, M. H., & Sears, C. L. (1996). Human intestinal epithelial cells swell and demonstrate actin rearrangement in response to the metalloprotease toxin of *Bacteroides fragilis*. *Infection and Immunity, 64*(12), 5022–5028.

Laharie, D., Asencio, C., Asselineau, J., Bulois, P., Bourreille, A., Moreau, J., et al. (2009). Association between entero-hepatic *Helicobacter* species and Crohn's disease: A prospective cross-sectional study. *Alimentary Pharmacology & Therapeutics, 30*(3), 283–293.

Lamb-Rosteski, J. M., Kalischuk, L. D., Inglis, G. D., & Buret, A. G. (2008). Epidermal growth factor inhibits *Campylobacter jejuni*-induced claudin-4 disruption, loss of epithelial barrier function, and *Escherichia coli* translocation. *Infection and Immunity, 76*(8), 3390–3398.

Lastovica, A. J. (2009). Clinical relevance of *Campylobacter concisus* isolated from pediatric patients. *Journal of Clinical Microbiology, 47*(7), 2360.

Laukoetter, M. G., Bruewer, M., & Nusrat, A. (2006). Regulation of the intestinal epithelial barrier by the apical junctional complex. *Current Opinion in Gastroenterology, 22*(2), 85–89.

Lauwaet, T., Oliveira, M. J., Callewaert, B., De Bruyne, G., Mareel, M., & Leroy, A. (2004). Proteinase inhibitors TPCK and TLCK prevent *Entamoeba histolytica* induced disturbance of tight junctions and microvilli in enteric cell layers in vitro. *International Journal for Parasitology, 34*(7), 785–794.

Leroy, A., Lauwaet, T., De Bruyne, G., Cornelissen, M., & Mareel, M. (2000). *Entamoeba histolytica* disturbs the tight junction complex in human enteric T84 cell layers. *FASEB Journal, 14*(9), 1139–1146.

Li, N., Russell, W. M., Douglas-escobar, M., Hauser, N., Lopez, M., & Neu, J. (2009). Live and heat-killed *Lactobacillus rhamnosus* GG: Effects on proinflammatory and anti-inflammatory cytokines/chemokines in gastrostomy-fed infant rats. *Pediatric Research, 66*(2), 203–207.

Liao, A. P., Petrof, E. O., Kuppireddi, S., Zhao, Y., Xia, Y., Claud, E. C., et al. (2008). *Salmonella* type III effector AvrA stabilizes cell tight junctions to inhibit inflammation in intestinal epithelial cells. *PLoS One, 3*(6), e2369.

Lima, A. A., Silva, T. M., Gifoni, A. M., Barrett, L. J., McAuliffe, I. T., Bao, Y., et al. (1997). Mucosal injury and disruption of intestinal barrier function in HIV-infected individuals with and without diarrhea and cryptosporidiosis in northeast Brazil. *American Journal of Gastroenterology, 92*(10), 1861–1866.

Lin, D., Edwards, A. S., Fawcett, J. P., Mbamalu, G., Scott, J. D., & Pawson, T. (2000). A mammalian PAR-3-PAR-6 complex implicated in Cdc42/Rac1 and aPKC signalling and cell polarity. *Nature Cell Biology, 2*, 540–547.

Lorrot, M., & Vasseur, M. (2007). How do the rotavirus NSP4 and bacterial enterotoxins lead differently to diarrhea? *Virology Journal, 4*, 31.

Ma, T. Y., Boivin, M. A., Ye, D., Pedram, A., & Said, H. M. (2005). Mechanism of TNF-alpha modulation of Caco-2 intestinal epithelial tight junction barrier: Role of myosin light-chain kinase protein expression. *American Journal of Physiology. Gastrointestinal and Liver Physiology, 288*, G422–G430.

MacCallum, A., Hardy, S. P., & Everest, P. H. (2005). *Campylobacter jejuni* inhibits the absorptive transport functions of Caco-2 cells and disrupts cellular tight junctions. *Microbiology, 151*(Pt 7), 2451–2458.

MacDonald, T. T., Hutchings, P., Choy, M. Y., Murch, S., & Cooke, A. (1990). Tumour necrosis factor-alpha and interferon-gamma production measured at the single cell level in normal and inflamed human intestine. *Clinical and Experimental Immunology, 81*(2), 301–305.

Mankertz, J., Tavalali, S., Schmitz, H., Mankertz, A., Riecken, E. O., Fromm, M., et al. (2000). Expression from the human occludin promoter is affected by tumor necrosis factor alpha and interferon gamma. *Journal of Cell Science, 113*(Pt 11), 2085–2090.

Markov, A. G., Veshnyakova, A., Fromm, M., Amasheh, M., & Amasheh, S. (2010). Segmental expression of claudin proteins correlates with tight junction barrier properties in rat intestine. *Journal of Comparative Physiology B, 180*(4), 591–598.

Matsuzawa, T., Kuwae, A., & Abe, A. (2005). Enteropathogenic *Escherichia coli* type III effectors EspG and EspG2 alter epithelial paracellular permeability. *Infection and Immunity, 73*, 6283–6289.

Matysiak-Budnik, T., Moura, I. C., Arcos-Fajardo, M., Lebreton, C., Ménard, S., Candalh, C., et al. (2008). Secretory IgA mediates retrotranscytosis of intact gliadin peptides via the transferrin receptor in celiac disease. *The Journal of Experimental Medicine, 205*, 143–154.

McNamara, B. P., Koutsouris, A., O'Connell, C. B., Nougayréde, J. P., Donnenberg, M. S., & Hecht, G. (2001). Translocated EspF protein from enteropathogenic *Escherichia coli* disrupts host intestinal barrier function. *The Journal of Clinical Investigation, 107*(5), 621–629.

Montalto, M., Cuoco, L., Ricci, R., Maggiano, N., Vecchio, F. M., & Gasbarrini, G. (2002). Immunohistochemical analysis of ZO-1 in the duodenal mucosa of patients with untreated and treated celiac disease. *Digestion, 65*, 227–233.

Morita, K., Furuse, M., Fujimoto, K., & Tsukita, S. (1999). Claudin multigene family encoding four-transmembrane domain protein components of tight junction strands. *Proceedings of the National Academy of Sciences of the United States of America, 96*, 511–516.

Morita, K., Sasaki, H., Furuse, M., & Tsukita, S. (1999). Endothelial claudin: Claudin-5/TMVCF constitutes tight junction strands in endothelial cells. *The Journal of Cell Biology, 147*, 185–194.

Mumy, K. L., Chen, X., Kelly, C. P., & McCormick, B. A. (2008). *Saccharomyces boulardii* interferes with Shigella pathogenesis by postinvasion signaling events. *American Journal of Physiology. Gastrointestinal and Liver Physiology, 294*(3), G599–G609.

Muza-Moons, M. M., Schneeberger, E. E., & Hecht, G. A. (2004). Enteropathogenic *Escherichia coli* infection leads to appearance of aberrant tight junctions strands in the lateral membrane of intestinal epithelial cells. *Cellular Microbiology, 6*(8), 783–793.

Narumiya, S., Ishizaki, T., & Watanabe, N. (1997). Rho effectors and reorganization of actin cytoskeleton. *FEBS Letters, 410*, 68–72.

Nava, P., López, S., Arias, C. F., Islas, S., & González-Mariscal, L. (2004). The rotavirus surface protein VP8 modulates the gate and fence function of tight junctions in epithelial cells. *Journal of Cell Science, 117*(Pt 23), 5509–5519.

Nitta, T., Hata, M., Gotoh, S., Seo, Y., Sasaki, H., Hashimoto, N., et al. (2003). Size-selective loosening of the blood-brain barrier in claudin-5-deficient mice. *The Journal of Cell Biology, 161*, 653–660.

Nusrat, A., Giry, M., Turner, J. R., Colgan, S. P., Parkos, C. A., Carnes, D., et al. (1995). Rho protein regulates tight junctions and perijunctional actin organization in polarized epithelia. *Proceedings of the National Academy of Sciences of the United States of America, 92*, 10629–10635.

Nusrat, A., von Eichel-Streiber, C., Turner, J. R., Verkade, P., Madara, J. L., & Parkos, C. A. (2001). *Clostridium difficile* toxins disrupt epithelial barrier function by altering membrane microdomain localization of tight junction proteins. *Infection and Immunity, 69*(3), 1329–1336.

Obiso, R. J., Jr., Azghani, A. O., & Wilkins, T. D. (1997). The Bacteroides fragilis toxin fragilysin disrupts the paracellular barrier of epithelial cells. *Infection and Immunity, 65*(4), 1431–1439.

Papini, E., Satin, B., Norais, N., de Bernard, M., Telford, J. L., Rappuoli, R., et al. (1998). Selective increase of the permeability of polarized epithelial cell monolayers by *Helicobacter pylori* vacuolating toxin. *The Journal of Clinical Investigation, 102*(4), 813–820.

Parashar, U. D., Hummelman, E. G., Bresee, J. S., Miller, M. A., & Glass, R. I. (2003). Global illness and deaths caused by rotavirus disease in children. *Emerging Infectious Diseases, 9*(5), 565–572.

Pizzuti, D., Bortolami, M., Mazzon, E., Buda, A., Guariso, G., D'Odorico, A., et al. (2004). Transcriptional downregulation of tight junction protein ZO-1 in active coeliac disease is reversed after a gluten-free diet. *Digestive and Liver Disease, 36*, 337–341.

Pothoulakis, C. (1996). Pathogenesis of *Clostridium difficile*-associated diarrhoea. *European Journal of Gastroenterology and Hepatology, 8*, 1041–1047.

Raddatz, D., Bockemuhl, M., & Ramadori, G. (2005). Quantitative measurement of cytokine mRNA in inflammatory bowel disease: Relation to clinical and endoscopic activity and outcome. *European Journal of Gastroenterology and Hepatology, 17*(5), 547–557.

Rahner, C., Mitic, L. L., & Anderson, J. M. (2001). Heterogeneity in expression and sub-cellular localization of claudins 2, 3, 4, and 5 in the rat liver, pancreas, and gut. *Gastroenterology, 120*, 411–422.

Rosenthal, R., Milatz, S., Krug, S. M., Oelrich, B., Schulzke, J. D., Amasheh, S., et al. (2010). The tight junction protein claudin-2 forms a paracellular water channel. *Journal of Cell Science* (in press).

Saitou, M., Furuse, M., Sasaki, H., Schulzke, J. D., Fromm, M., Takano, H., et al. (2000). Complex phenotype of mice lacking occludin, a component of tight junction strands. *Molecular Biology of the Cell, 11*, 4131–4142.

Sakaguchi, T., Köhler, H., Gu, X., McCormick, B. A., & Reinecker, H. C. (2002). *Shigella flexneri* regulates tight junction-associated proteins in human intestinal epithelial cells. *Cellular Microbiology, 4*(6), 367–381.

Sander, G. R., Cummins, A. G., Henshall, T., & Powell, B. C. (2005). Rapid disruption of intestinal barrier function by gliadin involves altered expression of apical junctional proteins. *FEBS Letters, 579*, 4851–4855.

Schauser, K., Olsen, J. E., & Larsson, L. I. (2005). *Salmonella typhimurium* infection in the porcine intestine: Evidence for caspase-3-dependent and -independent programmed cell death. *Histochemistry and Cell Biology, 123*(1), 43–50.

Schmitz, H., Barmeyer, C., Fromm, M., Runkel, N., Foss, H. D., Bentzel, C. J., et al. (1999). Altered tight junction structure contributes to the impaired epithelial barrier function in ulcerative colitis. *Gastroenterology, 116*(2), 301–309.

Schmitz, H., Fromm, M., Bentzel, C. J., Scholz, P., Detjen, K., Mankertz, J., et al. (1999). Tumor necrosis factor-alpha (TNFalpha) regulates the epithelial barrier in the human intestinal cell line HT-29/B6. *Journal of Cell Science, 112*(Pt 1), 137–146.

Schmitz, H., Rokos, K., Florian, P., Gitter, A. H., Fromm, M., Scholz, P., et al. (2002). Supernatants of HIV-infected immune cells affect the barrier function of human HT-29/B6 intestinal epithelial cells. *AIDS, 16*(7), 983–991.

Schulzke, J. D., Bentzel, C. J., Schulzke, I., Riecken, E. O., & Fromm, M. (1998). Epithelial tight junction structure in the jejunum of children with acute and treated celiac sprue. *Pediatric Research, 43*(4 Pt 1), 435–441.

Schulzke, J. D., Bojarski, C., Zeissig, S., Heller, F., Gitter, A. H., & Fromm, M. (2006). Disrupted barrier function through epithelial cell apoptosis. *Annals of the New York Academy of Sciences, 1072*, 288–299.

Schulzke, J. D., Gitter, A. H., Mankertz, J., Spiegel, S., Seidler, U., Amasheh, S., et al. (2005). Epithelial transport and barrier function in occludin-deficient mice. *Biochimica et Biophysica Acta, 1669*(1), 34–42.

Schulzke, J. D., Schulzke, I., Fromm, M., & Riecken, E. O. (1995). Epithelial barrier and ion transport in celiac sprue: Electrical measurements on intestinal aspiration biopsy specimens. *Gut, 37*, 777–782.

Schürmann, G., Brüwer, M., Klotz, A., Schmid, K. W., Senninger, N., & Zimmer, K. P. (1999). Transepithelial transport processes at the intestinal mucosa in inflammatory bowel disease. *International Journal of Colorectal Disease, 14*(1), 41–46.

Schumann, M., Richter, J. F., Wedell, I., Moos, V., Zimmermann-Kordmann, M., Schneider, T., et al. (2008). Mechanisms of epithelial translocation of the alpha(2)-gliadin-33mer in coeliac sprue. *Gut, 57*(6), 747–754.

Schwarz, B. T., Wang, F., Shen, L., Clayburgh, D. R., Su, L., Wang, Y., et al. (2007). LIGHT signals directly to intestinal epithelia to cause barrier dysfunction via cytoskeletal and endocytic mechanisms. *Gastroenterology, 132*, 2383–2394.

Scott, K. G., Meddings, J. B., Kirk, D. R., Lees-Miller, S. P., & Buret, A. G. (2002). Intestinal infection with *Giardia* spp. reduces epithelial barrier function in a myosin light chain kinase-dependent fashion. *Gastroenterology, 123*, 1179–1190.

Segain, J. P., Raingeard de la Blétière, D., Sauzeau, V., Bourreille, A., Hilaret, G., Cario-Toumaniantz, C., et al. (2003). Rho kinase blockade prevents inflammation via nuclear factor kappa B inhibition: Evidence in Crohn's disease and experimental colitis. *Gastroenterology, 124*(5), 1180–1187.

Söderholm, J. D., Streutker, C., Yang, P. C., Paterson, C., Singh, P. K., McKay, D. M., et al. (2004). Increased epithelial uptake of protein antigens in the ileum of Crohn's disease mediated by tumour necrosis factor alpha. *Gut, 53*(12), 1817–1824.

Sonoda, N., Furuse, M., Sasaki, H., Yonemura, S., Katahira, J., Horiguchi, Y., et al. (1999). *Clostridium perfringens* enterotoxin fragment removes specific claudins from tight junction strands: Evidence for direct involvement of claudins in tight junction barrier. *The Journal of Cell Biology, 147*(1), 195–204.

Stallmach, A., Giese, T., Schmidt, C., Ludwig, B., Mueller-Molaian, I., & Meuer, S. C. (2004). Cytokine/chemokine transcript profiles reflect mucosal inflammation in Crohn's disease. *International Journal of Colorectal Disease, 19*(4), 308–315.

Suzuki, K., Kokai, Y., Sawada, N., Takakuwa, R., Kuwahara, K., Isogai, E., et al. (2002). SS1 *Helicobacter pylori* disrupts the paracellular barrier of the gastric mucosa and leads to neutrophilic gastritis in mice. *Virchows Archiv, 440*(3), 318–324.

Takaishi, S., & Wang, T. C. (2007). Gene expression profiling in a mouse model of *Helicobacter*-induced gastric cancer. *Cancer Science, 98*(3), 284–293.

Tamura, A., Kitano, Y., Hata, M., Katsuno, T., Moriwaki, K., Sasaki, H., et al. (2008). Megaintestine in claudin-15-deficient mice. *Gastroenterology, 134*, 523–534.

Tatum, R., Zhang, Y., Salleng, K., Lu, Z., Lin, J. J., Lu, Q., et al. (2010). Renal salt wasting and chronic dehydration in claudin-7-deficient mice. *American Journal of Physiology. Renal Physiology, 298*, F24–F34.

Tripathi, A., Lammers, K. M., Goldblum, S., Shea-Donohue, T., Netzel-Arnett, S., Buzza, M. S., et al. (2009). Identification of human zonulin, a physiological modulator of tight junctions, as prehaptoglobin-2. *Proceedings of the National Academy of Sciences of the United States of America, 106*(39), 16799–16804.

Troeger, H., Epple, H. J., Schneider, T., Wahnschaffe, U., Ullrich, R., Burchard, G. D., et al. (2007). Effect of chronic *Giardia lamblia* infection on epithelial transport and barrier function in human duodenum. *Gut, 56*(3), 328–335.

Troeger, H., Loddenkemper, C., Schneider, T., Schreier, E., Epple, H. J., Zeitz, M., et al. (2009). Structural and functional changes of the duodenum in human norovirus infection. *Gut, 58*, 1070–1077.

Troeger, H., Richter, J. F., Beutin, L., Günzel, D., Dobrindt, U., Epple, H. J., et al. (2007). *Escherichia coli* alpha-haemolysin induces focal leaks in colonic epithelium: A novel mechanism of bacterial translocation. *Cellular Microbiology, 9*(10), 2530–2540.

Ukena, S. N., Singh, A., Dringenberg, U., Engelhardt, R., Seidler, U., Hansen, W., et al. (2007). Probiotic *Escherichia coli* Nissle 1917 inhibits leaky gut by enhancing mucosal integrity. *PLoS One, 2*(12), e1308.

Utech, M., Ivanov, A. I., Samarin, S. N., Bruewer, M., Turner, J. R., Mrsny, R. J., et al. (2005). Mechanism of IFN-gamma-induced endocytosis of tight junction proteins: Myosin II-dependent vacuolarization of the apical plasma membrane. *Molecular Biology of the Cell, 16*(10), 5040–5052.

Van Itallie, C. M., Fanning, A. S., & Anderson, J. M. (2003). Reversal of charge selectivity in cation or anion-selective epithelial lines by expression of different claudins. *American Journal of Physiology. Renal Physiology, 285*, F1078–F1084.

Van Itallie, C. M., Rahner, C., & Anderson, J. M. (2001). Regulated expression of claudin-4 decreases paracellular conductance through a selective decrease in sodium permeability. *The Journal of Clinical Investigation, 107*, 1319–1327.

Varon, C., Duriez, A., Lehours, P., Ménard, A., Layé, S., Zerbib, F., et al. (2009). Study of *Helicobacter pullorum* proinflammatory properties on human epithelial cells in vitro. *Gut, 58*(5), 629–635.

Vogelsang, H., Schwarzenhofer, M., & Oberhuber, G. (1998). Changes in gastrointestinal permeability in celiac disease. *Digestive Diseases, 16*, 333–336.

Wang, F., Graham, W. V., Wang, Y., Witkowski, E. D., Schwarz, B. T., & Turner, J. R. (2005). Interferon-gamma and tumor necrosis factor-alpha synergize to induce intestinal epithelial barrier dysfunction by up-regulating myosin light chain kinase expression. *American Journal of Pathology, 166*, 409–419.

Wang, F., Schwarz, B. T., Graham, W. V., Wang, Y., Su, L., Clayburgh, D. R., et al. (2006). IFN-gamma-induced TNFR2 expression is required for TNF-dependent intestinal epithelial barrier dysfunction. *Gastroenterology, 131*, 1153–1163.

Wang, W., Uzzau, S., Goldblum, S. E., & Fasano, A. (2000). Human zonulin, a potential modulator of intestinal tight junctions. *Journal of Cell Science, 113*(Pt 24), 4435–4440.

Wapenaar, M. C., Monsuur, A. J., van Bodegraven, A. A., Weersma, R. K., Bevova, M. R., Linskens, R. K., et al. (2008). Associations with tight junction genes PARD3 and MAGI2 in Dutch patients point to a common barrier defect for coeliac disease and ulcerative colitis. *Gut, 57*, 463–467.

Weng, X. H., Beyenbach, K. W., & Quaroni, A. (2005). Cultured monolayers of the dog jejunum with the structural and functional properties resembling the normal epithe-lium. *American Journal of Physiology. Gastrointestinal and Liver Physiology, 288*, G705–G717.

Wilde, C., & Aktories, K. (2001). The Rho-ADP-ribosylating C3 exoenzyme from *Clostridium botulinum* and related C3-like transferases. *Toxicon, 39*, 1647–1660.

Willemsen, L. E., Hoetjes, J. P., van Deventer, S. J., & van Tol, E. A. (2005). Abrogation of IFN-gamma mediated epithelial barrier disruption by serine protease inhibition. *Clinical and Experimental Immunology, 142*(2), 275–284.

Wine, E., Ossa, J. C., Gray-Owen, S. D., & Sherman, P. M. (2009). Adherent-invasive *Escherichia coli*, strain LF82 disrupts apical junctional complexes in polarized epithelia. *BMC Microbiology, 9*, 180.

Winkler, L., Gehring, C., Wenzel, A., Müller, S. L., Piehl, C., Krause, G., et al. (2009). Molecular determinants of the interaction between *Clostridium perfringens* enterotoxin fragments and claudin-3. *The Journal of Biological Chemistry, 284*(28), 18863–18872.

Wroblewski, L. E., Shen, L., Ogden, S., Romero-Gallo, J., Lapierre, L. A., Israel, D. A., et al. (2009). *Helicobacter pylori* dysregulation of gastric epithelial tight junctions by urease-mediated myosin II activation. *Gastroenterology, 136*(1), 236–246.

Ye, D., Ma, I., & Ma, T. Y. (2006). Molecular mechanism of tumor necrosis factor-alpha modulation of intestinal epithelial tight junction barrier. *American Journal of Physiology. Gastrointestinal and Liver Physiology, 290*(3), G496–G504.

Yu, A. S., Enck, A. H., Lencer, W. I., & Schneeberger, E. E. (2003). Claudin-8 expression in MDCK cells augments the paracellular barrier to cation permeation. *The Journal of Biological Chemistry, 278*, 17350–17359.

Zareie, M., Riff, J., Donato, K., McKay, D. M., Perdue, M. H., Soderholm, J. D., et al. (2005). Novel effects of the prototype translocating *Escherichia coli*, strain C25 on intestinal epithelial structure and barrier function. *Cellular Microbiology, 7*(12), 1782–1797.

Zeissig, S., Bojarski, C., Bürgel, N., Mankertz, J., Zeitz, M., Fromm, M., et al. (2004). Down-regulation of epithelial apoptosis and barrier repair in active Crohn's disease by tumour necrosis factor alpha antibody treatment. *Gut, 53*(9), 1295–1302.

Zeissig, S., Bürgel, N., Günzel, D., Richter, J. F., Mankertz, J., Wahnschaffe, U., et al. (2007). Changes in expression and distribution of claudin 2, 5 and 8 lead to discontinuous tight junctions and barrier dysfunction in active Crohn's disease. *Gut, 56*(1), 61–72.

Zolotarevsky, Y., Hecht, G., Koutsouris, A., Gonzalez, D. E., Quan, C., Tom, J., et al. (2002). A membrane-permeant peptide that inhibits MLC kinase restores barrier function in in vitro models of intestinal disease. *Gastroenterology, 123*, 163–172.

Zyrek, A. A., Cichon, C., Helms, S., Enders, C., Sonnenborn, U., & Schmidt, M. A. (2007). Molecular mechanisms underlying the probiotic effects of *Escherichia coli* Nissle 1917 involve ZO-2 and PKCzeta redistribution resulting in tight junction and epithelial barrier repair. *Cellular Microbiology, 9*(3), 804–816.

CHAPTER 10

Claudin Proteins and Neuronal Function

Jérôme Devaux,* Bozena Fykkolodziej,[†] and Alexander Gow[†,‡,§]

*Département Signalisation Neuronale, CRN2M, UMR, CNRS, Université de la Méditerranée—Université Paul Cézanne, IFR Jean Roche, Marseille, France

[†]Center for Molecular Medicine and Genetics, Wayne State University School of Medicine, Detroit, Michigan, USA

[‡]Carman and Ann Adams Department of Pediatrics, Wayne State University School of Medicine, Detroit, Michigan, USA

[§]Department of Neurology, Wayne State University School of Medicine, Detroit, Michigan, USA

I. OVERVIEW

The identification and characterization of the claudin family of tight junction (TJ) proteins in the late 1990s ushered in a new era for research into the molecular and cellular biology of intercellular junctions. Since that

Current Topics in Membranes, Volume 65 1063-5823/10 $35.00

DOI: 10.1016/S1063-5823(10)65010-7

time, TJs have been studied in the contexts of many diseases including deafness, male infertility, cancer, bacterial invasion, and liver and kidney disorders. In this review, we consider the role of claudins in the nervous system focusing on the mechanisms by which TJs in glial cells are involved in neuronal function. Electrophysiological evidence suggests that claudins may operate in the central nervous system (CNS) in a manner similar to polarized epithelia. We also evaluate hypotheses that TJs are the gatekeepers of an immune-privileged myelin compartment and that TJs emerged during evolution to form major adhesive forces within the myelin sheath. Finally, we consider the implications of disrupted CNS myelin TJs in the contexts of behavioral disorders (schizophrenia) and demyelinating or hypomyelinating diseases (multiple sclerosis and the leukodystrophies), and explore evidence of a possible mechanism governing affective disorder symptoms in patients with white matter abnormalities.

II. INTRODUCTION

TJs are evolutionarily conserved structural and physiological components of polarized epithelia that are essential for life (reviewed in Furuse & Tsukita, 2006; Southwood & Gow, 2001). The function and molecular composition of TJs have been unravelled over the last few decades and found to be comprised of intramembranous strands of protein polymers that form a dynamic meshwork at the apical edges of polarized epithelial cells. TJs play multiple roles in epithelia, including the generation of paracellular permeability barriers across epithelial sheets to regulate the flux of macromolecules into or out of the organism; the generation of macromolecular diffusion barriers within the plane of membranes to maintain cell polarity between the apical and basolateral domains; and the formation of ion selective paracellular pores to maintain electrochemical gradients and to regulate the composition of the epithelial microenvironment.

Genetic and biochemical studies show that the claudin family are principal mediators of the diverse morphological and physiological properties of TJs in higher eukaryotes (reviewed in Tsukita, Yamazaki, Katsuno, Tamura, & Tsukita, 2008), and many of these proteins are expressed in various cell types throughout the nervous system. In addition to their importance in elaborating and maintaining blood–brain barriers in the choroid plexus and vasculature, TJs also play crucial roles in motor and sensory systems. TJ proteins are rarely expressed by neurons, but one exception is olfactory neurons. The dendrites of these cells traverse the paracellular space of the nasal epithelium and localize claudins to intercalate with and maintain the epithelial barrier while carrying out their function to detecting odors (Brightman & Reese,

1969). The utilization of TJs in this manner is somewhat analogous to intestinal dendritic cells, which extend claudin-bearing projections between intestinal epithelial cells so as not to disrupt the TJ barrier while sampling intestinal flora (Rescigno et al., 2001).

TJs have been involved in nervous system function in early metazoan evolution with demonstrated importance at the blood–nerve barrier in *Drosophila* (Stork et al., 2008). In the mammalian CNS, TJs are key structural components of oligodendrocytes and support neuron function by increasing conduction velocity along small myelinated fibers (Devaux & Gow, 2008; Gow & Devaux, 2008). TJs are also present in PNS myelin elaborated by Schwann cells but likely perform different functions to TJs in CNS myelin (Devaux & Gow, 2008; Gow & Devaux, 2008). Of major focus in the current review, the CNS phenotype in *Claudin 11*-null mice demonstrates the importance of claudin 11 in myelin to maximizing nerve conduction.

The necessity for reliable conduction to plasticity and temporal processing in neural circuits implicates claudin 11 TJs not only broadly in brain function but also in behavioral disorders such as schizophrenia and other affective disorders associated with myelin abnormalities. A likely theme to emerge in the coming years will be the importance of TJs to neuron physiology because they form resistive barriers that maintain the composition of the extracellular microenvironment in the nervous system. This is consistent with the extensively characterized function of TJs in polarized epithelia in peripheral tissues.

III. FUNCTION OF TJS IN THE INVERTEBRATE NERVOUS SYSTEM

The claudin family of TJ proteins stems from 20–30 genes in mammals (reviewed in Southwood & Gow, 2001), with most of the expansion of this family occurring relatively recently during the evolution of marine and terrestrial vertebrates (Loh, Christoffels, Brenner, Hunziker, & Venkatesh, 2004). The genome of the invertebrate model organism *Drosophila* harbors six claudin family members expressed in different polarized epithelia, but two of these, *megatrachea* and *sinuous*, are expressed by peripheral nerve glial cells. These cells form several concentric layers around the nervous system, and generate an extracellular diffusion barrier (the blood–nerve barrier) that isolates neurons from the potassium-rich hemolymph bathing all organs in *Drosophila*.

Pleated septate junctions (pSJs) mediate the *Drosophila* blood–nerve barrier and are comprised of several adhesion molecule complexes which include *megatrachea* and *sinuous*. The major adhesion molecules forming pSJs are neurexin IV, contactin, and neuroglian, which are conserved in mammalian axoglial junctions (AJs; Fig. 1A) (reviewed in Banerjee, Sousa, & Bhat,

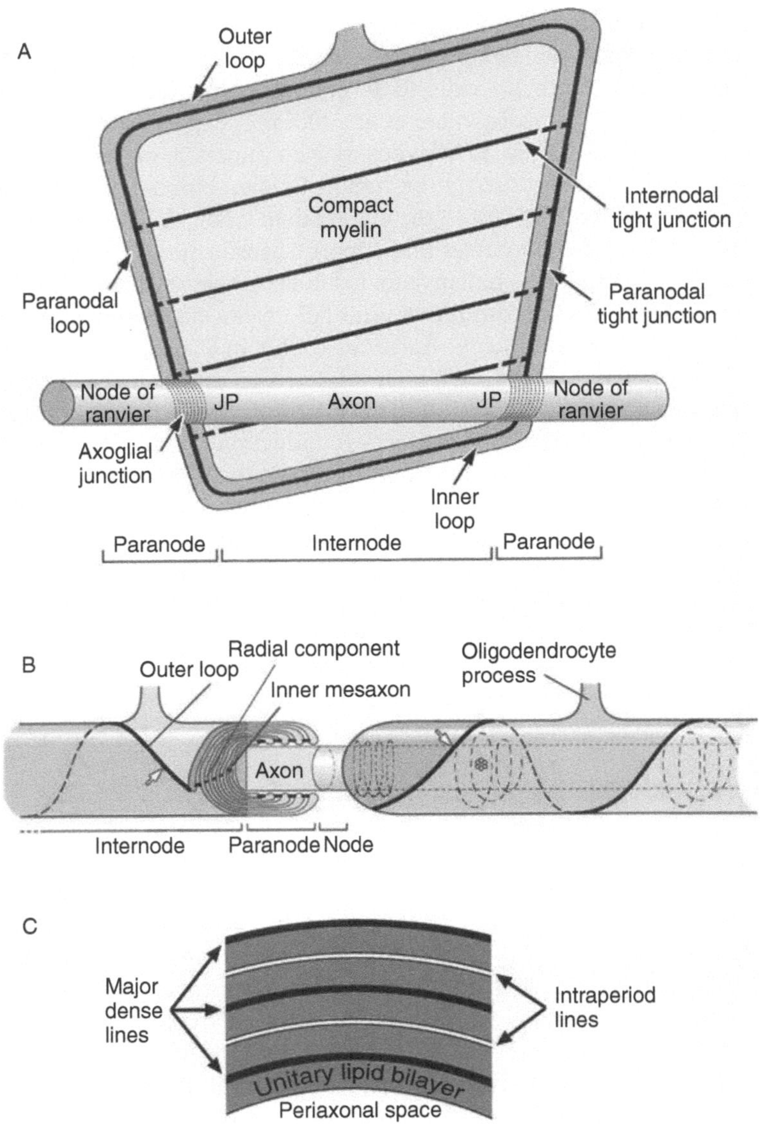

FIGURE 1 Organization of claudin 11 TJs in CNS myelinated fibers. (A) A CNS myelin sheath is unravelled to reveal the structure and organization of the multilayered myelin membrane that spirals around the axon. Myelin is a flat membrane envelope initially filled with cytoplasm on the inside which is extruded to the cell body as the sheath matures. Most of the membrane sheet compacts and the cytoplasmic membrane surface are closely apposed. A continuous channel of cytoplasm persists around the perimeter of the membrane. Autotypic TJs are also located around the perimeter and seal the edges of the membrane to block diffusion of macromolecules and increase the electrical resistance of the membrane. Veins of TJs run the

2006). However, orthologs of *megatrachea* and *sinuous* are not present in AJs, suggesting that structural and functional segregation of junctional elements has occurred during evolution (Nunes et al., 2006; Tepass, 2003).

Stork et al. (2008) have demonstrated that loss-of-function alleles of either *megatrachea* or *sinuous* results in disorganization of pSJs and infiltration of large molecular weight fluorescent dextrans into *Drosophila* nerves. Thus, these claudins are clearly involved in blood–nerve barrier function which likely serves to sequester axons from hemolymph that would abolish action potentials (APs) and paralyze the animals. However, the loss of any of the adhesive proteins causes dissolution of the pSJs, so it remains unclear if the claudins or the neurexin IV/contactin/neuroglian adhesion complex underlie the blood–nerve diffusion barrier in *Drosophila*. In contrast, the mammalian orthologs of this invertebrate adhesion complex do form a diffusion barrier at paranodal AJs in the absence of claudins.

Another important aspect of blood–nerve barrier biology in *Drosophila* remains unresolved and clouds the assignment of specific functions to *megatrachea* and *sinuous*. Canonical ultrastructural features of vertebrate TJs, including membrane kissing points and intramembranous TJ strands, have not been observed in pSJs using transmission or freeze-fracture electron microscopy. The absence of such direct evidence may stem from the difficulty of preserving sufficient ultrastructural detail in pSJs, which are dominated by dense arrays of septa between glial cells. Alternatively, *Drosophila* claudins may only perform supporting roles for the adhesive molecules in septa, which would suggest that they do not form TJs and are not functionally equivalent to vertebrate claudins.

---

length of the myelin internode through the compact region, which overlie each other on successive membrane layers in a radial alignment (the radial component). Membrane domains on the axon are also shown: nodes of Ranvier, which are short segments of bare axon between successive myelin sheaths in which sodium channels are clustered; axoglial junctions, which adheres the paranodal myelin to the axon in a continuous spiral; juxtaparanodal regions (JP), where potassium channels are clustered and the remaining axon internode (Axon). (B) Portions of two myelin sheaths are shown in their native conformation spirally wrapped around an axon and separated by a node of Ranvier (Node). To the left, the myelin sheath is cut to show the membrane spiral and the radial component. Claudin 11 TJs occlude the extracellular space at the inner mesaxon and outer loop (arrow), and between successive paranodal membrane layers (paranodal loops). In large myelinated fibers, Claudin 11 TJs also localize to spiral cytoplasmic channels (Schmidt–Lantermann incisures, not shown) within the compact region (asterisk). Reprinted from Gow (2008). (C) Schematic of a transverse section of compact myelin to show the major morphological features visible under the electron microscope. The major dense line and intraperiod line appear in myelin as alternating dark and light lines, although they actually spiral around the axon. The unitary lipid bilayer would comprise the envelope that is described in the legend of (A).

IV. FUNCTION OF TJS IN CNS MYELIN

A. Myelin Sheath Development and Structure

The mechanisms underlying myelin formation require complex sequential interactions between neurons and oligodendrocytes beginning with contact and recognition of axons by oligodendrocyte processes. Each process adheres to the axon and expands to form a large flat membrane sheet that wraps around the axon multiple times and elongates along the axon for several hundred micrometers to form a membrane tube (Fig. 1). Initially, the glial membrane is filled with cytoplasm, which is extruded as the membrane compacts around the axon and forms the characteristic stacked membrane spiral of the myelin sheath (Fig. 1B). Sodium channels are assembled in high-density clusters in short segments of axonal membrane, called nodes of Ranvier, between adjacent myelin sheaths to enable rapid saltatory conduction.

Dominant ultrastructural features of mature myelin sheaths that are visible under the electron microscope include the alternating major dense line (MDL) and intraperiod line (IPL). The MDL arises from the close apposition of cytoplasmic surfaces of the membrane and deposition of structural proteins to stabilize and possibly fuse the membrane surfaces. The IPL arises from the close apposition of extracellular surfaces of the membrane. Unlike the MDL, the membrane surfaces at the IPL remain separated by a distance that is likely determined by the extracellular domains of transmembrane proteins acting as spacers (Duncan, 1990; Stecca et al., 2000). The distance between successive layers of the MDL or IPL, known as the myelin period, is constant at approximately 16 nm in the CNS (reviewed in Kirschner, Ganser, & Caspar, 1984).

B. Major Structural Proteins in CNS Myelin

Three proteins—myelin basic protein (MBP), proteolipid protein 1 (PLP1), and claudin 11—comprise as much as 90% of the total protein content of CNS myelin. Importantly, the distributions of these proteins in myelin are distinct. MBP is a highly charged cytoplasmic extrinsic membrane protein that is localized to the MDL and stabilizes membrane apposition after compaction (Omlin, Webster, Pulkovits, & Cohen, 1982). PLP1 is an intrinsic membrane protein with four transmembrane domains. Its amino- and carboxyl-termini are exposed to the membrane cytoplasmic surface and may interact with MBP (Gow, Gragerov, Gard, Colman, & Lazzarini, 1997; Popot, Pham-Dinh, & Dautigny, 1991). Two domains of PLP1 protrude from the extracellular membrane surface and may form homomeric interactions in *trans* with PLP1 from juxtaposed lamellae to set the spacing of the IPL. Claudin 11 is also an integral membrane protein with a topology likely resembling PLP1. Unlike MBP and PLP1, which are evenly distributed throughout myelin sheaths, claudin 11 is confined to narrow proteinaceous veins which may only be a few molecules wide and extend the length of the sheath. These veins are reminiscent of a structural feature of CNS myelin, the radial component, which occupies a small fraction of the cross-sectional area of myelin.

C. TJs form the Radial Component in CNS Myelin

In pioneering observations, Peters (1961, 1964) identified the radial component in CNS myelin from ferricyanide- and permanganate-postfixed white matter tracts. The radial component appears as regularly spaced rod-like thickenings of the IPL in transmission electron micrographs and was hypothesized to comprise autotypic TJs (Fig. 2A). These morphological findings were confirmed by other groups (Schnapp & Mugnaini, 1978; Tabira, Cullen, Reier, & Webster, 1978). Early experiments indicated that the radial component conferred limited permeability of macromolecules and metal ions such as lanthanum into the myelin (Hirano, Becker, & Zimmerman, 1969; Tabira et al., 1978), which is consistent with the role that TJs play in polarized epithelia. Further attesting to its importance in myelin, the radial component is evolutionarily conserved in terrestrial vertebrates, amphibians, and fish (Shinowara, Beutel, & Revel, 1980; Tabira et al., 1978). Final evidence of the correspondence between the radial component and TJs (Fig. 2B–F) was provided by immunofluorescence labeling of white matter with anti-claudin 11 antibodies and ablation of the *Claudin 11* gene by homologous recombination in mice (Gow et al., 1999; Morita, Sasaki, Fujimoto, Furuse, & Tsukita, 1999). Importantly, the absence of claudin 11 in myelin is apparently without consequence to myelin ultrastructure other than the loss of the radial component (Devaux & Gow, 2008; Gow et al., 1999), indicating that claudin 11 is not crucial for myelin compaction.

D. TJs Contribute to Myelin Membrane Resistance

As a starting point for analyzing the function of claudin 11 TJs in myelin, it is reasonable to draw parallels with TJs in polarized epithelia, even though myelin TJs are autotypic rather than homotypic or heterotypic, and even though many of the cytoplasmic plaque proteins associated with most TJs (such as occludin and the MAGUK family of zonula occludens proteins,

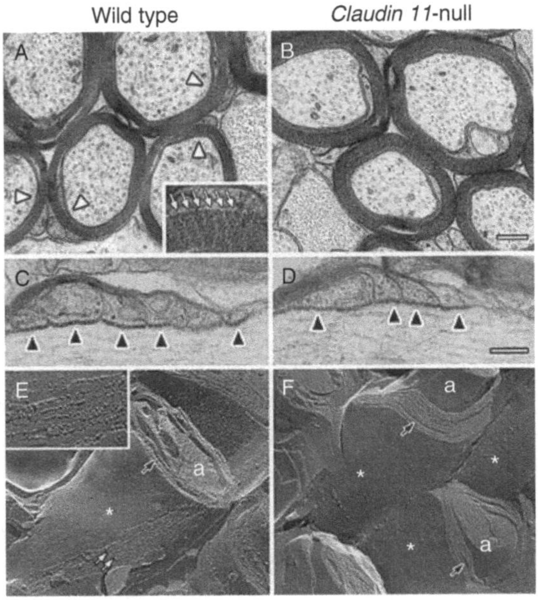

FIGURE 2 Tight junctions are absent in CNS myelin from *Claudin 11*-null mice. (A–D) Electron micrographs of transverse (A and B) and longitudinal (C and D) sections of optic nerve myelinated fibers from adult wild-type (A and C) and *Claudin 11*-null (B and D) mice. The optic nerves were postfixed with ferricyanide (Peters, 1961) to reveal the radial components (white arrowheads) in the myelin. Radial components appear as electron-lucent radial lines through compact myelin in wild-type mice (inset in A, white arrows), but are absent from *Claudin 11*-null mice. Overall myelin structure in the internode (A and B) and at paranodes (C and D) is unchanged by the absence of TJs, and axoglial junctions (also called transverse bands, black arrowheads in C and D) are morphologically normal. Scale bars = 0.2 μm. (E, F) Freeze fracture replicas of CNS myelinated fibers from adult wild-type (E) and *Claudin 11*-null (F) mice. Intramembranous strands (white arrows in E) run the length of the internode (see Fig. 1) and form the radial component within compact myelin (black arrow). The membranes are otherwise relatively smooth (asterisk). At high magnification (inset in E), the strands appear as discrete particles which represent monomers or multimeric complexes of claudin 11 in linear arrays. The strands are absent in *Claudin 11*-null mice (F) but the myelin membranes are compacted (black arrows in F) and otherwise morphologically normal. © Devaux and Gow (2008). Originally published in Devaux and Gow (2008), Gow et al. (1999).

ZO-1 to 3) are absent from myelin (Gow et al., 1999). Thus, myelin TJs may play occluding and resistive roles by sealing the edges of the membrane along the length of the sheath and at paranodes (Fig. 1B). In addition, they may maintain asymmetrical distributions of membrane proteins and lipids in paranodal regions of myelin that contact the axon and control extracellular fluid composition at the IPL, particularly when other structural myelin proteins are altered.

The absence of claudin 11 confers several phenotypes in mice, including tremors, gait abnormalities, and motor coordination deficits (Fig. 3). Examination of the electrophysiological properties of CNS myelinated tracts in optic nerve and spinal cord of *Claudin 11*-null mice (Devaux & Gow, 2008) reveals significant reductions in conduction and the excitability threshold of axons. These changes are more pronounced for small axons in optic nerve (those below 1 μm in diameter), which are normally ensheathed by relatively thin myelin sheaths. In contrast, large axons such as those in the ventral spinal cord, with correspondingly thicker myelin sheaths, are unaffected or only mildly affected by the absence of TJs. These data suggest that TJs contribute to the insulation of axons by significantly increasing the membrane resistance of thin myelin segments.

Computer simulations of myelin TJs corroborate these conclusions (Fig. 4D) and reveal possible mechanisms by which myelin resistance and axon excitability threshold are altered (Gow & Devaux, 2008). Analogous to polarized epithelia, TJs diminish the paracellular current pathway between membrane layers in the myelin sheath, and thereby decrease its capacitive charge during AP propagation. Thus, TJs diminish the time to charge the myelin capacitance, which favors the rapid regeneration and propagation of APs between nodes of Ranvier (Fig. 5A and B).

E. Axoglial Junctions Contribute to Axonal Conduction Independently of TJs

In addition to TJs, CNS myelin sheaths also contain AJs, which are septate-like adhesive junctions localized to paranodal regions of the sheath (Fig. 2C and D). These junctions function to: attach the lateral cytoplasmic loops of myelin to the surface of the axon (Fig. 1), specify the distance between the axonal and myelin membranes at paranodes, and define axonal membrane domains by lowering the lateral diffusion of axonal proteins such as ion channels. In addition, AJs are widely thought to form an impermeability barrier in the periaxonal space, although recent data suggest that AJs are more likely to act as space-filling struts that are relatively permeable even to small proteins (Devaux & Gow, 2008).

Three major components constitute AJs in mammals—caspr, contactin, and neurofascin 155—and the targeted deletion of any one of these components in mice abolishes AJs in CNS myelin (Bhat et al., 2001; Boyle et al., 2001; Pillai et al., 2009; Zonta et al., 2008). Consequences of the absence of AJs include conduction slowing, severe ataxia, and early death, which probably stem from dysfunctional nodes of Ranvier and large increases in the width of the periaxonal space at paranodes. Computer simulations indicate

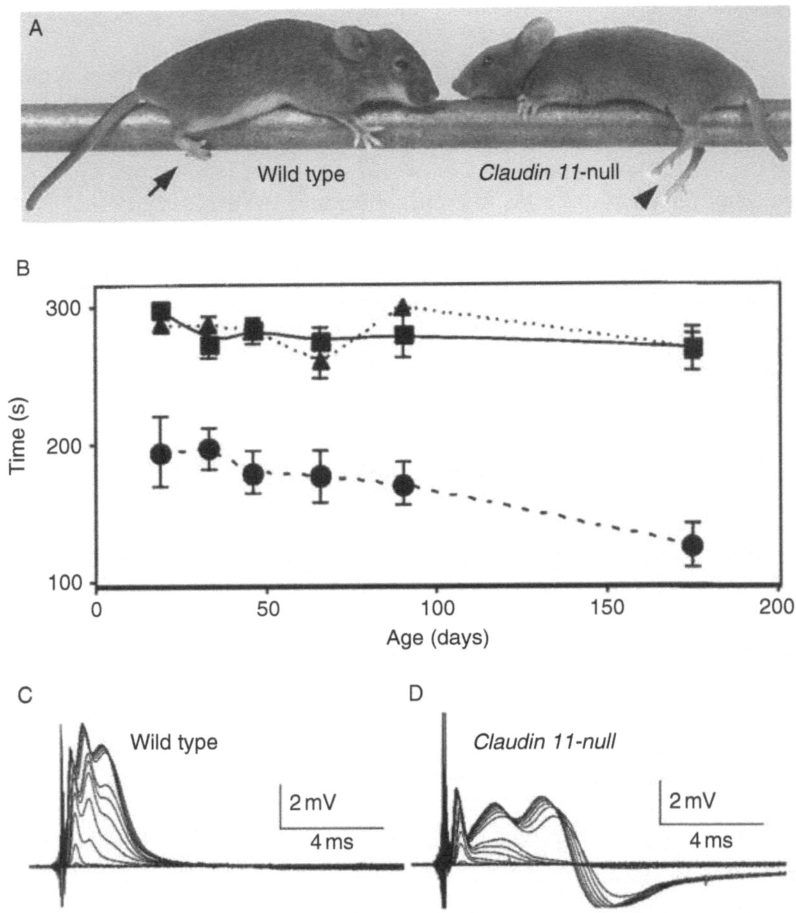

FIGURE 3 Neurological and central conduction deficits in *Claudin 11*-null mice. (A) Hind limb weakness is a lifelong phenotype of *Claudin 11*-null mice which does not enable these animals to grasp objects (arrowhead). Wild-type mice are able to perform this task (arrow). (B) The motor skills of wild-type (squares) and heterozygous (triangles) mice from weaning to adulthood are constant and indistinguishable. However, *Claudin 11*-null mice (circles) perform this task poorly at all ages and are progressively impaired. Error bars, ±S.E.M. Currently, it is unclear if this performance stems from central or peripheral deficits, because Claudin 11 is expressed by oligodendrocytes and epaxial muscles during development (Gow et al., 1999). (C, D) Compound APs (CAPs) elicited in the optic nerves of adult wild-type and *Claudin 11*-null mice by voltage pulses of increasing amplitude. CAPs show three components in the optic nerve, reflecting fiber populations with different conduction velocities and diameters. The latencies of the second and third CAP components are substantially increased in *Claudin 11*-null mice, which indicates that conduction velocity is decreased. In addition, fibers contributing to the second and third components require higher stimulation intensities for recruitment, indicating that the threshold of nodal excitability in these fibers is abnormal. In addition, CAPs from *Claudin 11*-null mice are followed by hyperpolarizing afterpotentials (D) that are not observed in wild-type nerves (C). Afterpotentials reflect the activation of internodal $K_v1.1/K_v1.2$ channels in the absence of myelin TJs; thus, internodal depolarization. © Devaux and Gow (2008). Originally published in Devaux and Gow (2008), Gow et al. (1999).

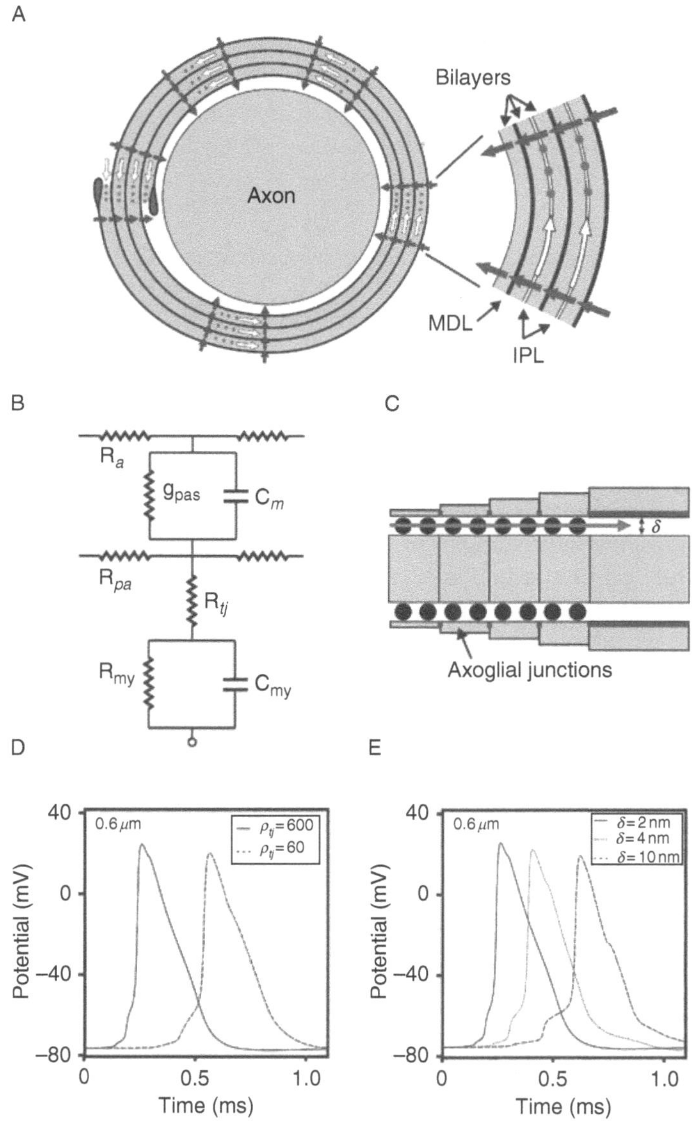

FIGURE 4 Claudin 11 TJs may form a resistive pathway in the myelin. (A) Schematic representation of a cross section of CNS axon ensheathed with four myelin wraps. The inset shows the myelin membrane bilayers in greater detail, including intraperiod lines (IPL) and major dense lines (MDL). Myelin can be represented in an electrical circuit as a cable surrounding the axon, where current can flow through the IPL (open arrows) between myelin layers or across the myelin membrane (closed arrows). Myelin TJs (gray dots) occlude the extracellular space between the membrane layers and increase the resistance of this compartment, thereby

that such increases can cause conduction slowing and conduction block, particularly in small diameter fibers (Fig. 4E). Interestingly, the absence of AJs does not cause the disassembly of TJs (Zonta et al., 2008), indicating that these intercellular junctions form independently and serve complementary roles in promoting myelin insulation of the internode and saltatory conduction in small myelinated fibers (Fig. 5C).

F. Are TJs Immune-Protective or Adhesive Components of Myelin?

Several hypotheses have been advanced to account for the presence of TJs in compact myelin. However, subsequent careful scrutiny has often brought into question the validity and veracity of the assumptions and conclusions leading to these hypotheses. For example, myelin TJs have been proposed as barriers to leukocyte migration which prevent them from coming into contact with myelin proteins that could be recognized as foreign antigens. In addition, myelin TJs have been suggested to be primary sites of adhesion that stabilize the multilamellar organization of compact myelin.

Early interest in myelin TJs revolved around the notion that these barriers might sequester myelin proteins away from immune cell surveillance, in similar fashion to the proposed function of TJs at the blood–testis barrier. The blood–testis barrier was originally thought to establish an immune-privileged

---

limiting the current path at the IPL. This is analogous to a paracellular barrier formed by TJs in polarized epithelia that increases the transepithelial resistance. (B) An equivalent electrical circuit of the myelin internode that is proposed to approximate the function of TJs. Thus, TJs may form a resistance (R_{tj}) in series with the myelin resistance (R_{my}) and capacitance (C_{my}). This resistive path decreases the capacitive charge of the myelin membrane and reduces the internodal delay during AP propagation. R_a and R_{pa} represent the resistances of the axoplasm and of the periaxonal space; g_{pas} and C_m represent the conductance and capacitance of the axonal membrane. (C) Schematic of a longitudinal section through the paranodal region, showing axoglial junctions (black dots) which occupy approximately 50% of the extracellular space at paranodes. Axoglial junctions maintain the width of the periaxonal space (δ) at paranodes and may form an axial resistance (R_{pa}; closed arrows) which prevents AP invasion of the juxtaparanodal and internodal regions. (D) Computer simulations of APs generated by a 0.6 μm diameter myelinated axon in the presence ($\rho_{tj} = 600\ \Omega\ cm^2$) and absence of claudin 11 TJs ($\rho_{tj} = 60\ \Omega\ cm^2$). Lowering TJ resistivity (ρ_{tj}) in the myelin increases the latency (reduces the conduction velocity) of APs in small diameter myelinated axons. (E) Computer simulations of APs generated in a 0.6 μm diameter myelinated axon with TJs and with functional axoglial junctions ($\delta = 2$ nm; extracellular occupancy 50%) or without axoglial junctions where the paranodes are detached from the axon ($\delta = 4$ or 10 nm). The absence of axoglial junctions also reduces conduction velocity. Surprisingly, simulations indicate that conduction block does not occur in larger fibers even when δ is >100 nm. Together, the simulations indicate that TJs and AJs are both required for normal conduction, particularly in small fibers; however, they are assembled independently in CNS myelin and function differently from each other. Printed from Gow & Devaux (2008) with permission from Cambridge University Press.

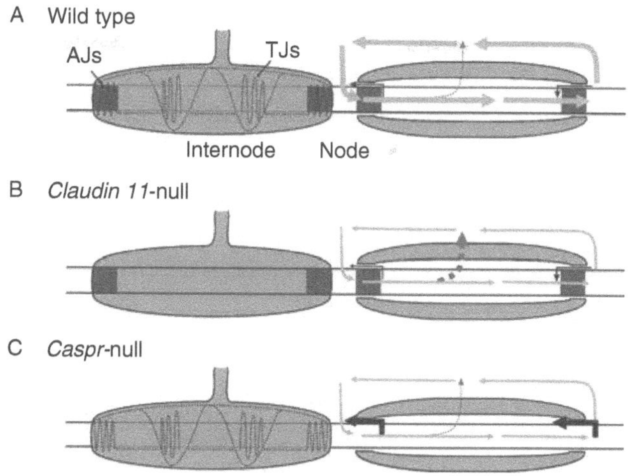

FIGURE 5 Roles of TJs and axoglial junctions (AJs) in axonal conduction. The schematics depict myelinated axons from (A) wild-type mice with TJs and AJs, (B) *Claudin 11*-null mice with AJs but no TJs, and (C) *Caspr*-null mice with TJs but no AJs. Current flow (right) through a myelin internode is also shown for each genotype. In wild-type animals, the majority of the axial current (looping arrows) flows to the next node because of the high resistance formed by the myelin around the axon. Small currents flow longitudinally through AJs at paranodes (black arrows) and radially through the myelin (dashed arrow). In *Claudin 11*-null mice, AJs are present; however, the loss of myelin TJs increases radial current flow through the IPL compartment of the myelin sheath (dashed arrows). This, in turn, decreases longitudinal current flow along the axon. In *Caspr*-null mice, the absence of AJs increases the distance between the axon and the myelin at paranodes, which in turn increases the current leak at paranodes and decreases longitudinal current flow along the axon. It is likely that similar changes occur in *Contactin*-null and *Neurofascin 155*-null mice. Thus, AJs and TJs play distinct but complementary functions in axonal conduction. (See Color Insert.)

compartment to preclude immune cell infiltration and interaction with novel antigens expressed in adolescence during spermiogenesis. Because myelination is also a late developmental event with respect to the perinatal establishment of immune tolerance, Mugnaini and Schnapp (1974) argued, by analogy, that myelin TJs were part of the blood–brain barrier and protected the CNS against demyelinating diseases such as multiple sclerosis (MS). Contemporary opinions on immunomodulation challenge this view (Pelletier & Byers, 1992; Yule, Mahi-Brown, & Tung, 1990), and an analysis of *Claudin 11*-null mice indicates that these mutants do not exhibit signs of immune cell infiltration, focal or disseminated demyelination, or spontaneous encephalomyelitis (Gow et al., 1999).

In a variety of studies, TJs also have been proposed to be adhesive structures that confer myelin stability. For example, hexachlorophene toxicity in rodents induces splitting of CNS myelin lamellae except in the vicinity of the radial component (Tabira et al., 1978). Rosenbluth and colleagues have investigated the adhesiveness of myelin TJs in optic nerves from *Plp1*-null mice (Rosenbluth, Nave, Mierzwa, & Schiff, 2006; Rosenbluth, Schiff, & Lam, 2009). Their ultrastructural analyses indicate, despite normal myelin compaction (also see Mobius, Patzig, Nave, & Werner, 2009), that the IPL in the mutants is disrupted by infiltration with hypotonic solutions, except at the radial component. Mobius and colleagues (Mobius, Patzig, Nave, & Werner, 2008; Mobius et al., 2009) also have asserted that these data provide evidence for the adhesiveness of TJs in myelin.

However, such interpretations do not exclude other explanations like tissue processing artifacts. Prior to hypotonic treatment, Rosenbluth et al. (2009) aldehyde-fixed the optic nerves, which undoubtedly cross-linked the claudin 11. Because 80–90% of the protein normally present at the IPL is missing in *Plp1*-null myelin, there is little wonder that the radial component remained intact in their study while swelling and splitting elsewhere at the surface of the virtually protein-free IPL was extensive.

Two studies have further explored the function of myelin TJs in the CNS of mice lacking both *Claudin 11* and *Plp1* genes (*Claudin 11:Plp1*-null). These double mutant mice exhibit significant clinical signs characterized by resting body tremors and ataxia with an onset around 2 weeks of age (A. Gow, unpublished data). At the ultrastructural level, Chow et al. (2005) found dramatic disruption of myelin in cross sections of optic nerve and spinal cord from double mutant mice with membrane whorls around many axons. This pathology was interpreted as indicating myelin instability and decompaction at the IPL in a similar fashion to *Plp1*-null myelin.

However as reported here, *Claudin 11:Plp1*-null double mutants exhibit more subtle disruption of myelin architecture than reported previously (Chow et al., 2005), and is characterized by essentially compacted but loosely wrapped myelin around the vast majority of axons (Fig. 6) with some fiber loss, intra-axonal accumulations of vesicles, redundant myelin sheaths, and an increased incidence of isolated cytoplasmic pockets within compact myelin. In general, the density of microtubules and neurofilaments is also increased in double mutant axons compared to controls (asterisks). The radial component, which is prevalent in control myelin (arrowheads), is absent in all *Claudin 11*-null genotypes and so too are TJs in noncompact regions of the myelin sheath such as the inner loop in the internode (Fig. 6C and D). Schmidt–Lantermann incisures are relatively rare in control rodent myelin because of the small size of axons (Blakemore, 1969) but are more frequent in the double mutants, suggesting that oligodendrocytes are compensating for deficiencies caused by structural changes in the mutant myelin (Fig. 6E and F).

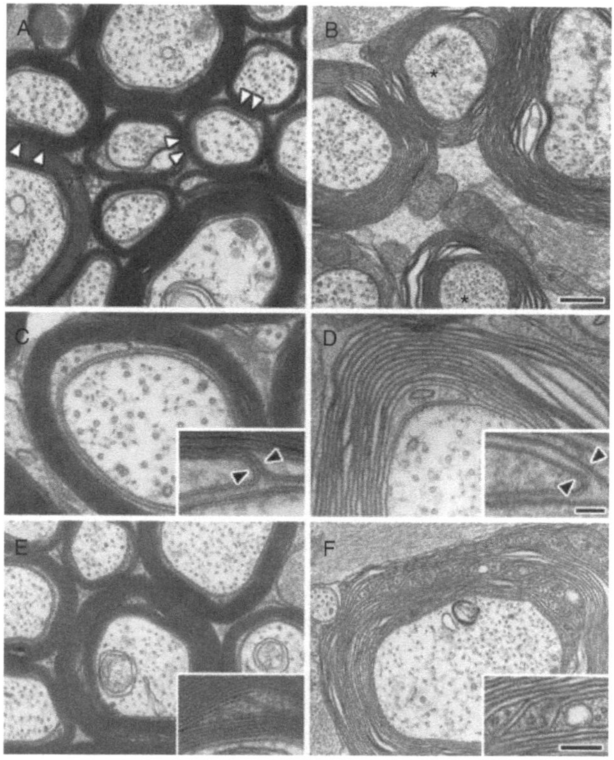

FIGURE 6 Ultrastructural abnormalities of CNS myelin in *Claudin 11:Plp1*-null mice. (A, B) Electron micrographs of optic nerve from adult *Claudin 11*-heterozygote control mice (A) and *Claudin 11:Plp1*-null double mutants (B). After glutaraldehyde fixation, the optic nerves were postfixed with osmium tetroxide and ferricyanide to reveal the radial component in almost all control myelin sheaths (arrowheads). Sheaths from double mutant optic nerves are multilamellar and largely compact, as indicated by the absence of cytoplasm in the internodal myelin. However, the membranes appear loosely wrapped around axons and sample preparation artifacts are present in most fibers. In general, the density of microtubules and neurofilaments is also increased in double mutant axons compared to controls (asterisk). (C, D) A characteristic feature of TJs at the inner loop in transverse sections of the myelin internode is the membrane kissing point at the junction between the periaxonal space and the intramyelinic compartment. The canonical pentalaminar appearance of the TJ is apparent in the inset (arrowheads) in *Claudin 11*-heterozygote mice (C). In *Claudin 11:Plp1*-null mice (D), the heptalaminar appearance of two closely apposed, but separate, membranes can be seen (arrowheads). The absence of a paracellular seal at the periaxonal-intramyelinic junction may provide access to the IPL for solutes from the periaxonal space and vice versa. Such access may disrupt electrostatic interactions that might keep the membrane surfaces of the IPL in close contact. (E, F) Schmidt–Lantermann incisures are less common in the rodent CNS than in larger mammals, likely because axons are much smaller (Blakemore, 1969). However, incisures in control mice (E) have similar organization and architecture to other mammals and frequently contain one or more microtubules (inset in E). Incisures are several-fold more common in *Claudin 11:Plp1*-null mice, suggesting that oligodendrocytes are compensating for the structural abnormalities of the myelin in these mutants. Scale bars: (A) and (B), 400 nm; (C) and (D), 100 or 33 (insets) nm; (E) and (F), 300 or 130 (insets) nm.

At high magnification (Fig. 7A), several abnormalities are apparent in the myelin from *Claudin 11:Plp1*-null double mutant mice. First, the myelin period is significantly greater and quite variable compared to controls (Fig. 7B). Second, the membrane surfaces at the IPL are much further

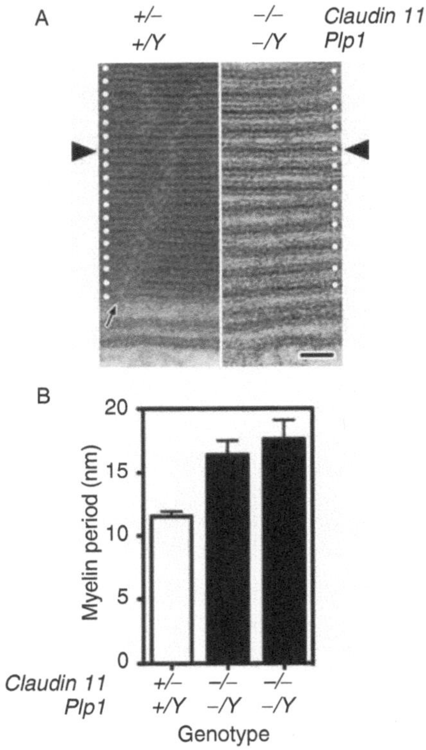

FIGURE 7 Abnormal myelin period, IPL and MDL in CNS myelin from *Claudin 11:Plp1*-null mice. (A) Electron micrographs of internodal myelin from optic nerves of adult *Claudin 11*-heterozygote and *Claudin 11:Plp1*-null mice at high magnification. The MDL from each mouse has been aligned at the black arrowheads and white dots on either edge show the spacing of the myelin lamellae for each animal. In the control myelin, the MDL is evenly spaced in each layer and the IPL appears as a narrow line at this magnification. The radial component, visualized by postfixing with ferricyanide, can be seen coursing diagonally through the myelin (black arrow). The myelin from the double mutant mouse is devoid of cytoplasm, but the lamellae do not appear completely compacted and several abnormalities are visible. Thus, the myelin period is abnormally large and variable; the IPL is too widely spaced; and the width of the MDL may be increased. (B) Myelin period in adult optic nerve from one *Claudin 11*-heterozygote (control) and two *Claudin 11:Plp1*-null double mutant mice was determined by averaging three measurements for each of 30 myelinated fibers of different diameters (mean ± S.D.). These data reflect the micrographs in (A), that the myelin period is larger and more variable in the double mutants. Scale bar in (A) 25 nm.

apart than for the control, which probably accounts for the majority of the increased period. However, it is unclear if the lack of PLP1 at the IPL allows the membranes to move apart passively, or whether the absence of claudin 11 enables solutes to penetrate deep into the internode and cause generalized swelling. The latter explanation seems likely in light of previous investigations and the compressed appearance of internodal membranes in *Plp1*-null myelin which contain TJs (Klugmann et al., 1997; Rosenbluth, Stoffel, & Schiff, 1996; Stecca et al., 2000).

Finally and irrespective of the mechanism, the myelin from *Claudin 11: Plp1*-null mice appears as a spiral of pairs of lipid bilayers fused at the MDL, the width of which may be slightly greater than normal (compare MDLs in the vicinity of the black arrowheads). These data suggest that claudin 11 or, more likely PLP1, may contribute to stabilizing the cytoplasmic membrane apposition either independently of, or in conjunction with, MBP. The PLP1-specific domain has a significant net positive charge to interact with negatively charged lipid headgroups, which is similar to the proposed function of the cytoplasmic tail of the P_0 glycoprotein in PNS myelin (Martini, Mohajeri, Kasper, Giese, & Schachner, 1995).

From the perspective of myelin TJ function, the ultrastructural defects in *Claudin 11:Plp1*-null double mutant myelin reveal that oligodendrocytes can still assemble and largely compact a multilamellar membrane around axons even with as much as 60% of the total myelin protein missing and a near protein-free IPL. Such a remarkable feature of myelin lipid self-assembly into spiral membranes in the presence of MBP has also been demonstrated using purified components (Mateu, Luzzati, London, Gould, & Vosseberg, 1973; Riccio, Fasano, Borenshtein, Bleve-Zacheo, & Kirschner, 2000). Accordingly, these data imply that there may be little need for strongly adhesive proteins in myelin and cast substantial doubt on recent schemes that propose claudin 11 TJs as major adhesive structures.

V. POTENTIAL CLINICAL RELEVANCE OF CNS MYELIN TJS TO NEUROLOGICAL DISEASES

A. *Schizophrenia*

Schizophrenia is a complex mental disorder with genetic, neurological, and environmental causes. Schizophrenia symptoms are widely thought to arise from altered brain connectivity, particularly stemming from abnormal neurotransmitter metabolism (Benes & Berretta, 2000, 2001). However, functional magnetic resonance imaging, tractography, and postmortem molecular analyses suggest white matter may be involved, with reduced myelin volume or integrity in some patients (Andreasen et al., 1994; Davis et al., 2003; Flynn et al., 2003). Consistent with this idea, schizophrenia symptoms are relatively common in several diseases with known myelin abnormalities including MS, metachromatic leukodystrophy, and X-linked adrenoleukodystrophy (Davis et al., 2003; Denier et al., 2007; Feinstein, du Boulay, & Ron, 1992; Hyde, Ziegler, & Weinberger, 1992; Kopala et al., 2000; Walterfang, Wood, Velakoulis, Copolov, & Pantelis, 2005). Because CNS axons require adequate myelination to maintain neurophysiological function, it is plausible that myelin dysfunction may negatively impact brain connectivity and executive functions in forebrain cortex.

Microarray analyses from several groups have revealed lower expression levels of several myelin-specific genes, including *Claudin 11*, in postmortem tissue from affected regions of schizophrenia brain while expression of other myelin genes remains unaffected (Katsel, Davis, & Haroutunian, 2005; McCullumsmith et al., 2007). These correlative data suggest that myelin function may be important for normal cognition and that white matter perturbations may play a role in mental disorders. Our observations in *Claudin 11*-null mice have lead us to postulate that reduced *CLAUDIN 11* gene expression may slow the propagation of APs in small myelinated axons (< 1 μm diameter) and thereby affect brain function in humans (Devaux & Gow, 2008).

In vivo imaging implicates abnormal corpus callosum function in the development of schizophrenia (Walterfang et al., 2005) and this interhemispheric connective tract is largely (45–70%) comprised of axons below 1 μm diameter (Aboitiz, Scheibel, Fisher, & Zaidel, 1992). *CLAUDIN 11* dysfunction may significantly increase transit times along this white matter tract and lead to callosal disconnectivity. If so, the disruption of myelin TJs may have broader implications and may impact more general brain processes. To date, mutations in the *CLAUDIN 11* gene have not been implicated in familial forms of schizophrenia, although several single-nucleotide polymorphisms have been identified in the vicinity. Nevertheless, examining schizophrenia-related behavior and callosal connectivity in *Claudin 11*-null mice will shed light on the possibility that TJ function is linked to schizophrenia and possibly other affective disorders.

B. Multiple Sclerosis

MS is the most common human demyelinating disease affecting CNS axons. MS is widely thought to stem from autoimmune attack, inflammatory demyelination, and axonal damage; however, recent studies have challenged this notion and suggest that immune involvement is secondary to the primary defect (Trapp & Nave, 2008). Irrespective of the primary cause, many MS patients present circulating T-cells and antibodies reactive against myelin proteins (Cross, Trotter, & Lyons, 2001; O'Connor, Bar-Or, & Hafler, 2001).

The abundance of claudin 11 in myelin has led to studies investigating whether claudin 11 is targeted by autoreactive T-cells or antibodies in MS patients. In one study, 50% of patients with relapsing–remitting MS harbored higher levels of anti-claudin 11 antibodies than controls (Bronstein, Lallone, Seitz, Ellison, & Myers, 1999), while a second study concluded that claudin 11-reactive T-cells were absent from MS patients (Tiwari-Woodruff et al., 2001). These data suggest that a humoral response against myelin TJs may play a role in the development of relapsing-remitting MS. Nevertheless, claudin 11 induces experimental autoimmune encephalomyelitis (EAE) and optic neuritis in SJL/J mice with T-cell infiltration and demyelination (Kaushansky, Eisenstein, Oved, & Ben-Nun, 2008; Kaushansky, Hemo, Eisenstein, & BenNun, 2007; Kaushansky et al., 2006; Stevens, Chen, Seitz, Sercarz, & Bronstein, 1999) and EAE can be adoptively transferred to naïve mice by injecting claudin 11 reactive T-cells. Thus, the pathogenic actions of anti-claudin 11 antibodies in relapsing–remitting MS remain to be demonstrated and passive cotransfer of antibodies and T-cells reactive against claudin 11 will help in determining whether anti-claudin 11 antibodies may destabilize myelin TJs and affect conduction.

VI. ROLE OF CLAUDIN 11 IN CNS DEVELOPMENT

Other novel functions of claudin 11 in oligodendrocyte biology have recently emerged. Yeast-two-hybrid screens suggest that claudin 11 interacts with several transmembrane proteins, including: OAP-1, a member of the tetraspanin superfamily; $K_v3.1$, a voltage-gated potassium channel subunit; and integrin β1 (Tiwari-Woodruff et al., 2001, 2006). Overexpression of claudin 11 or OAP-1 in oligodendrocyte cell lines enhances their proliferation, while deletion of *Claudin 11* or $K_v3.1$ gene reduces proliferation and migration of oligodendrocyte precursors *in vitro*. However, the relevance of these data to oligodendrocyte biology is controversial, because loss-of-function alleles of $K_v3.1$ and *Claudin 11* do not confer overt defects in oligodendrocyte proliferation, migration, or myelination *in vivo*. Nevertheless, it is intriguing that claudin 11 may be involved in oligodendrocyte differentiation during development, in contrast to the notion that TJs are only expressed by differentiated cells.

VII. TJS IN PNS MYELIN

Although there is strong correspondence between the distribution of TJs in myelin elaborated by oligodendrocytes and Schwann cells (Hall & Williams, 1971; Mugnaini, Osen, Schnapp, & Friedrich, 1977; Revel & Hamilton, 1969), it is unlikely that claudins function similarly in the CNS and PNS for several reasons. First and foremost, the absence of claudin 11 in CNS myelin causes conduction slowing in myelinated axons (Devaux & Gow, 2008; Gow & Devaux, 2008; Gow et al., 2004, 1999), but there is little evidence that conduction in the PNS is perturbed in *Claudin 19*-null mice (Miyamoto et al., 2005) or in patients with *CLAUDIN 19* mutations (Konrad et al., 2006; Lee et al., 2006). Second, TJs in CNS myelin are important for normal conduction in axons below 1 μm diameter with thin myelin sheaths. In general, myelinated axons in the PNS are greater than 1 μm in diameter (Waxman & Bennett, 1972) and the myelin is sufficiently thick so as not to benefit from the resistive properties of TJs. Finally, claudin 11 TJs in CNS myelin are continuous around the edges of the sheath and form a resistive barrier, but claudin 19 TJs are discontinuous in PNS myelin which would minimize the effectiveness of a barrier to current flow.

Irrespective of any differences in claudin function between CNS and PNS myelin, TJs may fulfill important roles in Schwann cells. Poliak, Matlis, Ullmer, Scherer, and Peles (2002) suggest that TJs may coordinate the assembly and maintenance of microarchitectural features in myelinated fibers, such as gap junctions at Schmidt–Lantermann incisures. Gap junctions and TJ proteins colocalize in PNS myelin and incorporate cytoplasmic plaque anchoring proteins such as ZO-1. Perhaps, TJs reduce the width of the extracellular space between myelin lamellae at the IPL to mediate coupling of connexin hemichannels at apposed membrane surfaces. Indeed, a role for claudin 11 TJs in maintaining gap junction microdomains has been established for basal cells in the stria vascularis of the cochlea (Gow et al., 2004).

Acknowledgments

This work has been supported by grants to J. D. from the Association Française contre les Myopathies and the National Multiple Sclerosis Society (RG3839A1/T) and to A. G. from NIDCD, NIH (DC006262), NINDS, NIH (NS43783), the National Multiple Sclerosis Society (RG2891), and the William and Marie Carls Foundation, Detroit, Michigan.

References

Aboitiz, F., Scheibel, A. B., Fisher, R. S., & Zaidel, E. (1992). Fiber composition of the human corpus callosum. *Brain Research, 598*(1–2), 143–153.

Andreasen, N. C., Arndt, S., Swayze, V., 2nd., Cizadlo, T., Flaum, M., O'Leary, D., et al. (1994). Thalamic abnormalities in schizophrenia visualized through magnetic resonance image averaging. *Science, 266*(5183), 294–298.

Banerjee, S., Sousa, A. D., & Bhat, M. A. (2006). Organization and function of septate junctions: An evolutionary perspective. *Cell Biochemistry and Biophysics, 46*(1), 65–77.
Benes, F. M., & Berretta, S. (2000). Amygdalo-entorhinal inputs to the hippocampal formation in relation to schizophrenia. *Annals of the New York Academy of Sciences, 911*, 293–304.
Benes, F. M., & Berretta, S. (2001). GABAergic interneurons: Implications for understanding schizophrenia and bipolar disorder. *Neuropsychopharmacology, 25*(1), 1–27.
Bhat, M. A., Rios, J. C., Lu, Y., Garcia-Fresco, G. P., Ching, W., Martin, M. S., et al. (2001). Axon–glia interactions and the domain organization of myelinated axons requires neurexin iv/caspr/paranodin. *Neuron, 30*(2), 369–383.
Blakemore, W. F. (1969). Schmidt–Lanterman incisures in the central nervous system. *Journal of Ultrastructural Research, 29*, 496–498.
Boyle, M. E., Berglund, E. O., Murai, K. K., Weber, L., Peles, E., & Ranscht, B. (2001). Contactin orchestrates assembly of the septate-like junctions at the paranode in myelinated peripheral nerve. *Neuron, 30*(2), 385–397.
Brightman, M. W., & Reese, T. S. (1969). Junctions between intimately apposed cell membranes in the vertebrate brain. *The Journal of Cell Biology, 40*(3), 648–677.
Bronstein, J. M., Lallone, R. L., Seitz, R. S., Ellison, G. W., & Myers, L. W. (1999). A humoral response to oligodendrocyte-specific protein in MS: A potential molecular mimic. *Neurology, 53*(1), 154–161.
Chow, E., Mottahedeh, J., Prins, M., Ridder, W., Nusinowitz, S., & Bronstein, J. M. (2005). Disrupted compaction of CNS myelin in an OSP/Claudin-11 and PLP/DM20 double knockout mouse. *Molecular and Cellular Neurosciences, 29*(3), 405–413.
Cross, A. H., Trotter, J. L., & Lyons, J. (2001). B cells and antibodies in CNS demyelinating disease. *Journal of Neuroimmunology, 112*(1–2), 1–14.
Davis, K. L., Stewart, D. G., Friedman, J. I., Buchsbaum, M., Harvey, P. D., Hof, P. R., et al. (2003). White matter changes in schizophrenia: Evidence for myelin-related dysfunction. *Archives of General Psychiatry, 60*(5), 443–456.
Denier, C., Orgibet, A., Roffi, F., Jouvent, E., Buhl, C., Niel, F., et al. (2007). Adult-onset vanishing white matter leukoencephalopathy presenting as psychosis. *Neurology, 68*(18), 1538–1539.
Devaux, J. J., & Gow, A. (2008). Tight junctions potentiate the insulative properties of small CNS myelinated axons. *The Journal of Cell Biology, 183*(5), 909–921.
Duncan, I. D. (1990). Dissection of the phenotype and genotype of the X-linked myelin mutants. In: I. D. Duncan, R. P. Skoff & D. R. Colman (Eds.), Vol. 605, *Myelination and dysmyelination* (pp. 110–121). New York: New York Academy of Sciences.
Feinstein, A., du Boulay, G., & Ron, M. A. (1992). Psychotic illness in multiple sclerosis. A clinical and magnetic resonance imaging study. *British Journal of Psychiatry, 161*, 680–685.
Flynn, S. W., Lang, D. J., Mackay, A. L., Goghari, V., Vavasour, I. M., Whittall, K. P., et al. (2003). Abnormalities of myelination in schizophrenia detected in vivo with MRI, and post-mortem with analysis of oligodendrocyte proteins. *Molecular Psychiatry, 8*(9), 811–820.
Furuse, M., & Tsukita, S. (2006). Claudins in occluding junctions of humans and flies. *Trends in Cell Biology, 16*(4), 181–188.
Gow, A. (2008). Major components of myelin in the mammalian central and peripheral nervous systems. In B. Kalman & T. H. III Brannagan (Eds.), *Neuroimmunology in clinical practice* (pp. 12–25). Malden: Blackwell Publishing.
Gow, A., Davies, C., Southwood, C. M., Frolenkov, G., Chrustowski, M., Ng, L., et al. (2004). Deafness in *Claudin 11*-null mice reveals the critical contribution of basal cell tight junctions to stria vascularis function. *Journal of Neuroscience, 24*(32), 7051–7062.
Gow, A., & Devaux, J. J. (2008). A model of tight junction function in CNS myelinated axons. *Neuron Glial Biology, 4*(4), 307–317.

Gow, A., Gragerov, A., Gard, A., Colman, D. R., & Lazzarini, R. A. (1997). Conservation of topology, but not conformation, of the proteolipid proteins of the myelin sheath. *Journal of Neuroscience, 17*, 181–189.

Gow, A., Southwood, C. M., Li, J. S., Pariali, M., Riordan, G. P., Brodie, S. E., et al. (1999). CNS myelin and sertoli cell tight junction strands are absent in *Osp/Claudin 11*-null mice. *Cell, 99*(6), 649–659.

Hall, S. M., & Williams, P. L. (1971). The distribution of electron-dense tracers in peripheral nerve fibres. *Journal of Cell Science, 8*, 541–555.

Hirano, A., Becker, N. H., & Zimmerman, H. M. (1969). Isolation of the periaxonal space of the central myelinated nerve fiber with regard to the diffusion of peroxidase. *The Journal of Histochemistry and Cytochemistry, 17*, 512–516.

Hyde, T. M., Ziegler, J. C., & Weinberger, D. R. (1992). Psychiatric disturbances in metachromatic leukodystrophy. Insights into the neurobiology of psychosis. *Archives of Neurology, 49*(4), 401–406.

Katsel, P., Davis, K. L., & Haroutunian, V. (2005). Variations in myelin and oligodendrocyte-related gene expression across multiple brain regions in schizophrenia: A gene ontology study. *Schizophrenia Research, 79*(2–3), 157–173.

Kaushansky, N., Eisenstein, M., Oved, J. H., & Ben-Nun, A. (2008). Activation and control of pathogenic T cells in OSP/claudin-11-induced EAE in SJL/J mice are dominated by their focused recognition of a single epitopic residue (OSP58M). *International Immunology, 20*(11), 1439–1449.

Kaushansky, N., Hemo, R., Eisenstein, M., & BenNun, A. (2007). OSP/claudin-11-induced EAE in mice is mediated by pathogenic T cells primarily governed by OSP192Y residue of major encephalitogenic region OSP179-207. *European Journal of Immunology, 37*(7), 2018–2031.

Kaushansky, N., Zhong, M. C., Kerlero de Rosbo, N., Hoeftberger, R., Lassmann, H., & Ben-Nun, A. (2006). Epitope specificity of autoreactive T and B cells associated with experimental autoimmune encephalomyelitis and optic neuritis induced by oligodendrocyte-specific protein in SJL/J mice. *Journal of Immunology, 177*(10), 7364–7376.

Kirschner, D. A., Ganser, A. L., & Caspar, D. L. D. (1984). Diffraction studies of molecular organization and membrane interactions in myelin. In P. Morell (Ed.), *Myelin* (pp. 51–95). (2nd ed.). New York: Plenum.

Klugmann, M., Schwab, M. H., Puhlhofer, A., Schneider, A., Zimmermann, F., Griffiths, I. R., et al. (1997). Assembly of CNS myelin in the absence of proteolipid protein. *Neuron, 18*(1), 59–70.

Konrad, M., Schaller, A., Seelow, D., Pandey, A. V., Waldegger, S., Lesslauer, A., et al. (2006). Mutations in the tight-junction gene claudin 19 (CLDN19) are associated with renal magnesium wasting, renal failure, and severe ocular involvement. *American Journal of Human Genetics, 79*(5), 949–957.

Kopala, L. C., Tan, S., Shea, C., Orlik, H., Vandorpe, R., & Honer, W. G. (2000). Adrenoleukodystrophy associated with psychosis. *Schizophrenia Research, 45*(3), 263–265.

Lee, N. P., Tong, M. K., Leung, P. P., Chan, V. W., Leung, S., Tam, P. C., et al. (2006). Kidney claudin-19: Localization in distal tubules and collecting ducts and dysregulation in polycystic renal disease. *FEBS Letters, 580*(3), 923–931.

Loh, Y. H., Christoffels, A., Brenner, S., Hunziker, W., & Venkatesh, B. (2004). Extensive expansion of the claudin gene family in the teleost fish, Fugu rubripes. *Genome Research, 14*(7), 1248–1257.

Martini, R., Mohajeri, M. H., Kasper, S., Giese, K. P., & Schachner, M. (1995). Mice doubly deficient in the genes for P0 and myelin basic protein show that both proteins contribute to the formation of the major dense line in peripheral nerve myelin. *Journal of Neuroscience, 15*(6), 4488–4495.

Mateu, L., Luzzati, V., London, Y., Gould, R. M., & Vosseberg, F. G. (1973). X-ray diffraction and electron microscope study of the interactions of myelin components. The structure of a lamellar phase with a 150 to 180 A repeat distance containing basic proteins and acidic lipids. *Journal of Molecular Biology, 75*(4), 697–709.

McCullumsmith, R. E., Gupta, D., Beneyto, M., Kreger, E., Haroutunian, V., Davis, K. L., et al. (2007). Expression of transcripts for myelination-related genes in the anterior cingulate cortex in schizophrenia. *Schizophrenia Research, 90*(1–3), 15–27.

Miyamoto, T., Morita, K., Takemoto, D., Takeuchi, K., Kitano, Y., Miyakawa, T., et al. (2005). Tight junctions in Schwann cells of peripheral myelinated axons: A lesson from claudin-19-deficient mice. *The Journal of Cell Biology, 169*(3), 527–538.

Mobius, W., Patzig, J., Nave, K. A., & Werner, H. B. (2008). Phylogeny of proteolipid proteins: Divergence, constraints, and the evolution of novel functions in myelination and neuroprotection. *Neuron Glial Biology, 4*(2), 111–127.

Mobius, W., Patzig, J., Nave, K. A., & Werner, H. B. (2009). Phylogeny of proteolipid proteins: Divergence, constraints, and the evolution of novel functions in myelination and neuroprotection. *Neuron Glial Biology, 1–17.*

Morita, K., Sasaki, H., Fujimoto, K., Furuse, M., & Tsukita, S. (1999). Claudin-11/OSP-based tight junctions of myelin sheaths in brain and Sertoli cells in testis. *The Journal of Cell Biology, 145*(3), 579–588.

Mugnaini, E., Osen, K. K., Schnapp, B., & Friedrich, V. L. Jr., (1977). Distribution of Schwann cell cytoplasm and plasmalemmal vesicles (caveole) in peripheral myelin sheaths. An electron microscopic study with thin sections and freeze-fracturing. *Journal of Neurocytology, 6*, 647–668.

Mugnaini, E., & Schnapp, B. (1974). Possible role of zonula occludens of the myelin sheath in demyelinating conditions. *Nature, 251*, 725–727.

Nunes, F. D., Lopez, L. N., Lin, H. W., Davies, C., Azevedo, R. B., Gow, A., et al. (2006). Distinct subdomain organization and molecular composition of a tight junction with adherens junction features. *Journal of Cell Science, 119*(Pt 23), 4819–4827.

O'Connor, K. C., Bar-Or, A., & Hafler, D. A. (2001). The neuroimmunology of multiple sclerosis: Possible roles of T and B lymphocytes in immunopathogenesis. *Journal of Clinical Immunology, 21*(2), 81–92.

Omlin, F., Webster, H.de.F., Pulkovits, C. G., & Cohen, S. R. (1982). Immunocytochemical localization of BP in major dense line regions of central and peripheral myelin. *The Journal of Cell Biology, 95*, 242–248.

Pelletier, R. M., & Byers, S. W. (1992). The blood–testis barrier and Sertoli cell junctions: Structural considerations. *Microscopy Research and Technique, 20*, 3–33.

Peters, A. (1961). A radial component of central myelin sheaths. *Journal of Biophysical and Biochemical Cytology, 11*, 733–735.

Peters, A. (1964). Further observations on the structure of myelin sheaths in the central nervous system. *Journal of Cell Biology, 20*, 281–296.

Pillai, A. M., Thaxton, C., Pribisko, A. L., Cheng, J. G., Dupree, J. L., & Bhat, M. A. (2009). Spatiotemporal ablation of myelinating glia-specific neurofascin (Nfasc NF155) in mice reveals gradual loss of paranodal axoglial junctions and concomitant disorganization of axonal domains. *Journal of Neuroscience Research, 87*(8), 1773–1793.

Poliak, S., Matlis, S., Ullmer, C., Scherer, S. S., & Peles, E. (2002). Distinct claudins and associated PDZ proteins form different autotypic tight junctions in myelinating Schwann cells. *The Journal of Cell Biology, 159*(2), 361–372.

Popot, J.-L., Pham-Dinh, D., & Dautigny, A. (1991). Major myelin proteolipid: The 4-alpha-helix topology. *The Journal of Membrane Biology, 120*(3), 233–246.

Rescigno, M., Urbano, M., Valzasina, B., Francolini, M., Rotta, G., Bonasio, R., et al. (2001). Dendritic cells express tight junction proteins and penetrate gut epithelial monolayers to sample bacteria. *Nature Immunology, 2*(4), 361–367.

Revel, J.-P., & Hamilton, D. W. (1969). The double nature of the intermediate dense line in peripheral nerve myelin. *Anatomical Record, 163*, 7–16.

Riccio, P., Fasano, A., Borenshtein, N., Bleve-Zacheo, T., & Kirschner, D. A. (2000). Multilamellar packing of myelin modeled by lipid-bound MBP. *Journal of Neuroscience Research, 59*(4), 513–521.

Rosenbluth, J., Nave, K. A., Mierzwa, A., & Schiff, R. (2006). Subtle myelin defects in PLP-null mice. *Glia, 54*(3), 172–182.

Rosenbluth, J., Schiff, R., & Lam, P. (2009). Effects of osmolality on PLP-null myelin structure: Implications re axon damage. *Brain Research, 1253*, 191–197.

Rosenbluth, J., Stoffel, W., & Schiff, R. (1996). Myelin structure in proteolipid protein (PLP)-null mouse spinal cord. *The Journal of Comparative Neurology, 371*, 336–344.

Schnapp, B., & Mugnaini, E. (1978). Membrane architecture of myelinated fibers as seen by freeze-fracture. In S. G. Waxman (Ed.), *Physiology and pathobiology of axons* (pp. 83–123). New York: Raven.

Shinowara, N. L., Beutel, W. B., & Revel, J.-P. (1980). Comparative analysis of junctions in the myelin sheath of central and peripheral axons of fish, amphibians and mammals: A freeze-fracture study using complementary replicas. *Journal of Neurocytology, 9*, 15–38.

Southwood, C. M., & Gow, A. (2001). Functions of OSP/claudin 11-containing parallel tight junctions: Implications from the knockout mouse. In J. M. Anderson & M. Cereijido (Eds.), *Tight junctions* (pp. 719–741). (2nd ed.). New York: CRC Press.

Stecca, B., Southwood, C. M., Gragerov, A., Kelley, K. A., Friedrich, V. L., Jr., Gow, A. (2000). The evolution of lipophilin genes from invertebrates to tetrapods: DM-20 cannot replace PLP in CNS myelin. *Journal of Neuroscience, 20*(11), 4002–4010.

Stevens, D. B., Chen, K., Seitz, R. S., Sercarz, E. E., & Bronstein, J. M. (1999). Oligodendrocyte-specific protein peptides induce experimental autoimmune encephalomyelitis in SJL/J mice. *Journal of Immunology, 162*(12), 7501–7509.

Stork, T., Engelen, D., Krudewig, A., Silies, M., Bainton, R. J., & Klambt, C. (2008). Organization and function of the blood–brain barrier in *Drosophila*. *Journal of Neuroscience, 28*(3), 587–597.

Tabira, T., Cullen, M. J., Reier, P. J., & Webster, H. (1978). An experimental analysis of interlamellar tight junctions in amphibian and mammalian C.N.S. myelin. *Journal of Neurocytology, 7*(4), 489–503.

Tepass, U. (2003). Claudin complexities at the apical junctional complex. *Nature Cell Biology, 5*(7), 595–597.

Tiwari-Woodruff, S., BeltranParrazal, L., Charles, A., Keck, T., Vu, T., & Bronstein, J. (2006). K+ channel K(V)3.1 associates with OSP/claudin-11 and regulates oligodendrocyte development. *American Journal of Physiology. Cell Physiology, 291*(4), C687–C698.

Tiwari-Woodruff, S. K., Buznikov, A. G., Vu, T. Q., Micevych, P. E., Chen, K., Kornblum, H. I., et al. (2001). OSP/claudin-11 forms a complex with a novel member of the tetraspanin super family and beta1 integrin and regulates proliferation and migration of oligodendrocytes. *The Journal of Cell Biology, 153*(2), 295–305.

Trapp, B. D., & Nave, K. A. (2008). Multiple sclerosis: An immune or neurodegenerative disorder? *Annual Review of Neuroscience, 31*, 247–269.

Tsukita, S., Yamazaki, Y., Katsuno, T., Tamura, A., & Tsukita, S. (2008). Tight junction-based epithelial microenvironment and cell proliferation. *Oncogene, 27*(55), 6930–6938.

Walterfang, M., Wood, S. J., Velakoulis, D., Copolov, D., & Pantelis, C. (2005). Diseases of white matter and schizophrenia-like psychosis. *Australian and New Zealand Journal of Psychiatry, 39*(9), 746–756.

Waxman, S. G., & Bennett, M. V. (1972). Relative conduction velocities of small myelinated and non-myelinated fibres in the central nervous system. *Nature—New Biology, 238*(85), 217–219.

Yule, T. D., Mahi-Brown, C. A., & Tung, K. S. (1990). Role of testicular autoantigens and influence of lymphokines in testicular autoimmune disease. *Journal of Reproductive Immunology, 18*(1), 89–103.

Zonta, B., Tait, S., Melrose, S., Anderson, H., Harroch, S., Higginson, J., et al. (2008). Glial and neuronal isoforms of Neurofascin have distinct roles in the assembly of nodes of Ranvier in the central nervous system. *The Journal of Cell Biology, 181*(7), 1169–1177.

CHAPTER 11

Claudin is Skin Deep☆

Kursad Turksen*,†,‡,§ and Tammy-Claire Troy*

*Regenerative Medicine Program, Sprott Centre for Stem Cell Research, Ottawa Hospital Research Institute, Ottawa, ON, Canada

†Department of Cellular and Molecular Medicine, Faculty of Medicine, University of Ottawa, Ottawa Hospital, Ottawa, ON, Canada

‡Department of Medicine, Division of Endocrinology, Ottawa Hospital, Ottawa, ON, Canada

§Department of Medicine, Division of Dermatology, Ottawa Hospital, Ottawa, ON, Canada

I. OVERVIEW

The skin acts as the interface between the organism and the outside environment, functioning as a physical and selective barrier against chemical and biological invasion (*outside-to-inside*), as well as retaining body solutes and maintaining thermoregulation (*inside-to-outside*). This complex role is achieved through an epidermal differentiation program that gives rise to a tightly regulated and selective epidermal permeability barrier (EPB). Dysfunction of this barrier during development, or at any time during the lifespan, leads to death under severe circumstances or to a propensity for chronic skin conditions or diseases. In spite of its important role, the molecular

☆We pay homage to "*Beauty is skin deep*" by Elaine Fuchs (Fuchs, 1998).

Current Topics in Membranes, Volume 65 1063-5823/10 $35.00

DOI: 10.1016/S1063-5823(10)65011-9

mechanisms behind the formation and function of the EPB are still poorly understood. The recent discovery of the Claudin (Cldn) family of tight junction (TJ) proteins and their structural–functional characterization has uncovered their essential role in the EPB and other tissue–organ barriers in the body. This chapter summarizes our emerging understanding of Cldns and their role in TJs and in the EPB in normal development and certain skin diseases.

II. INTRODUCTION

Skin is one of the largest organs of the body and is strategically located at the interface between the organism and the outside world. In its mature form, the epidermis protects the organism against microorganism invasion and UV irradiation, inhibits water loss, regulates body temperature, and plays a critical role in the body's immune system. Despite these critical functions, the molecular basis and mechanisms governing the regulation of mammalian epidermal differentiation with the concomitant formation of the EPB are still poorly understood. Generally stated, it is known that a functional EPB is assembled from lipids and proteins into a structure dubbed by some as "*bricks and mortar*." While an interesting analogy, such a static structure or view fails to provide a molecular mechanism for the selectivity in permeability that this barrier displays both outside-to-inside and inside-to-outside. However, the recent discovery and characterization of the Cldn family of TJ proteins have uncovered their essential role in the formation, function, and maintenance of the EPB and other tissue–organ barriers in the body (Angelow, Ahlstrom, & Yu, 2008; Findley & Koval, 2009; Segre, 2003; Van Itallie & Anderson, 2006). Most importantly, Cldn-family-based TJs provide essential new understanding of at least some of the molecular basis behind selective permeability *in vivo*. Progress during the last few years includes the differential expression and localization of different Cldns in TJs and TJ fibrils using both *in vitro* and *in vivo* models. Gain-of-function and loss-of-function approaches in cells, mouse models, and human specimens have provided evidence that Cldn-containing TJs are responsible for the EPB (Furuse et al., 2002; Troy, Arabzadeh, Lariviere, Enikanolaiye, & Turksen, 2009). This chapter summarizes these data and our emerging understanding of Cldns and their role in TJs and in the EPB in normal development and certain skin diseases.

III. THE EPIDERMIS AND EPIDERMAL TERMINAL DIFFERENTIATION

The mammalian epidermis is formed during development through a highly orchestrated process involving cell fate selection, lineage commitment, mesenchymal–epithelial interactions, stratification, and terminal differentiation.

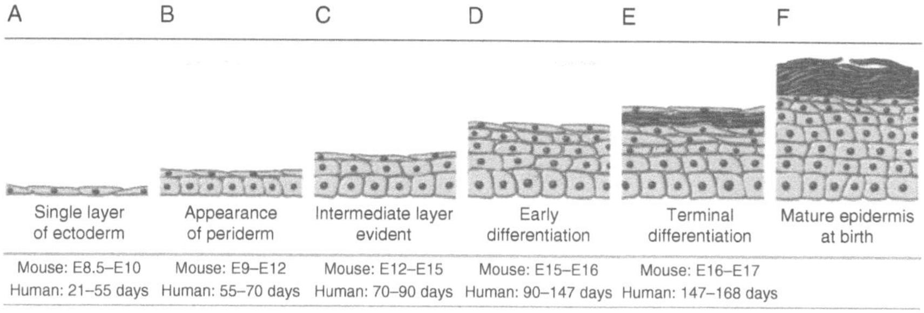

FIGURE 1 Schematic representation of the development of the mature epidermis and the formation of the EPB (adapted from Byrne et al., 2003). During development, the epidermis is derived from a single layer of ectoderm (A), which becomes covered by a layer of periderm (B). Then differentiation of the epidermis commences with the appearance of the intermediate layer of cells between the basal cells and the periderm (C), which undergoes progressive differentiation to form the spinous and granular layers of cells (D, E) as well as the stratum corneum of the mature epidermis (F).

Morphogenesis of the mature epidermis (Fig. 1) is completed just before birth, with the appearance of the granular and cornified layers (Byrne, Tainsky, & Fuchs, 1994; Kopan & Fuchs, 1989; Mack, Anand, & Maytin, 2005; Weiss & Zelickson, 1975a, 1975b). Formation of the insoluble cornified envelopes of the stratum corneum occurs just before birth (Byrne et al., 1994; Elias, 2005, 2007; Fuchs & Horsley, 2008; Hardman, Moore, Ferguson, & Byrne, 1999; Hardman, Sisi, Banbury, & Byrne, 1998; Mack et al., 2005; Madison, 2003; Nemes & Steinert, 1999; Presland & Dale, 2000; Proksch, Brandner, & Jensen, 2008; Steinert & Marekov, 1999; Turksen & Troy, 2002), thus preparing the mammalian newborn to switch from an aquatic environment to survival in a terrestrial one.

The epidermal differentiation program is a self-renewal process that is "*la raison d'être*" of epidermal cells and, like Sisyphus in pushing the rock up the hill in perpetuity, epidermal cells are "*condemned*" to repeat the process indefinitely for the normal functioning of the EPB. In this precisely orchestrated homeostatic process, epidermal cells in the proliferative basal compartment, in response to discrete signals, become irreversibly committed to terminal differentiation and move upwards away from the basal layer (Byrne, 1997; Fuchs, 2007; Fuchs & Byrne, 1994; Fuchs & Horsley, 2008; Turksen & Troy, 1998). This process can be very accurately traced by the expression profile of epidermal structural proteins, keratins, and other proteins involved in terminal differentiation (Candi, Schmidt, & Melino, 2005; Fuchs, 1995; Gu & Coulombe, 2007; Steinert, 1995). Briefly, in response to differentiation signals, epidermal cells in the basal layer, which express keratin 5 and keratin 14 (K5/K14), commit to terminal differentiation and give rise to the differentiated layers of the skin by leaving the basal layer, shutting down K5/K14, and inducing K1/K10 expression (Stoler, Kopan, Duvic, & Fuchs, 1988). As cells progress to the spinous layer, they begin to change shape and to express involucrin, a protein involved in the eventual formation of the insoluble cornified envelope (Banks-Schlegel & Green, 1981). Granular cells, which directly follow the spinous cells, are characterized by electron-dense keratohyalin granules containing filaggrin, a protein that facilitates the aggregation of keratin filaments (Dale, 1985). As the terminally differentiating cells transit from the granular layer to the cornified layer, they undergo a destruction of their organelles to form the cornified envelope. The stratum corneum—the actual barrier between the external and internal environments of the body (Mack et al., 2005; Marshall, Hardman, & Byrne, 2000; Marshall, Hardman, Nield, & Byrne, 2001; Segre, 2003)—is finally formed through the sequential expression, processing, and deposition of numerous proteins of the epidermal terminal differentiation program, including cornified envelopes (for review see Candi et al., 2005; Mack et al., 2005) (Fig. 1).

Extensive biochemical studies over the years have led to the identification of the structural components and their assembly in contributing to the insoluble stratum corneum, a structure that is rich in lipid and protein and likened to, or reminiscent of, a "*bricks and mortar*" structure (Kalinin, Kajava, & Steinert, 2002; Nemes & Steinert, 1999; Presland & Dale, 2000). However, such a static model, which comprises approximately 30 structurally important (but most likely redundant) molecules (Segre, 2003), individually or collectively cannot easily account for the known inside-to-outside and outside-to-inside selectivity of the barrier (Segre, 2003, 2006); nor do they explain its dynamics in response to fluctuating environmental conditions. Interestingly, the established role of TJs in selective barrier formation and functioning did provide a potential molecular basis for such a permeability barrier in the epidermis, but exactly how was not clear until recently.

IV. TIGHT JUNCTIONS AND THE TIGHT JUNCTION PERMEABILITY BARRIER

A. *TJs and the Discovery of Cldns*

The first identification of TJs was accomplished in 1963 using transmission electron microscopy (TEM), whereby Farquhar and Palade described structures localized at the apical end of epithelial cells that appear morphologically as "close contacts"—without extracellular gaps or intramembranous material—of the plasma membranes of neighboring polarized epithelial cells (Farquhar & Palade, 1963). Further to the initial observation with TEM, freeze fracture analysis indicated that TJ structures appeared as continuous networks of strands—or fibrils—that surround the apical end of polarized cells. Based on further characterization, it became evident that TJs are responsible for the control of paracellular transport between epithelial cells (*gate function*) as well as the maintenance of apical/basolateral polarity by preventing the diffusion of membrane lipids and/or proteins from one surface domain to another (*fence function*).

In spite of extensive functional studies, elucidation of the molecular composition of TJs lagged behind that of adherence and gap junctions. TJs are now believed to be formed by interacting transmembrane molecules that span the extracellular space and connect neighboring membranes. Early studies resulted in the identification of occludin as the first integral membrane protein identified as being associated with TJs (Furuse et al., 1993). To date, peripheral protein members of the MAGUK family (e.g., ZO-1, ZO-2, and ZO-3) as well as lesser known molecules such as symplekin, cingulin, and 7H6 antigen are now all known to be associated with TJs. Since its identification, many investigators queried the role of occludin in TJs, but occludin could not provide a convincing or rational molecular mechanism by which the TJ heterogeneity observed in various epithelia could be explained. This was further emphasized when occludin-deficient embryonic stem (ES) cells and mice were characterized as exhibiting no impairment in the formation of TJ strands or barrier function (Saitou et al., 1998; Schulzke et al., 2005).

The dispensability of occludin in TJ formation and barrier function, both in ES cells and knockout mice, accelerated the search for additional integral transmembrane proteins that might fulfill these roles. In the 1990s, Tsukita's laboratory isolated two novel $\sim$22 kDa integral membrane proteins that they named from a Latin word "*claudere*" (to close) as Cldn1 and Cldn2, and provided evidence that they are responsible for TJ formation (Furuse, Fujita, Hiiragi, Fujimoto, & Tsukita, 1998). Additional studies led fairly quickly to the identification of 23 distinct Cldns in mammalian cells. With a few notable exceptions (e.g., Cldn16 and Cldn23), most Cldns encode $\sim$20–27 kDa integral membrane proteins and are structurally analogous with four transmembrane domains, two extracellular loops, one intracellular loop, a short N terminus as well as a COOH intracellular tail of varying length. Recent analyses have led to the classification of Cldns into two groups according to their degree of sequence similarity, these are the so-called "classic Cldns" (1–10, 14, 15, 17, 19) and the "nonclassic Cldns" (11–13, 16, 18, 20–24) (Krause et al., 2008). Numerous studies are currently being done to address the structure–function relationships of Cldns. One tentative molecular model of homophilic interactions addresses the different functional contributions of the two extracellular loops of TJs (Krause et al., 2008).

The cytoplasmic tail is generally considered to be the most divergent structural component among Cldn family members (Turksen & Troy, 2004), which also possesses a number of sites that provide clues towards its role in cell polarity and differentiation. These include potential phosphorylation sites (D'Souza, Agarwal, & Morin, 2005; Fujibe et al., 2004; Simard, Di Pietro, Young, Plaza, & Ryan, 2006; Tanaka, Kamata, & Sakai, 2005), a palmitoylation site (Findley & Koval, 2009; Van Itallie, Gambling, Carson, & Anderson, 2005), and a PDZ-binding sequence (Itoh et al., 1999). In fact, most members of the Cldn family have PDZ-binding domains at their COOH terminus that enable recruitment of other TJ-associated proteins [e.g., proteins with a PDZ domain such as the MAGUK family proteins (ZO-1, ZO-2, and ZO-3) (Itoh et al., 1999)]. Other proteins containing PDZ domains [e.g., multi-PDZ domain scaffolding proteins, PATJ (Roh et al., 2002), and MUPP-1 (Poliak, Matlis, Ullmer, Scherer, & Peles, 2002)] also selectively recognize and bind this sequence, and it is predicted that other Cldn tail domain-interacting proteins (CTIPs) remain to be identified. However, it seems likely that as yet unidentified novel molecules interact with Cldns in regulating gene expression and epidermal differentiation. Posttranslational modifications within the tail domain, including phosphorylation and palmitoylation are also thought to regulate Cldn activities, including their targeting to the membrane and insertion into TJs (Van Itallie et al.) to regulate paracellular permeability (Simard et al., 2006) and differentiation (Arabzadeh, Troy, & Turksen, 2006). Phosphorylation of a number of Cldns has been demonstrated to be required for their assembly into TJs [e.g., for Cldn1, Cldn4 (Banan et al., 2005), and Cldn16 (Ikari et al., 2008)]; however, to date most Cldns have not been subjected to exhaustive analyses.

It is beyond the scope of this chapter to review the differential expression of different Cldn family members in diverse TJ-containing tissues and organs; a topic addressed throughout this volume. Suffice it to say that Cldns are responsible for TJ fibril formation and, together with Cldns in neighboring cells and the cytoplasmic proteins they recruit, provide a tissue-specific permeability barrier to fit the physiological needs of the specific organ. Much remains to be elucidated about the mechanism of assembly of Cldns into TJs, including interactions between various Cldns in TJs and the molecular basis of the observed barrier heterogeneity. For the rest of this chapter, we will discuss Cldns in the epidermis.

B. TJs and the Epidermis

As introduced earlier, the epidermis fulfills the crucial role of being a dynamic and selective barrier that harbors gatekeeper function and possesses both inside-to-outside and outside-to-inside function between the organism and its environment (Fig. 2). Although the existence of this two-way selective permeability is now widely accepted (Elias and Friend, 1975; Elias, McNutt, & Friend, 1977), until a decade ago the existence of TJs in the epidermis was not commonly acknowledged; even though, quite ironically, morphologically

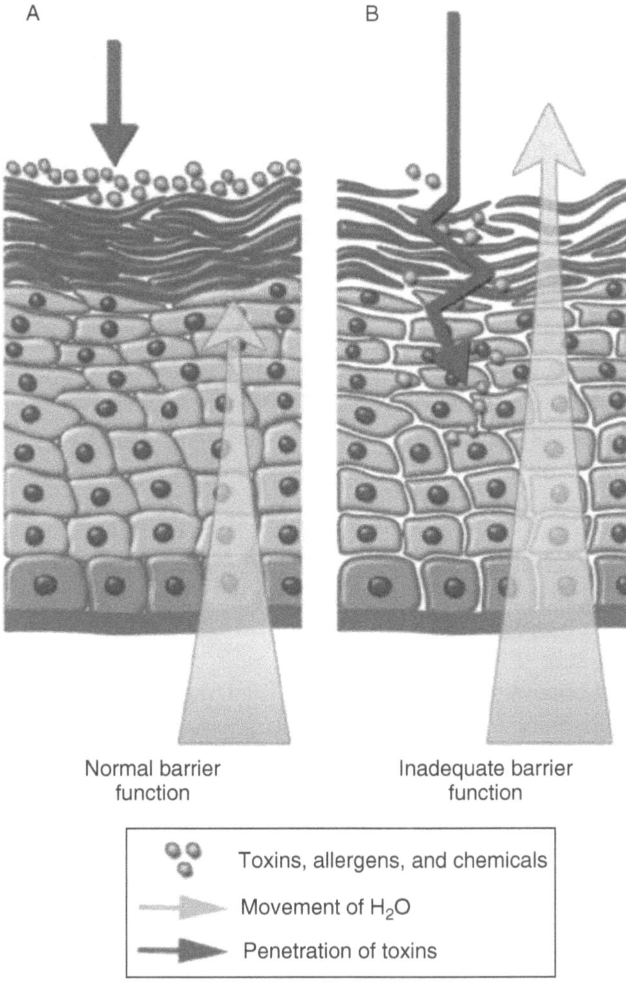

FIGURE 2 The epidermis fulfills the crucial role of being a dynamic and selective barrier that harbors gatekeeper function and possesses both inside-to-outside and outside-to-inside function between the organism and its environment. When the barrier is intact (A), TEWL is restricted and penetration by toxins, allergens and/or chemicals is inhibited. When the barrier is inadequate (B), both inside-to-outside and outside-to-inside function between the organism and its environment is defective (adapted from Proksch et al., 2009). (See Color Insert.)

recognizable epidermal TJs were documented in the early 1970s by TEM (Hashimoto, 1971). Other contemporaneous studies demonstrated the existence of a functional barrier in the epidermis [e.g., the electron-dense tracer lanthanum stopped at the granular layer of the human epidermis, which contained TJs (Hashimoto, 1971)]. Furthermore, the existence of outside-to-inside permeability has very elegantly been demonstrated by means of a dye penetration assay whereby Byrne and colleagues established the patterned formation of the permeability barrier during development (Hardman et al., 1998, 1999). This functional barrier has been confirmed and correlated with other developmental changes in the epidermis, including the expression of TJ molecules and the formation of TJs (Troy, Li, O'Malley, & Turksen, 2007; Troy, Rahbar, Arabzadeh, Cheung, & Turksen, 2005). Yet, doubt remained about the presence and role of epidermal TJs. After the identification of Cldns, Franke's group (Brandner et al., 2002; Franke, 2009; Franke, Rickelt, Barth, & Pieperhoff, 2009; Langbein et al., 2002, 2003; Schluter, Moll, Wolburg, & Franke, 2007; Schluter, Wepf, Moll, & Franke, 2004) used systematic ultrastructural studies to readdress the question of the existence of TJs in the mammalian epidermis. The group reported canonical TJ-like structures in the granular layers of the epidermis, as generally observed in simple epithelia, but also reported for the first time "new TJ-like structures" that appeared to be unique to the stratified epithelium. The suprabasal layers of the epidermis were found to contain TJ protein-containing junctions (including Cldns) of variable sizes, characterized by a 10–30-nm dense lamina interposed between the two adjacent membranes. They also reported the presence of structures that they named "*sandwich junctions*" or "*juncturae structae*." In addition, they identified another TJ protein-containing structure; a very small structure that they termed "*stud junctions*" or "*puncta occludentia*" as the smallest identifiable TJ-like unit that occurs in most strata (Fig. 3). Collectively, these observations compellingly confirmed and extended earlier observations to solidify the concept of structurally recognizable Cldn-containing TJs as functional and indispensable components of the permeability barrier of the epidermis.

C. Cldns and Epidermal Development and Function in Health and Disease

Numerous studies have described the developmental formation of the EPB (Arabzadeh et al., 2006; Byrne & Hardman, 2005; Hardman et al., 1998; Troy et al., 2005, 2007, 2009; Turksen & Troy, 2002). Thus, it is well recognized that a disruption or delay in its formation before birth may have severe consequences in the survival of the organism (Cartlidge, 2000;

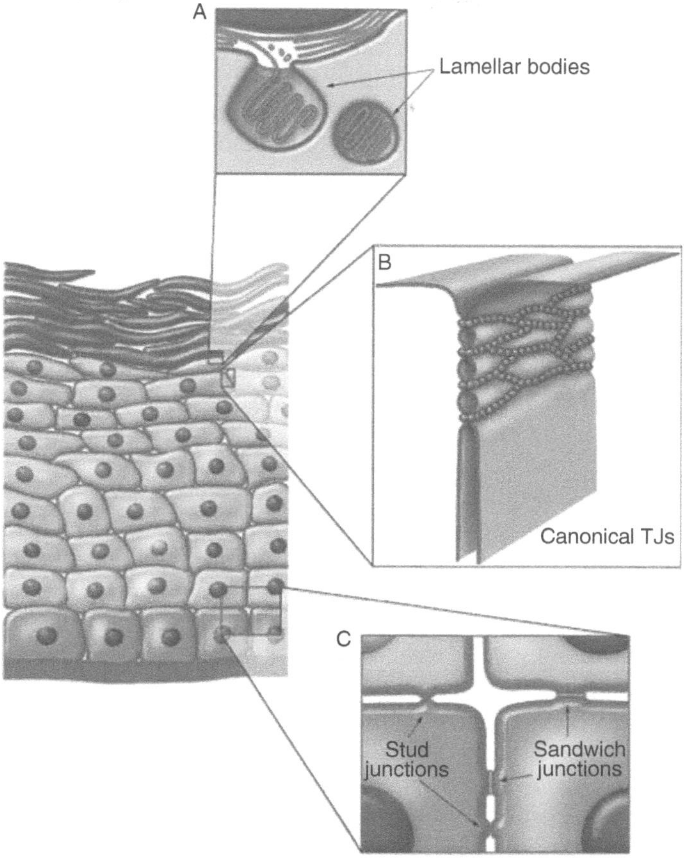

FIGURE 3 Pictorial demonstration of the components of the permeability barrier of the epidermis (adapted from Morita and Miyachi, 2003; Proksch et al., 2008). At the interface to the stratum corneum, the contents of lamellar bodies are secreted to contribute to the formation of the stratum corneum (A). Location of the canonical epithelial TJs in the granular layer is indicated (B). Location and distribution of noncanonical TJs (sandwich junctions and stud junctions) in all layers of the epidermis (C). (See Color Insert.)

Harpin & Rutter, 1983). For example, human premature infants are often born with a poorly developed skin barrier, consequently resulting in high transepidermal water loss (TEWL; leading to poor temperature control and difficulty in maintaining fluid balance), percutaneous absorption (leading to infection and sepsis), and associated trauma. The seriousness of these consequences has led to significant interest in understanding what role Cldns may play in normal epidermal development and in pathological conditions.

During development as well as in normal homeostasis, epidermal cells differentially express Cldns in a cell- and differentiation-specific manner; moreover, their expression/localization is modulated in injury and disease. Gain-of-function and loss-of-function experiments in animal models extended functional evidence for the critical role of Cldn-containing TJs in the formation and function of the epidermis (Furuse et al., 2002; Turksen & Troy, 2002). Tsukita's group demonstrated that Cldn1 knockout mice do not form a functional outside-to-inside permeability barrier [intradermally injected small molecules (600 Da) penetrated the mutant but not the wild-type epidermis (Furuse et al., 2002)] and die shortly after birth with epidermal differentiation defects and severe TEWL (a measure of inside-to-outside function) (Furuse et al., 2002). Moreover, our group established that depending on the level of forced overexpression of normal or mutant forms of Cldn6 in its endogenous site of expression (the suprabasal compartment), the epidermis is more or less severely compromised (Arabzadeh et al., 2006; Troy et al., 2005, 2009; Turksen & Troy, 2002). For example, severe EPB dysfunction manifested in extreme TEWL and neonatal lethality when native Cldn6 was expressed at high levels (Turksen & Troy, 2002); lower levels of expression resulted in less-severe EPB dysfunction that normalized postnatally (Troy et al., 2005). Overexpression of a mutant form of Cldn6 lacking its entire cytoplasmic tail domain (Inv-Cldn6-CΔ187 mice) exhibited no apparent prenatal epidermal developmental defects, but an abnormal postnatal lifelong epidermal hyperproliferation was observed (Arabzadeh et al., 2006). Higher-level overexpression of a different mutant lacking only the C-terminal half of the tail domain of Cldn6 (Inv-Cldn6-CΔ196 mice) resulted in a lethal barrier dysfunction with marked hyperproliferative squamous invaginations/cysts replacing hair follicles (Troy & Turksen, 2007), while lower-level expression resulted in an aging-related skin barrier defect manifested by an intrinsic propensity for injury, inefficient repair, and chronic dermatitis (Troy et al., 2009). Expression of a mutant Cldn6 with a shorter tail deletion (also known as removing the PDZ domain and a putative PKA phosphorylation site; Inv-Cldn6-CΔ206 mice) resulted in another distinct developmental defect in epidermal differentiation in which EPB formation delays followed by a robust repair response occurred for postnatal epidermal maturation (Enikanolaiye et al., 2010). It is notable that formation of a skin barrier with functional TEWL characteristics indistinguishable from the wild type occurred more rapidly than, or prior to, the complete morphological maturation of the epidermis in postnatal Inv-Cldn6-CΔ206 mice. This indicates an ability to disconnect aspects of the two processes, an observation that is interesting from a developmental standpoint, but may also be therapeutically important (see also below). Collectively, these and other studies underscore at least three notions: (1) Cldn homeostasis is crucial to the development, function, and maintenance of the epidermis; (2) domains and motifs in the Cldn cytoplamic tail domain participate (presumably with a plethora of other molecules many of which remain to be identified) to regulate those processes (see also above); and (3) skin conditions and diseases—some of which are life threatening—arise from alterations in Cldn homeostasis and function in mice. These findings provide the basis for much of the research involving Cldns today and are beginning to be applied to understanding the pathogenesis of many human diseases.

It is worth noting that disruptions in the expression/localization profile of one Cldn often, and perhaps always, leads to modifications in the expression/localization profiles of other Cldns. Upon injury or in response to differentiation abnormalities, the spatial expression of the suprabasal Cldns (Cldn6, Cldn11, Cldn12, and Cldn18) generally expands or shrinks to encompass all the cells of the perturbed suprabasal zone with some concomitant loss in cell membrane association (Arabzadeh, Troy, & Turksen, 2007, 2008; ; Troy et al., 2005, 2009; Turksen & Troy, 2002). This was true, for example, in the immature Inv-Cldn6-CΔ206 epidermis, where localization and expression normalized to a strictly membranous association in a suprabasal zone comparable in thickness to that of the wild type as the epidermis matured by postnatal day 10 (Enikanolaiye et al., 2010). Moreover, Cldn1 undergoes more dramatic alterations in response to epidermal homeostasis dysregulation. In the developing epidermis, Cldn1 is first restricted to the stratifying layers and matures to occupy as well the basal layer upon the completion of barrier formation at E17.5 (Troy & Turksen, 2007). However, in response to TPA-induced acute injury as well as loss of cell polarity as observed in tumorigenesis, Cldn1 expression is downregulated in both the basal layer and the immediate suprabasal layers of the epidermis (Arabzadeh et al., 2007, 2008), changes were also observed in the intrinsic aging process of the Inv-Cldn6-CΔ196 transgenic epidermis (Troy et al., 2009) and in the delayed epidermal maturation that we now report in Inv-Cldn6-CΔ206 transgenic mice (Enikanolaiye et al., 2010). Notably, Cldn1 expression normalized with the normalization of epidermal differentiation markers and epidermal maturation (*vide infra*). While the upstream and downstream consequences of changes in Cldn expression profiles remain to be fully elucidated, it is clear that in spite of their TJ localization, considerable homologies, and sequence similarities, specific Cldns play very distinct roles presumably via their specific binding partners and targets (see Table I).

An area of considerable interest, but for which few concrete examples yet exist, is a role for Cldns in hereditary skin diseases. Currently, the only epidermal example is neonatal ichthyosis–sclerosing cholangitis (NISCH) syndrome, a rare autosomal recessive ichthyosis disorder characterized by scalp hypotrichosis, scarring alopecia, ichthyosis, and sclerosing cholangitis.

TABLE I

TJs and Cldns in the Various Compartments of the Epidermis

Epidermal compartment	Tight junction	Lamellated and sandwich junctions	Stud junction	Cldn1	Cldn6	Cldn11	Cldn12	Cldn18
Basal layer	−	+	+	+	−	−	−	−
Suprabasal layers	−	+	+	+	+	+	+	+
Granular layer	+	+	+	+	+	+	+	+
Stratum corneum	−	+	+	−	−	−	−	−

Described for the first time in 2002, NISCH is caused by a mutation in the gene coding for Cldn1; four patients carrying the same Cldn1 gene mutation have been described to date (Feldmeyer et al., 2006). Changes in Cldn expression levels or structure–function are thought to be good candidates for a number of other skin diseases in which patients exhibit permeability barrier defects; yet thus far no other specific examples have been reported. It is known, however, that Cldn expression is altered in diverse epithelia in a variety of inflammatory diseases (Heller, Fromm, Gitter, Mankertz, & Schulzke, 2008; Troy & Turksen, 2007) including enteropathies such as ulcerative colitis and Crohn's disease—another inflammatory disease with barrier defects. Among the skin diseases, psoriasis is a chronic inflammatory skin disease affecting approximately 2% of the western population, with significantly increased occurrence in individuals with Crohn's disease (Watson et al., 2007). Initial studies investigated the expression of Cldns in the skin of healthy volunteers and patients with chronic plaque psoriasis. Immunolabeling for Cldn1, Cldn3, Cldn4, and Cldn7 demonstrated that the epidermal Cldn expression profile is altered, at least in part by interleukin-1β (Kirschner et al., 2009; Watson et al., 2007).

Another area of considerable interest is the elucidation of what role Cldn modulations may play in the changes observed in the EPB following external injuries (such as UVB exposure or tumorigenesis). Recent studies document a correlation between changes in the expression patterns of TJ-related molecules, including Cldn1, and epidermal permeability in UVB-irradiated mouse skin (Yamamoto et al., 2008). Three days after irradiation, Cldn1 protein levels were elevated and *in vivo* permeability assays suggested that barrier function was perturbed, thereby providing evidence that changes in the expression patterns of Cldn1, and other TJ-associated molecules, are associated with perturbed barrier function following UVB irradiation. Regarding tumorigenesis, a two-stage chemical carcinogenesis model resulted in a reduction in the number of suprabasal epidermal layers positive for Cldn6, Cldn11, Cldn12, and Cldn18 during disease progression, especially in the lower strata of the expanded suprabasal zone of the murine epidermis (Arabzadeh et al., 2007). In addition, a variably reduced cell membrane association of those differentiation-specific Cldns was observed, especially within the infiltrating epidermal structures. In contrast, Cldn1 (which is normally expressed in all the living layers of the epidermis) remained restricted to the cell membrane throughout the tumorigenesis protocol. However, commencing 2 weeks after treatment there was a marked decrease in the number of Cldn1-positive basal cells, and the zone of Cldn1-null epidermal cells was expanded up into the lower stratified epidermis throughout the progression of the tumorigenesis protocol. Changes in Cldn expression are also beginning to be documented in human skin tumor biopsies (Morita, Tsukita, & Miyachi, 2004; Troy & Turksen, unpublished observations). Thus, it seems unequivocal that Cldns are deregulated in response to external injury. However, much remains to be learned about what changes occur first, which have the most serious consequences for the maintenance of normal function, and whether these changes can be reversed in a therapeutic program to treat, reverse, and/or cure skin cancer or UV-damaged skin.

V. CONCLUSIONS—CHALLENGES FOR THE FUTURE

As briefly summarized, Cldn-containing TJs (both typical and atypical) are critically important in the formation and function of the epidermis and the EPB. Accumulating data indicate that Cldns are developmentally regulated, differentially expressed during epidermogenesis, and that their expression/localization is modulated in various skin conditions and diseases. A growing number of knockout and transgenic mouse models expressing native or mutated forms of various Cldns support the strongly held hypothesis that Cldn dysregulation is not merely a consequence but rather a cause of epidermal pathologies. However, a great deal remains to be uncovered about what transcription factors or other regulatory molecules lie upstream of Cldns, what regulates their expression at the transcriptional or posttranscriptional levels, how structurally and dynamically (e.g., via signaling pathways) they contribute to the formation and maintenance of the epidermis and the EPB, and how Cldn dysregulation alters their roles in TJs and the EPB. These queries will no doubt be addressed in the plight of researchers to uncover the molecular mechanisms governing the regulation of Cldns in development, epidermal differentiation, barrier formation, and disease. This is but the first step in demonstrating that the significance of Cldns in these processes is more than just skin deep.

Acknowledgments

We thank Jane Aubin for continued valuable discussions and her critical review of this manuscript. K. T. acknowledges J. D. Salinger for Holden Caulfield's experiences. Our work on epidermal differentiation and Cldns has been supported by grants from the Canadian Institutes of Health Research (CIHR), and its Institutes of Musculoskeletal Health and Arthritis (IMHA) and of Aging (IA). We are grateful to the artistic talents of Lauren O'Malley who prepared the diagrams used in this review.

References

Angelow, S., Ahlstrom, R., & Yu, A. S. (2008). Biology of Claudins. *American Journal of Physiology Renal Physiology, 295(4)*, F867–F876.

Arabzadeh, A., Troy, T. C., & Turksen, K. (2006). Role of the cldn6 cytoplasmic tail domain in membrane targeting and epidermal differentiation in vivo. *Molecular and Cellular Biology, 26*, 5876–5887.

Arabzadeh, A., Troy, T. C., & Turksen, K. (2007). Changes in the distribution pattern of Claudin tight junction proteins during the progression of mouse skin tumorigenesis. *BMC Cancer, 7*, 196.

Arabzadeh, A., Troy, T. C., & Turksen, K. (2008). Claudin expression modulations reflect an injury response in the murine epidermis. *The Journal of Investigative Dermatology, 128*, 237–240.

Banan, A., Zhang, L. J., Shaikh, M., Fields, J. Z., Choudhary, S., Forsyth, C. B., et al. (2005). Theta isoform of protein kinase C alters barrier function in intestinal epithelium through modulation of distinct claudin isotypes: A novel mechanism for regulation of permeability. *The Journal of Pharmacology and Experimental Therapeutics, 313*, 962–982.

Banks-Schlegel, S., & Green, H. (1981). Involucrin synthesis and tissue assembly by keratinocytes in natural and cultured human epithelia. *The Journal of Cell Biology, 90*, 732–737.

Brandner, J. M., Kief, S., Grund, C., Rendl, M., Houdek, P., Kuhn, C., et al. (2002). Organization and formation of the tight junction system in human epidermis and cultured keratinocytes. *European Journal of Cell Biology, 81*, 253–263.

Byrne, C. (1997). Regulation of gene expression in developing epidermal epithelia. *Bioessays, 19*, 691–698.

Byrne, C., & Hardman, M. J. (2005). Whole-mount assays for gene induction and barrier formation in the developing epidermis. *Methods in Molecular Biology, 289*, 127–136.

Byrne, C., Hardman, M., & Nield, K. (2003). Covering the limb—formation of the integument. *Journal of Anatomy, 202*, 113–123.

Byrne, C., Tainsky, M., & Fuchs, E. (1994). Programming gene expression in developing epidermis. *Development, 120*, 2369–2383.

Candi, E., Schmidt, R., & Melino, G. (2005). The cornified envelope: A model of cell death in the skin. *Nature Reviews. Molecular Cell Biology, 6*, 328–340.

Cartlidge, P. (2000). The epidermal barrier. *Seminars in Neonatology, 5*, 273–280.

Dale, B. A. (1985). Filaggrin, the matrix protein of keratin. *American Journal of Dermatopathology, 7*, 65–68.

D'Souza, T., Agarwal, R., & Morin, P. J. (2005). Phosphorylation of claudin-3 at threonine 192 by cAMP-dependent protein kinase regulates tight junction barrier function in ovarian cancer cells. *Journal of Biological Chemistry, 280*, 26233–26240.

Elias, P. M. (2005). Stratum corneum defensive functions: An integrated view. *The Journal of Investigative Dermatology, 125*, 183–200.
Elias, P. M. (2007). The skin barrier as an innate immune element. *Seminars in Immunopathology, 29*, 3–14.
Elias, P. M., & Friend, D. S. (1975). The permeability barrier in mammalian epidermis. *The Journal of Cell Biology, 65*, 180–191.
Elias, P. M., McNutt, N. S., & Friend, D. S. (1977). Membrane alterations during cornification of mammalian squamous epithelia: A freeze-fracture, tracer, and thin-section study. *Anatomical Record, 189*, 577–594.
Enikanolaiye, A., Lariviere, N., Troy, T. C., Arabzadeh, A., Atasoy, E., & Turksen, K. (2010). Involucrin-claudin-6 tail deletion mutant (C{Delta}206) transgenic mice: a model of delayed epidermal permeability barrier formation and repair. *Disease Models and Mechanisms, 3*, 167–180.
Farquhar, M. G., & Palade, G. E. (1963). Junctional complexes in various epithelia. *The Journal of Cell Biology, 17*, 375–409.
Feldmeyer, L., Huber, M., Fellmann, F., Beckmann, J. S., Frenk, E., & Hohl, D. (2006). Confirmation of the origin of NISCH syndrome. *Human Mutation, 27*, 408–410.
Findley, M. K., & Koval, M. (2009). Regulation and roles for claudin-family tight junction proteins. *IUBMB Life, 61*, 431–437.
Franke, W. W. (2009). Discovering the molecular components of intercellular junctions—a historical view. *Cold Spring Harbor Perspectives in Biology, 1*, a003061.
Franke, W. W., Rickelt, S., Barth, M., & Pieperhoff, S. (2009). The junctions that don't fit the scheme: Special symmetrical cell-cell junctions of their own kind. *Cell and Tissue Research, 338*, 1–17.
Fuchs, E. (1995). Keratins and the skin. *Annual Review of Cell Developmental Biology, 11*, 123–153.
Fuchs, E. (1998). Beauty is skin deep: The fascinating biology of the epidermis and its appendages. *Harvey Lectures, 94*, 47–77.
Fuchs, E. (2007). Scratching the surface of skin development. *Nature, 445*, 834–842.
Fuchs, E., & Byrne, C. (1994). The epidermis: Rising to the surface. *Current Opinion in Genetics and Development, 4*, 725–736.
Fuchs, E., & Horsley, V. (2008). More than one way to skin. *Genes and Development, 22*, 976–985.
Fujibe, M., Chiba, H., Kojima, T., Soma, T., Wada, T., Yamashita, T., et al. (2004). Thr203 of claudin-1, a putative phosphorylation site for MAP kinase, is required to promote the barrier function of tight junctions. *Experimental Cell Research, 295*, 36–47.
Furuse, M., Fujita, K., Hiiragi, T., Fujimoto, K., & Tsukita, S. (1998). Claudin-1 and -2: Novel integral membrane proteins localizing at tight junctions with no sequence similarity to occludin. *The Journal of Cell Biology, 141*, 1539–1550.
Furuse, M., Hata, M., Furuse, K., Yoshida, Y., Haratake, A., Sugitani, Y., et al. (2002). Claudin-based tight junctions are crucial for the mammalian epidermal barrier: A lesson from claudin-1-deficient mice. *The Journal of Cell Biology, 156*, 1099–1111.
Furuse, M., Hirase, T., Itoh, M., Nagafuchi, A., Yonemura, S., & Tsukita, S. (1993). Occludin: A novel integral membrane protein localizing at tight junctions. *The Journal of Cell Biology, 123*, 1777–1788.
Gu, L. H., & Coulombe, P. A. (2007). Keratin function in skin epithelia: A broadening palette with surprising shades. *Current Opinion in Cell Biology, 19*, 13–23.
Hardman, M. J., Moore, L., Ferguson, M. W., & Byrne, C. (1999). Barrier formation in the human fetus is patterned. *The Journal of Investigative Dermatology, 113*, 1106–1113.

Hardman, M. J., Sisi, P., Banbury, D. N., & Byrne, C. (1998). Patterned acquisition of skin barrier function during development. *Development, 125*, 1541–1552.

Harpin, V. A., & Rutter, N. (1983). Barrier properties of the newborn infant's skin. *The Journal of Pediatrics, 102*, 419–425.

Hashimoto, K. (1971). Intercellular spaces of the human epidermis as demonstrated with lanthanum. *The Journal of Investigative Dermatology, 57*, 17–31.

Heller, F., Fromm, A., Gitter, A. H., Mankertz, J., & Schulzke, J. D. (2008). Epithelial apoptosis is a prominent feature of the epithelial barrier disturbance in intestinal inflammation: Effect of pro-inflammatory interleukin-13 on epithelial cell function. *Mucosal Immunology, 1*(Suppl. 1), S58–S61.

Ikari, A., Ito, M., Okude, C., Sawada, H., Harada, H., Degawa, M., et al. (2008). Claudin-16 is directly phosphorylated by protein kinase A independently of a vasodilator-stimulated phosphoprotein-mediated pathway. *Journal of Cellular Physiology, 214*, 221–229.

Itoh, M., Furuse, M., Morita, K., Kubota, K., Saitou, M., & Tsukita, S. (1999). Direct binding of three tight junction-associated MAGUKs, ZO-1, ZO-2, and ZO-3, with the COOH termini of claudins. *The Journal of Cell Biology, 147*, 1351–1363.

Kalinin, A. E., Kajava, A. V., & Steinert, P. M. (2002). Epithelial barrier function: Assembly and structural features of the cornified cell envelope. *Bioessays, 24*, 789–800.

Kirschner, N., Poetzl, C., von den Driesch, P., Wladykowski, E., Moll, I., Behne, M. J., et al. (2009). Alteration of tight junction proteins is an early event in psoriasis: Putative involvement of proinflammatory cytokines. *American Journal of Pathology, 175*, 1095–1106.

Kopan, R., & Fuchs, E. (1989). A new look into an old problem: Keratins as tools to investigate determination, morphogenesis, and differentiation in skin. *Genes and Development, 3*, 1–15.

Krause, G., Winkler, L., Mueller, S. L., Haseloff, R. F., Piontek, J., & Blasig, I. E. (2008). Structure and function of claudins. *Biochimica et Biophysica Acta, 1778*, 631–645.

Langbein, L., Grund, C., Kuhn, C., Praetzel, S., Kartenbeck, J., Brandner, J. M., et al. (2002). Tight junctions and compositionally related junctional structures in mammalian stratified epithelia and cell cultures derived therefrom. *European Journal of Cell Biology, 81*, 419–435.

Langbein, L., Pape, U. F., Grund, C., Kuhn, C., Praetzel, S., Moll, I., et al. (2003). Tight junction-related structures in the absence of a lumen: Occludin, claudins and tight junction plaque proteins in densely packed cell formations of stratified epithelia and squamous cell carcinomas. *European Journal of Cell Biology, 82*, 385–400.

Mack, J. A., Anand, S., & Maytin, E. V. (2005). Proliferation and cornification during development of the mammalian epidermis. *Birth Defects Research. PartC, Embryo Today, 75*, 314–329.

Madison, K. C. (2003). Barrier function of the skin: "la raison d'etre" of the epidermis. *The Journal of Investigative Dermatology, 121*, 231–241.

Marshall, D., Hardman, M. J., & Byrne, C. (2000). SPRR1 gene induction and barrier formation occur as coordinated moving fronts in terminally differentiating epithelia. *The Journal of Investigative Dermatology, 114*, 967–975.

Marshall, D., Hardman, M. J., Nield, K. M., & Byrne, C. (2001). Differentially expressed late constituents of the epidermal cornified envelope. *Proceedings of the National Academy of Sciences of the United States of America, 98*, 13031–13036.

Morita, K., & Miyachi, Y. (2003). Tight junctions in the skin. *Journal of Dermatological Science, 31*, 81–89.

Morita, K., Tsukita, S., & Miyachi, Y. (2004). Tight junction-associated proteins (occludin, ZO-1, claudin-1, claudin-4) in squamous cell carcinoma and Bowen's disease. *British Journal of Dermatology, 151*, 328–334.

Nemes, Z., & Steinert, P. M. (1999). Bricks and mortar of the epidermal barrier. *Experimental and Molecular Medicine, 31*, 5–19.

Poliak, S., Matlis, S., Ullmer, C., Scherer, S. S., & Peles, E. (2002). Distinct claudins and associated PDZ proteins form different autotypic tight junctions in myelinating Schwann cells. *The Journal of Cell Biology, 159*, 361–372.

Presland, R. B., & Dale, B. A. (2000). Epithelial structural proteins of the skin and oral cavity: Function in health and disease. *Critical Reviews in Oral Biology and Medicine, 11*, 383–408.

Proksch, E., Brandner, J. M., & Jensen, J. M. (2008). The skin: An indispensable barrier. *Experimental Dermatology, 17*, 1063–1072.

Proksch, E., Folster-Holst, R., Brautigam, M., Sepehrmanesh, M., Pfeiffer, S., & Jensen, J. M. (2009). Role of the epidermal barrier in atopic dermatitis. *Journal der Deutschen Dermatologischen Gesellschaft, 7*, 899–910.

Roh, M. H., Makarova, O., Liu, C. J., Shin, K., Lee, S., Laurinec, S., et al. (2002). The Maguk protein, Pals1, functions as an adapter, linking mammalian homologues of Crumbs and Discs Lost. *The Journal of Cell Biology, 157*, 161–172.

Saitou, M., Fujimoto, K., Doi, Y., Itoh, M., Fujimoto, T., Furuse, M., et al. (1998). Occludin-deficient embryonic stem cells can differentiate into polarized epithelial cells bearing tight junctions. *The Journal of Cell Biology, 141*, 397–408.

Schluter, H., Moll, I., Wolburg, H., & Franke, W. W. (2007). The different structures containing tight junction proteins in epidermal and other stratified epithelial cells, including squamous cell metaplasia. *European Journal of Cell Biology, 86*, 645–655.

Schluter, H., Wepf, R., Moll, I., & Franke, W. W. (2004). Sealing the live part of the skin: The integrated meshwork of desmosomes, tight junctions and curvilinear ridge structures in the cells of the uppermost granular layer of the human epidermis. *European Journal of Cell Biology, 83*, 655–665.

Schulzke, J. D., Gitter, A. H., Mankertz, J., Spiegel, S., Seidler, U., Amasheh, S., et al. (2005). Epithelial transport and barrier function in occludin-deficient mice. *Biochimica et Biophysica Acta, 1669*, 34–42.

Segre, J. (2003). Complex redundancy to build a simple epidermal permeability barrier. *Current Opinion in Cell Biology, 15*, 776–782.

Segre, J. A. (2006). Epidermal barrier formation and recovery in skin disorders. *The Journal of Clinical Investigation, 116*, 1150–1158.

Simard, A., Di Pietro, E., Young, C. R., Plaza, S., & Ryan, A. K. (2006). Alterations in heart looping induced by overexpression of the tight junction protein Claudin-1 are dependent on its C-terminal cytoplasmic tail. *Mechanisms of Development, 123*, 210–227.

Steinert, P. M. (1995). A model for the hierarchical structure of the human epidermal cornified cell envelope. *Cell Death and Differentiation, 2*, 33–40.

Steinert, P. M., & Marekov, L. N. (1999). Initiation of assembly of the cell envelope barrier structure of stratified squamous epithelia. *Molecular Biology of the Cell, 10*, 4247–4261.

Stoler, A., Kopan, R., Duvic, M., & Fuchs, E. (1988). Use of monospecific antisera and cRNA probes to localize the major changes in keratin expression during normal and abnormal epidermal differentiation. *The Journal of Cell Biology, 107*, 427–446.

Tanaka, M., Kamata, R., & Sakai, R. (2005). EphA2 phosphorylates the cytoplasmic tail of Claudin-4 and mediates paracellular permeability. *The Journal of Biological Chemistry, 280*, 42375–42382.

Troy, T. C., & Turksen, K. (2007). The targeted overexpression of a Claudin mutant in the epidermis of transgenic mice elicits striking epidermal and hair follicle abnormalities. *Molecular Biotechnology, 36*, 166–174.

Troy, T. C., Arabzadeh, A., Lariviere, N. M., Enikanolaiye, A., & Turksen, K. (2009). Dermatitis and aging-related barrier dysfunction in transgenic mice overexpressing an epidermal-targeted claudin 6 tail deletion mutant. *PLoS ONE, 4*, e7814.

Troy, T. C., Li, Y., O'Malley, L., & Turksen, K. (2007). The temporal and spatial expression of Claudins in epidermal development and the accelerated program of epidermal differentiation in K14-CaSR transgenic mice. *Gene Expression Patterns, 7*, 423–430.

Troy, T. C., Rahbar, R., Arabzadeh, A., Cheung, R. M., & Turksen, K. (2005). Delayed epidermal permeability barrier formation and hair follicle aberrations in Inv-Cldn6 mice. *Mechanisms of Development, 122*, 805–819.

Turksen, K., & Troy, T. C. (1998). Epidermal cell lineage. *Biochemistry and Cell Biology, 76*, 889–898.

Turksen, K., & Troy, T. C. (2002). Permeability barrier dysfunction in transgenic mice overexpressing claudin 6. *Development, 129*, 1775–1784.

Turksen, K., & Troy, T. C. (2004). Barriers built on claudins. *Journal of Cell Science, 117*, 2435–2447.

Van Itallie, C. M., & Anderson, J. M. (2006). Claudins and epithelial paracellular transport. *Annual Review of Physiology, 68*, 403–429.

Van Itallie, C. M., Gambling, T. M., Carson, J. L., & Anderson, J. M. (2005). Palmitoylation of claudins is required for efficient tight-junction localization. *Journal of Cell Science, 118*, 1427–1436.

Watson, R. E., Poddar, R., Walker, J. M., McGuill, I., Hoare, L. M., Griffiths, C. E., et al. (2007). Altered claudin expression is a feature of chronic plaque psoriasis. *The Journal of Pathology, 212*, 450–458.

Weiss, L. W., & Zelickson, A. S. (1975a). Embryology of the epidermis: Ultrastructural aspects. II. Period of differentiation in the mouse with mammalian comparisons. *Acta Dermato-Venereologica, 55*, 321–329.

Weiss, L. W., & Zelickson, A. S. (1975b). Embryology of the epidermis: ultrastructural aspects. III. Maturation and primary appearance of dendritic cells in the mouse with mammalian comparisons. *Acta Dermato-Venereologica, 55*, 431–442.

Yamamoto, T., Kurasawa, M., Hattori, T., Maeda, T., Nakano, H., & Sasaki, H. (2008). Relationship between expression of tight junction-related molecules and perturbed epidermal barrier function in UVB-irradiated hairless mice. *Archives of Dermatological Research, 300*, 61–68.

CHAPTER 12

The Involvement of Tight Junction Protein Claudin-1 in Hepatitis C Virus Entry

Christopher Davis, Helen J. Harris, and Jane A. McKeating
Institute for Biomedical Research, University of Birmingham, Birmingham, United Kingdom

I. OVERVIEW

Viruses exploit normal cellular processes to accomplish every step in their life cycle and these processes can differ in diverse cell types. Virus entry into a host cell is defined by specific interaction(s) with cell surface proteins or receptors that confer host and cellular tropism. Recent advances in the development of in vitro systems to study Hepatitis C virus (HCV) replication have demonstrated a role for tetraspanin CD81, scavenger receptor B I (SR-BI) and the tight junction proteins Claudin-1 and Occludin in viral entry, suggesting a multi-step internalization process. SR-BI and CD81 bind HCV encoded glycoproteins,

Current Topics in Membranes, Volume 65

1063-5823/10 $35.00
DOI: 10.1016/S1063-5823(10)65012-0

suggesting a classical role for these molecules as receptors. In contrast, there is limited evidence for TJ protein association with HCV, which may reflect an indirect role for these proteins in virus internalization.

II. INTRODUCTION

Hepatitis C virus (HCV), the sole member of the *Hepacivirus* genus in the Flaviviridae, poses a global health burden with an estimated 170 million infected individuals. HCV is classified into seven genotypes which differ from each other in the order of 31–33% at the nucleotide level (Simmonds et al., 2005). The acute phase of infection is often subclinical and the majority of individuals develop persistent infection with progressive liver pathology, frequently culminating in fibrosis and hepatocellular carcinoma (Shepard, Finelli, & Alter, 2005). HCV infection is the leading indication for liver transplantation in many parts of the world (Brown, 2005). Infection is also associated with a variety of extrahepatic syndromes, including cryoglobulinemia, glomerulonephritis, and central nervous system abnormalities (Galossi, Guarisco, Bellis, & Puoti, 2007; Hoofnagle, 2002). Current estimates suggest that 40–74% of chronically infected individuals will develop at least one of these conditions.

The virus has a positive stranded RNA genome of around 9.4 kb in length, encapsulated within an envelope. The RNA can directly interact with the 40s ribosomal subunit via an internal ribosomal entry site at the 5′ end of the genome, allowing translation of the viral RNA into a single polyprotein and circumventing the usual translation processes of cellular messenger RNA (Brown, Zhang, Ping, & Lemon, 1992; Pestova, Shatsky, Fletcher, Jackson, & Hellen, 1998). The polyprotein is co- and posttranslationally cleaved by cellular and viral proteases into 10 individual proteins. There are three structural proteins: Core (C), which forms the viral capsid, and the two envelope glycoproteins, E1 and E2. Processing of the polyprotein releases six nonstructural proteins (NS2–NS5B) that are essential for replicating the viral RNA. The tenth protein, p7, belongs to a group of ion channel forming viral proteins called viroporins (Gonzalez & Carrasco, 2003); the isolated p7 protein has been shown to act very similarly to influenza-encoded M2, inducing the formation of ion channels. It is still debated whether p7 is a structural or nonstructural protein (Steinmann et al., 2007; StGelais et al., 2009).

At present, neither therapeutic nor preventive vaccines are available and the only current therapy is a combination of pegylated interferon α-2a and Ribavirin (Manns, Wedemeyer, & Cornberg, 2006). Unfortunately, the rates of sustained virological response are 46% for individuals infected with genotype 1, 76% for those infected with genotypes 2 or 3, and 77% for those infected with genotype 3. Given the low response rates, toxicity, and cost of treatment, new therapies are urgently required. One major target for antiviral therapy is the entry process; this has been successful for other infections, notably HIV (Rusconi, Scozzafava, Mastrolorenzo, & Supuran, 2007). Thus, there is a growing interest in defining the HCV entry process.

For reasons that have not fully been determined, it is extremely difficult to culture HCV *in vitro* directly from the peripheral blood or liver of infected patients (Farquhar & McKeating, 2008). To overcome this, a number of tools have been developed to study HCV entry that have proved invaluable. These include soluble forms of the HCV glycoproteins, retroviral pseudotypic particles bearing HCV glycoproteins, and, most recently, a strain of HCV that can replicate and assemble infectious particles in cultured cells (HCVcc).

III. SOLUBLE HCV GLYCOPROTEIN E2

One of the first tools developed to study the interaction of HCV envelope proteins and cellular receptors was a soluble form of the envelope glycoprotein E2 (Selby, Glazer, Masiarz, & Houghton, 1994). sE2 protein can be produced in large quantities and was used as a surrogate model to study virus–cell surface interactions. It must be noted that HCV E1E2 glycoproteins exist as a heterodimer on the particle surface and sE2 association with host cell proteins may not recapitulate the true interaction(s) and function of native particle-associated glycoproteins (Deleersnyder et al., 1997; Flint et al., 2000; Op De Beeck et al., 2004). However, two recent publications demonstrate that sE2 can inhibit HCV infection of hepatoma cells, suggesting that its interaction(s) with the cell surface can recapitulate some aspects of virus–receptor engagement (Krey et al., 2010; Whidby et al., 2009).

IV. HCV PSEUDOPARTICLES

To address the possible shortcomings of sE2 and to measure HCV glycoprotein-mediated entry events, we and others developed retroviral particles that incorporate HCV glycoproteins, termed HCV pseudoparticles (HCVpp) (Bartosch, Dubuisson, & Cosset, 2003; Drummer, Maerz, & Poumbourios, 2003; Hsu et al., 2003). Retrovirus particles bud from the plasma membrane and incorporate host cell or exogenous proteins into their envelope. HCVpp can be designed to express diverse glycoproteins (Lavillette et al., 2005; McKeating et al., 2004) and entry into target cells is dependent on expression of both E1E2 glycoproteins. HCVpp infect hepatocytes and hepatoma-derived cell lines, suggesting that virus glycoprotein–receptor interactions may in part define HCV tropism for the liver (Bartosch, Dubuisson, et al.; Hsu et al.). This system has allowed investigators to define the role of cellular proteins in the HCV internalization process (reviewed in Budkowska (2009)).

V. THE JFH-1 STRAIN OF HCV

One of the key milestones in HCV research was the development of a HCV strain with the ability to replicate, assemble, and release infectious particles from cultured cells (HCVcc) *in vitro* (Lindenbach et al., 2005; Wakita et al., 2005; Zhong et al., 2006). This strain was cloned from a patient with fulminant hepatitis and is termed Japanese Fulminant Hepatitis-1 (JFH-1). JFH-1 particles are infectious for chimpanzees as well as for UpA-SCID chimeric mice, which contain transplanted human hepatocytes (Wakita et al.; Zhong et al.). Virus recovered from JFH-1cc-infected animals can replicate *in vitro* (Lindenbach et al., 2006). Chimeric JFH-1 viruses expressing the structural proteins from diverse HCV genotypes enable experiments to study the role of glycoprotein diversity in virus entry (Gottwein et al., 2007; Gottwein et al., 2009; Lindenbach et al., 2006; Scheel et al., 2008). The JFH-1 strain of virus allows studies on the full life cycle from entry to egress, not only in hepatocyte-derived cell lines but in cultures of primary human hepatocytes (Molina et al., 2008; Ploss et al., 2010). Serum-derived HCV from an infected patient or experimentally infected animal shows a different density profile from HCVcc propagated *in vitro* (Lindenbach et al., 2006), reinforcing earlier reports that plasma-derived HCV associates with lipoproteins (Nielsen, Bassendine, Burt, Bevitt, & Toms, 2004; Thomssen, Bonk, & Thiele, 1993). Recent studies, demonstrating that infectious particle assembly and/or secretion are linked to the lipoprotein synthesis machinery of the hepatocyte, suggest an important role for particle-associated lipoproteins in the HCV life cycle (Popescu & Dubuisson, 2009). We recently reported that particle density associates with the sensitivity of virus to neutralization by glycoprotein-specific antibodies, with more dense particles with lower lipid content being more sensitive to neutralization (Grove et al., 2008).

VI. DISCOVERY OF RECEPTORS

All viruses must transverse the barrier of the cellular membrane and this is achieved by interaction with cell surface proteins or receptors that mediate viral binding and entry (Fig. 1). Virus–receptor interactions are highly specific and confer host and cellular tropism. HCV entry is a complex process involving a number of host factors, where initial adsorption of the virus may be facilitated by low affinity attachment factors like glycosaminoglycans (Barth et al., 2003),

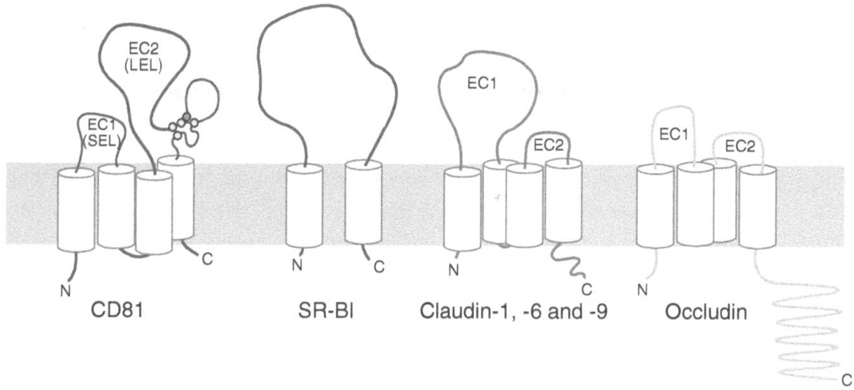

FIGURE 1 Hepatitis C virus entry factors. The diagram shows a schematic representation of the known entry factors for HCV. CD81, Claudin-1, -6, and -9, and Occludin share a similar overall topology of four transmembrane domains with two extracellular loop regions, denoted as EC1 and EC2. SR-BI has only two transmembrane domains and one extracellular loop.

lectins (Cormier, Durso, et al., 2004; Lozach et al., 2003; Pohlmann et al., 2003), and low-density lipoprotein (LDL) receptor (Agnello, Abel, Elfahal, Knight, & Zhang, 1999; Molina et al., 2007; Owen, Huang, Ye, & Gale, 2009). Four entry factors tetraspanin CD81 (Bartosch, Vitelli, et al., 2003; Cormier, Tsamis, et al., 2004; Flint et al., 2006; Hsu et al., 2003; Pileri et al., 1998), scavenger receptor class B type I (SR-BI) (Catanese et al., 2010; Grove et al., 2007; Scarselli et al., 2002), and tight junction proteins claudin-1 and occludin (Evans et al., 2007; Liu et al., 2009; Meertens et al., 2008; Ploss et al., 2009; Yang et al., 2008; Zheng et al., 2007) are essential for clathrin-dependent endocytosis of virus particles (Cai et al., 2007; Zhang et al., 2004) (Fig. 1). Ectopic expression of human CD81, SR-BI, Claudin-1, and Occludin render cell lines permissive for HCV entry, suggesting that additional host factors are broadly expressed and conserved amongst different mammalian species (Ploss et al.). Current evidence suggests that a species-specific interaction of HCV with these entry factors may restrict virus entry into nonhuman cells and may contribute to the narrow species tropism observed for HCV. Further information on the role of the individual proteins will be discussed below.

VII. CD81

CD81 is member of the tetraspanin superfamily of type III transmembrane proteins that contain four transmembrane domains and conserved signature amino acid residues, including a conserved CCG motif and 4–6 cysteine residues that form critical disulfide bonds within the second large extracellular loop (LEL or EC2). Tetraspanins associate with a complex array of tetraspanin and nontetraspanin proteins at cholesterol-enriched microdomains and exert a diverse array of biological functions including cell–cell adhesion, cell migration, and proliferation (Hemler, 2005). The first evidence for the involvement of CD81 in HCV entry was gained by screening a human cDNA library expressed in mouse fibroblasts to bind sE2 (Pileri et al., 1998). More recent studies have demonstrated the importance of the CD81–E2 interaction in the viral life cycle by inhibiting HCV infectivity with monoclonal antibodies specific for CD81 (Zhang et al., 2004). HCV sE2 interacts with a series of discontinuous amino acid residues in the LEL (Cormier, Tsamis, et al., 2004; Drummer, Wilson, & Poumbourios, 2005; Flint et al., 2006; Pileri et al.). siRNA silencing of CD81 expression in hepatoma cells prevented HCVpp and HCVcc entry while the introduction of CD81 into HepG2 cells, which are a CD81 negative hepatocyte cell line, them permissive for HCVpp and HCVcc infection, demonstrating the essential role of CD81 in viral entry (Lavillette et al., 2005; Lindenbach et al., 2005; Zhang et al.). A soluble form of CD81-LEL binds and neutralizes HCVpp and HCVcc infectivity (McKeating et al., 2004; Zhang et al.). Kitadokoro et al. (2001) reported on the dimeric structure of CD81-LEL, suggesting that motifs within this domain drive CD81 dimerization. Disruption of dimer formation, via a series of specific mutations within CD81-LEL, associated with reduced sE2 binding (Drummer et al.). Consistent with these observations, dimeric forms of CD81-LEL are 10-fold more efficient at binding sE2 compared to monomeric CD81-LEL (Nakajima, Cocquerel, Kiyokawa, Fujimoto, & Levy, 2005). The kinetics of the interaction(s) between E2 and E1E2 with CD81-LEL were studied by surface plasma resonance, demonstrating a lower dissociation rate for E1E2–CD81 than E2–CD81 binding. Although CD81 is a highly specific receptor for HCV E1E2, its ubiquitous expression (Hemler, 2005) led investigators to look for additional entry factors.

VIII. SCAVENGER RECEPTOR CLASS B TYPE I

The initial evidence for a role of SR-BI in HCV entry came from the observation that HepG2 cells, which do not express CD81, bound HCV sE2, this was shown to be a specific interaction with SR-BI (Scarselli et al., 2002), a scavenger protein involved with lipoprotein metabolism. Lipoproteins comprise a cholesterol ester core surrounded by lipids and apoproteins. These structures transport triglycerides and cholesterol around the body and are categorized as high-density (HDL), low-density (LDL), or very-low-density lipoproteins (VLDL) (Krieger, 1999). The major role of SR-BI is the selective uptake of cholesterol from HDL particles bound to SR-BI via the apoprotein ApoAI, resulting in the selective transfer of cholesterol to the membrane or endocytosis of the HDL particle (Fidge, 1999). SR-BI has been shown to interact with a variety of ligands including VLDL, LDL, HDL, and chemically modified forms, such as oxidized (oxLDL) and acetylated LDL (Krieger, 1997). HCV infection can be inhibited by oxLDL and serum amyloid-α (Cai et al., 2007; Lavie et al., 2006; von Hahn et al., 2006). In contrast, HDL has been shown to enhance HCV entry and reduce the efficacy of neutralizing antibodies (Voisset et al., 2005). This enhancement can be prevented by blocking the movement of cholesterol from bound HDL into the cell, suggesting that the cholesterol content of the membrane plays an important role in HCV entry. SR-BI has been shown to interact with the hypervariable region of E2, as deletion of this region ablates sE2-SR-BI interaction(s) (Scarselli et al.). Silencing of SR-BI expression or treatment of cells with antibodies specific to SR-BI reduced HCV entry (Dreux et al., 2009; Grove et al., 2007; Lavillette et al., 2005; von Hahn et al.), whereas overexpression of SR-BI enhanced virus entry implying that SR-BI expression in Huh7 cells is limiting (Grove et al., 2008). Taken together these facts demonstrate an important role for SR-BI in HCV viral entry.

IX. CLAUDINS

Since many cell lines express CD81 and SR-BI and are nonpermissive for HCVpp entry (Bartosch, Dubuisson, et al., 2003; Hsu et al., 2003), other factors were predicted to be required for viral entry. A cDNA library derived from a permissive hepatocellular carcinoma cell line was cloned into nonpermissive 293 T cells, which express both CD81 and SR-BI. Permissive clones were identified to express Claudin-1 (Evans et al., 2007). Further work by Zheng et al. demonstrated that HCV could infect the Claudin-1 negative cell line Bel7402, due to the expression of Claudin-9, where silencing Claudin-9 ablated HCVpp entry. A screen of several Claudin family members showed that Claudin-9 and the closely related Claudin-6, which are relatively poorly expressed in the liver (Mee et al., 2008) could mediate HCVpp entry into 293-T cells, whereas none of the other Claudin family members showed receptor activity (Zheng et al., 2007).

The genesis of chimeric proteins between receptor active and inactive Claudin-1 and -7 molecules identified Claudin-1 EC1 as the minimal requirement for HCV entry into 293 T cells (Evans et al., 2007). Exchanging residues in Claudin-1 and Claudin-7 EC1 at the two positions where they differ (codons 32 and 48) resulted in a loss and a gain of receptor activity, respectively (Fig. 2) (Evans et al.). These positions were also identified by Liu et al. (2009) who found further involvement of residues

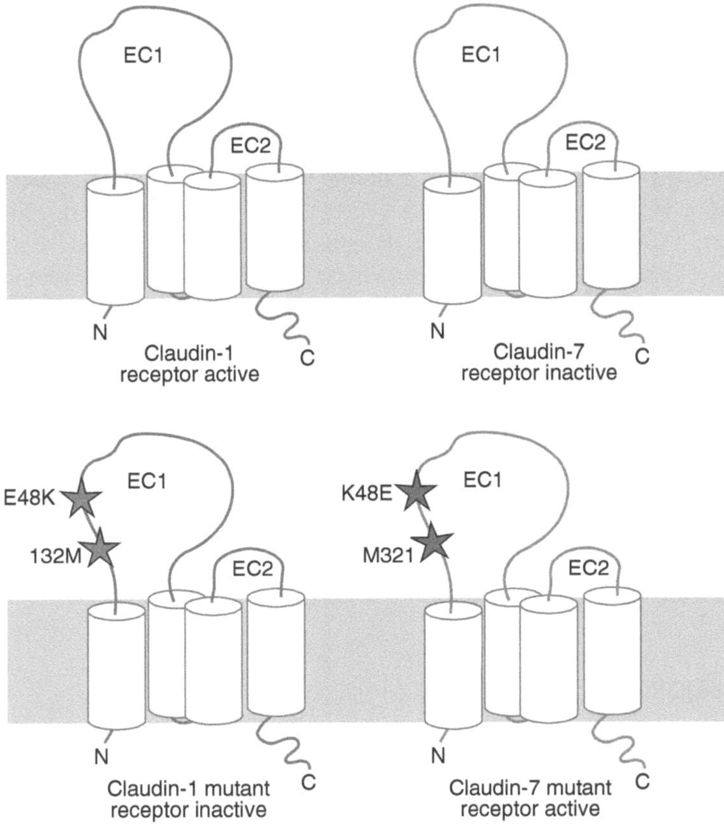

FIGURE 2 Schematic representation of receptor active and inactive Claudin proteins. The diagrams show a representation of the Claudin-1 and Claudin-7 constructs created by Evans et al., with their receptor activities detailed. The upper row shows wild type Claudin-1 and -7 and the lower row their mutations, with amino acid changes and positions highlighted.

F148 and R158 in Claudin-1 EC2 loop (Liu et al.). A more extensive analysis by Cukierman et al. demonstrated that mutation of a conserved EC1 motif among all Claudins, W30-GLW51-C54-C64, as well as mutation of codon D38 ablated receptor activity (Cukierman, Meertens, Bertaux, Kajumo, & Dragic, 2009). Mutational analysis of Claudin-9 demonstrated that residues S38 and V45 are essential for receptor activity (Zheng et al., 2007). Although no clear mechanism for the receptor inactivity was shown in these publications, the authors suggest that cis-interaction(s) between Claudin molecules on opposing cells may be important for HCV entry, with receptor inactive Claudin-1 mutants showing reduced localization at cell–cell contacts (Liu et al.).

To date, a direct interaction between the HCV particle and Claudin-1, -6, or -9 has not been demonstrated, suggesting an indirect role for Claudins in the HCV internalization process. We and others have reported that Claudin-1 associates with CD81 (Cukierman et al., 2009; Harris et al., 2008; Kovalenko, Yang, & Hemler, 2007), suggesting a role for the coreceptor complex in virus entry. Recent data from our laboratory demonstrates a relationship between receptor active Claudins and their association and organization with CD81 at the plasma membrane by fluorescent resonance energy transfer (FRET) and imaging methodologies. Mutation of residues 32 and 48 in Claudin-1 EC1 ablates CD81 association and HCV receptor activity (Fig. 2). Furthermore, mutation of the same residues in the receptor inactive Claudin-7 molecule enabled CD81 complex formation and virus entry (Harris et al., 2010). Modulation of Claudin-1–CD81 complexes by cholesterol depletion or ligation with specific antibodies inhibits HCV entry, demonstrating an essential role for Claudin–CD81 complexes in HCV infection (Krieger et al., 2010).

X. OCCLUDIN

Ectopic expression of CD81, SR-BI, and Claudin-1 in nonhuman cells failed to confer HCV entry. This led Ploss et al. (2009) to repeat the cDNA library screening process in the nonpermissive mouse cell line (NIH3T3) engineered to express human CD81, SR-BI, and Claudin-1. This led to the identification of Occludin as a fourth entry factor. The authors demonstrated that expression of Occludin in the nonpermissive renal carcinoma cell line, 786-O, or in TZM cells, a HeLa cell derivative, allowed viral entry, and siRNA-mediated silencing of Occludin expression in Huh-7.5 hepatoma cells significantly reduced infection of both HCVcc and HCVpp. Finally, the authors demonstrate that rodent SR-BI and Claudin-1 proteins could confer HCV entry since NIH3T3 or CHO cells expressing human CD81 and Occludin were permissive for HCVpp entry (Ploss et al.).

Using an independent approach, Liu et al. (2009) demonstrated a role for Occludin in HCV entry by siRNA silencing proteins known to associate with Claudin-1, including Occludin, ZO-1, JAM-A, and Coxsackie and Adenovirus receptor (CAR). The authors demonstrated that reduction of Claudin-1, Occludin, or ZO-1 expression inhibited HCVpp entry. Silencing Claudin-1 or Occludin did not alter CD81 or SR-BI expression, however, silencing ZO-1 led to a significantly reduced Claudin-1 expression. Chimeric proteins expressing extracellular loops of Claudin-1 or Occludin demonstrated that proteins expressing Claudin-1 EC1 allowed HCV entry. This result is consistent with previous chimeric Claudin-1/7 data from Evans et al.

(2007) and the observation that the C-terminus of Claudin-1 is dispensable for receptor activity. Similar to Claudin-1, there is minimal data for an interaction of HCV particles with Occludin (Benedicto et al., 2008).

Occludin was first identified by Furuse et al. (1993) as an integral membrane protein located at tight junctions. Structurally it is similar to the Claudins, with four transmembrane domains and two cytoplasmic regions at the N- and C-terminal, and is involved in the maintenance of tight junctions. Studies on knockout mice demonstrated that tight junctions could still form in the absence of Occludin with no gross phenotypic abnormalities (Saitou et al., 2000), suggesting that Occludin is not essential for tight junction integrity. The C-terminal region of Occludin interacts with the guanylate kinase region of ZO-1 (Feldman, Mullin, & Ryan, 2005; Haskins, Gu, Wittchen, Hibbard, & Stevenson, 1998; Paris, Tonutti, Vannini, & Bazzoni, 2008). As part of the regulation of tight junctions, Occludin has been shown to co-internalize with Claudin-1 in a Clathrin-mediated manner (Matsuda, Kubo, Furuse, & Tsukita, 2004; Shen, Weber, & Turner, 2008), as well as being induced by stress to internalize via caveolae-mediated processes (Stamatovic, Keep, Wang, Jankovic, & Andjelkovic, 2009).

XI. TIGHT JUNCTION PROTEINS AND VIRUS ENTRY

Tight junctions are responsible for the regulation of movement of solutes, ions, and water across endothelia or epithelia (Fig. 4). They also form a barrier preventing the access of pathogens to the underlying tissue (Guttman & Finlay, 2009; Krause et al., 2009; Van Itallie, Betts, Smedley, McClane, & Anderson, 2008). Up to 30 proteins have been identified with distinct functions in maintaining barrier function of tight junctions (Shen et al., 2008). Several viruses have evolved to exploit tight junction proteins to enter and infect target cells, examples include adenovirus, coxsackievirus-B (CVB), reovirus, rotavirus, and HCV (Guttman & Finlay).

CVB entry is facilitated by decay-accelerating factor (DAF), CAR, and Occludin (Coyne & Bergelson, 2006; Coyne, Shen, Turner, & Bergelson, 2007). In polarized Caco-2 cells, CVB engagement of apically localized DAF activates the tyrosine kinase, AbI, which in turn induces Rac GTPase, leading to a relocalization of the virus–DAF complex to the tight junction. Once at the tight junction the virus can interact with CAR, which induces a conformational change in the virus that primes particle internalization via a caveola-dependent manner (Coyne & Bergelson; Coyne et al.). Treatment of cells with an siRNA specific for Occludin reduced CVB entry and trapped virus particles at tight junctions. Thus, Occludin is believed to play a role in

CVB internalization at the tight junction. In contrast, CVB entry into non-polarized HeLa cells is caveolin-independent and seems to involve a different series of CVB–cell interactions (Patel, Coyne, & Bergelson, 2009).

Parallels between CVB and HCV entry are evident; both involve particle interaction with attachment receptors that may precede virus engagement of tight junction proteins. HCV entry is dependent on protein kinase A (PKA) activation that is known to activate Ras as part of the cAMP-PKA pathway (Farquhar et al., 2008). Ras is involved in proliferation, differentiation, and more specifically cytoskeletal reorganization (Chiaradonna, Balestrieri,

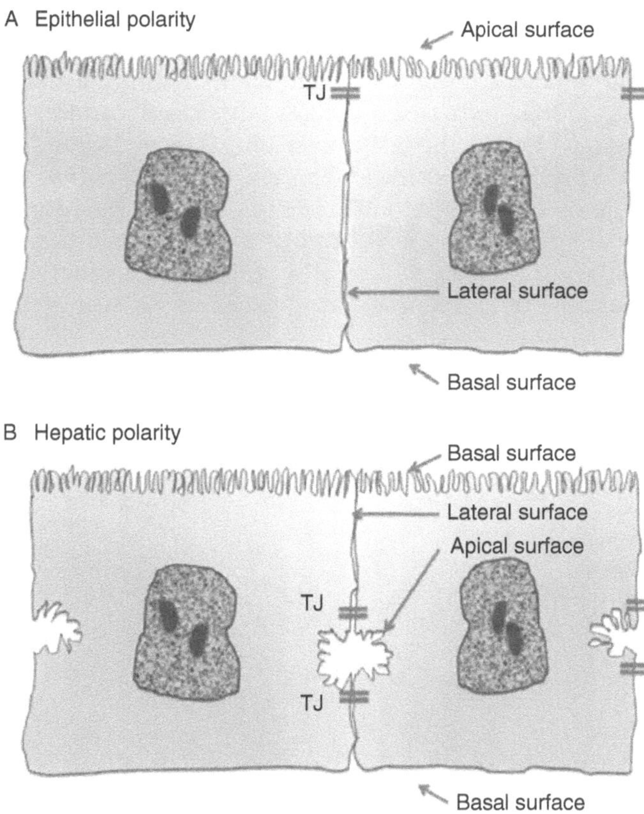

FIGURE 3 Schematic representation of polarized epithelial and hepatic cells. Two types of polarization are shown. (A) Simple epithelial polarity: Epithelial cells form a polarized continuous apical surface providing a barrier between the extracellular environment and the deeper tissue of the body. The tight junctions between neighboring cells are in close proximity to the apical surface. (B) Hepatic polarity: Hepatocytes form an apical surface between cells which generates a tubular structure, the bile canaliculus. The tight junctions seal this tube and isolate its contents.

Gaglio, & Vanoni, 2008). Although DAF has been shown to move with CVB to the tight junction of polarized cells, the same has not been shown for HCV and CD81 or SR-BI. Although similarities exist between the entry processes for these two viruses, more work is required to ascertain whether HCV enters polarized cells in a CVB-like manner or by another, distinct, mechanism.

XII. EFFECTS OF CELL POLARIZATION ON HCV ENTRY

Viruses that utilize tight junction proteins as part of their entry strategy mostly infect epithelial surfaces which show a simple planar polarity. These cells present apical and basal surfaces with their tight junctions located very close to the apical surface (Furuse & Tsukita, 2006; Guttman & Finlay, 2009). In contrast hepatocytes, the target of HCV, have a more complex polarity with at least two basal surfaces, the apical surface forming an enclosed channel or tube, the bile canaliculi (Decaens, Durand, Grosse, & Cassio, 2008) (Fig. 3). We recently reported that polarization of HepG2–CD81 cells limits HCV infection. Promoting HepG2 polarity with PKA inhibitors or human oncostatin M led to a significant reduction in HCVpp entry. Similarly, reducing HepG2 polarity with vascular endothelial growth factor or phorbol ester, a PKA activator, increased HCV entry (Mee et al.,

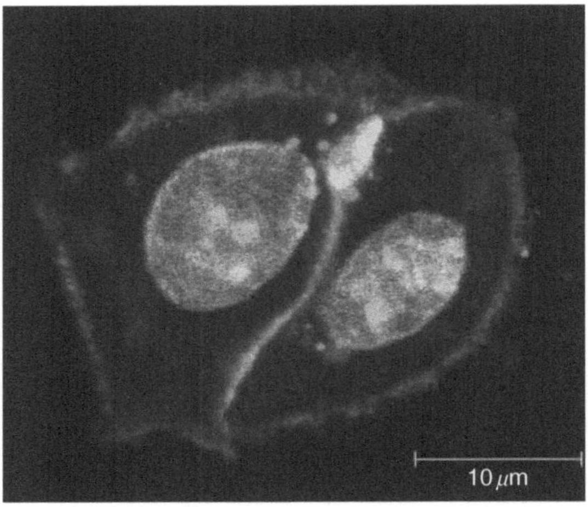

FIGURE 4 Confocal image of polarized HepG2 cells expressing fluorophore-labeled Claudin-1. HepG2 cells engineered to express AcGFP-labeled Claudin-1 (g.CLDN1) were allowed to polarize, fixed, and stained for ZO-1 localization (red), which identifies the tight junction. The g.CLDN1 colocalizes with ZO-1 at the tight junction (yellow color) and is apparent at the basal and lateral plasma membrane surfaces. (See Color Insert.)

2009, 2010). One noticeable effect of increasing HepG2 polarity was a clear redistribution of Claudin-1 to apically located tight junctions, with a concomitant reduction in basolateral pools of Claudin-1 (Mee et al., 2009).

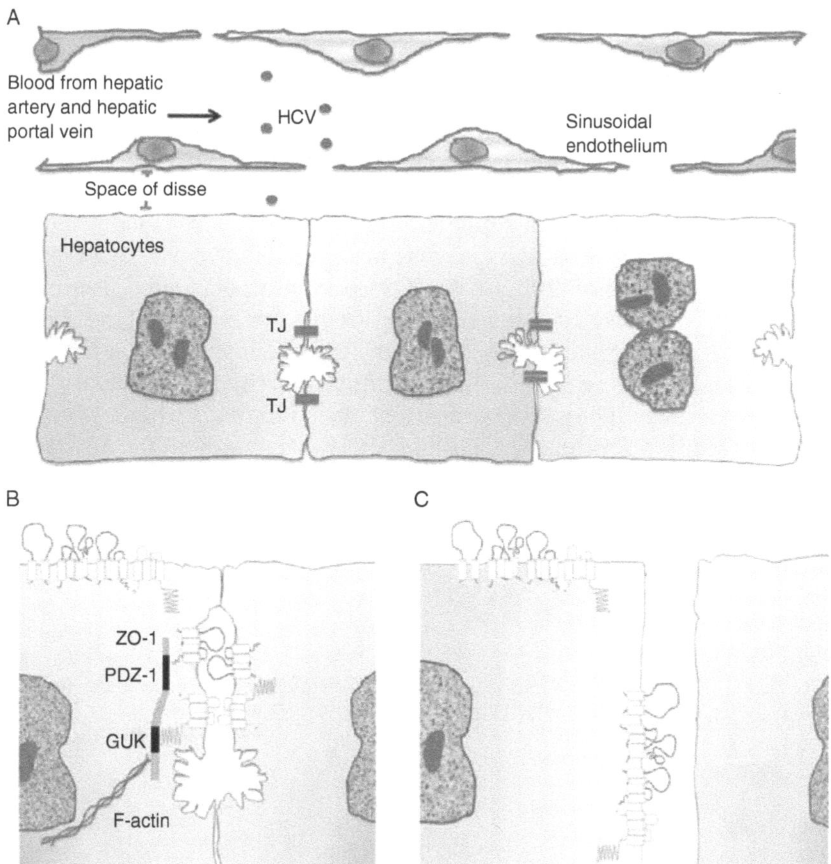

FIGURE 5 Schematic representation of the hepatic environment. HCV enters the liver in the blood carried by the hepatic artery. (A) The fenestrated sinusoidal endothelial layer which the virus has to traverse to gain access to its primary target cell, the hepatocyte. (B) In the case of polarized hepatocytes, although all receptors are detectable at the basal surface, the receptor proteins associated with the tight junctions may be inaccessible due to the closing of the space between neighboring cells. (C) Depolarization of the hepatocytes, due to an underlying condition or to the actions of VEGF, PKC inhibition, etc., alters the presentation of the putative entry factors. In this state all of the receptor proteins will be available to allow viral engagement. (See Color Insert.)

Staining of human liver tissue demonstrates that Claudin-1 is predominantly detected at tight junctions; however, basal and lateral hepatocyte surfaces stain for Claudin-1 (Reynolds et al., 2008). Expression of fluorophore-labeled Claudin-1 in HepG2 cells demonstrates discrete pools of Claudin-1 at the tight junction and basal/lateral surfaces (Fig. 4) (Mee et al., 2009). To investigate the presence and location of Claudin-1–CD81 receptor complexes in polarized HepG2 cells, we transduced them to stably express AcGFP- and DsRED-tagged versions of Claudin-1 and CD81. Claudin-1 was found to associate with basolateral pools of CD81, whereas, CD81 was largely excluded from the tight junction and exhibited minimal association with Claudin-1 at this location.

Since HCV enters the liver via the sinusoidal blood, the virus will encounter receptors expressed on the sinusoidal or basal surface of the hepatocyte (Fig. 5). The location of Claudin-1–CD81 complexes at the basolateral surface of polarized hepatoma cells supports a model where virus engagement of Claudin-1–CD81 at the basal membrane may initiate the particle internalization process. The role of occludin in HCV entry into polarized cells is poorly defined and is hindered by a lack of antibodies to extracellular-expressed epitopes. HCV internalizes via a clathrin-dependent process and fusion is believed to occur within the early endosomes (Blanchard et al., 2006; Tscherne et al., 2006). At present it is unknown whether any of the viral receptors, including CD81 and Claudin-1, are endocytosed with HCV and further research on the trafficking and endocytic routing of receptor complexes and virus particles (Coller et al., 2009) in polarized hepatocytes is required to fully appreciate the complex entry process of HCV in the liver.

Acknowledgments

We thank our lab colleagues and collaborators for helpful discussions over the years and acknowledge our funding from NIAID, MRC, and The Wellcome Trust.

References

Agnello, V., Abel, G., Elfahal, M., Knight, G. B., & Zhang, Q. X. (1999). Hepatitis C virus and other flaviviridae viruses enter cells via low density lipoprotein receptor. *Proceedings of the National Academy of Sciences of the United States of America, 96*, 12766–12771.

Barth, H., Schafer, C., Adah, M. I., Zhang, F., Linhardt, R. J., Toyoda, H., et al. (2003). Cellular binding of hepatitis C virus envelope glycoprotein E2 requires cell surface heparan sulfate. *The Journal of Biological Chemistry, 278*, 41003–41012.

Bartosch, B., Dubuisson, J., & Cosset, F. L. (2003). Infectious hepatitis C virus pseudo-particles containing functional E1-E2 envelope protein complexes. *The Journal of Experimental Medicine, 197*, 633–642.

Bartosch, B., Vitelli, A., Granier, C., Goujon, C., Dubuisson, J., et al. (2003). Cell entry of hepatitis C virus requires a set of co-receptors that include the CD81 tetraspanin and the SR-B1 scavenger receptor. *The Journal of Biological Chemistry, 278*, 41624–41630.

Benedicto, I., Molina-Jimenez, F., Barreiro, O., Maldonado-Rodriguez, A., Prieto, J., et al. (2008). Hepatitis C virus envelope components alter localization of hepatocyte tight junction-associated proteins and promote occludin retention in the endoplasmic reticulum. *Hepatology, 48*, 1044–1053.

Blanchard, E., Belouzard, S., Goueslain, L., Wakita, T., Dubuisson, J., et al. (2006). Hepatitis C virus entry depends on clathrin-mediated endocytosis. *Journal of Virology, 80*, 6964–6972.

Brown, R. S. (2005). Hepatitis C and liver transplantation. *Nature, 436*, 973–978.

Brown, E. A., Zhang, H., Ping, L. H., & Lemon, S. M. (1992). Secondary structure of the 5′ nontranslated regions of hepatitis C virus and pestivirus genomic RNAs. *Nucleic Acids Research, 20*, 5041–5045.

Budkowska, A. (2009). Mechanism of cell infection with hepatitis C virus (HCV)—a new paradigm in virus-cell interaction. *Polish Journal of Microbiology, 58*, 93–98.

Cai, Z., Cai, L., Jiang, J., Chang, K. S., van der Westhuyzen, D. R., et al. (2007). Human serum amyloid A protein inhibits hepatitis C virus entry into cells. *Journal of Virology, 81*, 6128–6133.

Catanese, M. T., Ansuini, H., Graziani, R., Huby, T., Moreau, M., et al. (2010). Role of scavenger receptor class B type I in hepatitis C virus entry: Kinetics and molecular determinants. *Journal of Virology, 84*, 34–43.

Chiaradonna, F., Balestrieri, C., Gaglio, D., & Vanoni, M. (2008). RAS and PKA pathways in cancer: New insight from transcriptional analysis. *Frontiers in Bioscience, 13*, 5257–5278.

Coller, K. E., Berger, K. L., Heaton, N. S., Cooper, J. D., Yoon, R., et al. (2009). RNA interference and single particle tracking analysis of hepatitis C virus endocytosis. *PLoS Pathogens, 5*, e1000702.

Cormier, E. G., Durso, R. J., Tsamis, F., Boussemart, L., Manix, C., et al. (2004). L-SIGN (CD209L) and DC-SIGN (CD209) mediate transinfection of liver cells by hepatitis C virus. *Proceedings of the National Academy of Sciences of the United States of America, 101*, 14067–14072.

Cormier, E. G., Tsamis, F., Kajumo, F., Durso, R. J., Gardner, J. P., et al. (2004). CD81 is an entry coreceptor for hepatitis C virus. *Proceedings of the National Academy of Sciences of the United States of America, 101*, 7270–7274.

Coyne, C. B., & Bergelson, J. M. (2006). Virus-induced Abl and Fyn kinase signals permit coxsackievirus entry through epithelial tight junctions. *Cell, 124*, 119–131.

Coyne, C. B., Shen, L., Turner, J. R., & Bergelson, J. M. (2007). Coxsackievirus entry across epithelial tight junctions requires occludin and the small GTPases Rab34 and Rab5. *Cell Host & Microbe, 2*, 181–192.

Cukierman, L., Meertens, L., Bertaux, C., Kajumo, F., & Dragic, T. (2009). Residues in a highly conserved Claudin-1 motif are required for hepatitis C virus entry and mediate the formation of cell-cell contacts. *Journal of Virology, 83*, 5477–5484.

Decaens, C., Durand, M., Grosse, B., & Cassio, D. (2008). Which in vitro models could be best used to study hepatocyte polarity? *Biology of the Cell, 100*, 387–398.

Deleersnyder, V., Pillez, A., Wychowski, C., Blight, K., Xu, J., et al. (1997). Formation of native hepatitis C virus glycoprotein complexes. *Journal of Virology, 71*, 697–704.

Dreux, M., Dao Thi, V. L., Fresquet, J., Guerin, M., Julia, Z., et al. (2009). Receptor complementation and mutagenesis reveal SR-BI as an essential HCV entry factor and functionally imply its intra- and extra-cellular domains. *PLoS Pathogens, 5*, e1000310.

Drummer, H. E., Maerz, A., & Poumbourios, P. (2003). Cell surface expression of functional hepatitis C virus E1 and E2 glycoproteins. *FEBS Letters, 546*, 385–390.

Drummer, H. E., Wilson, K. A., & Poumbourios, P. (2005). Determinants of CD81 dimerization and interaction with hepatitis C virus glycoprotein E2. *Biochemical and Biophysical Research Communications, 328*, 251–257.

Evans, M. J., von Hahn, T., Tscherne, D. M., Syder, A. J., Panis, M., et al. (2007). Claudin-1 is a hepatitis C virus co-receptor required for a late step in entry. *Nature, 446*, 801–805.

Farquhar, M. J., & McKeating, J. A. (2008). Primary hepatocytes as targets for hepatitis C virus replication. *Journal of Viral Hepatitis, 15*, 849–854.

Farquhar, M. J., Harris, H. J., Diskar, M., Jones, S., Mee, C. J., et al. (2008). Protein kinase A-dependent step(s) in hepatitis C virus entry and infectivity. *Journal of Virology, 82*, 8797–8811.

Feldman, G. J., Mullin, J. M., & Ryan, M. P. (2005). Occludin: Structure, function and regulation. *Advanced Drug Delivery Reviews, 57*, 883–917.

Fidge, N. H. (1999). High density lipoprotein receptors, binding proteins, and ligands. *Journal of Lipid Research, 40*, 187–201.

Flint, M., Dubuisson, J., Maidens, C., Harrop, R., Guile, G. R., et al. (2000). Functional characterization of intracellular and secreted forms of a truncated hepatitis C virus E2 glycoprotein. *Journal of Virology, 74*, 702–709.

Flint, M., von Hahn, T., Zhang, J., Farquhar, M., Jones, C. T., et al. (2006). Diverse CD81 proteins support hepatitis C virus infection. *Journal of Virology, 80*, 11331–11342.

Furuse, M., & Tsukita, S. (2006). Claudins in occluding junctions of humans and flies. *Trends in Cell Biology, 16*, 181–188.

Furuse, M., Hirase, T., Itoh, M., Nagafuchi, A., Yonemura, S., et al. (1993). Occludin: A novel integral membrane protein localizing at tight junctions. *Journal of Cell Biology, 123*, 1777–1788.

Galossi, A., Guarisco, R., Bellis, L., & Puoti, C. (2007). Extrahepatic manifestations of chronic HCV infection. *Journal of Gastrointestinal and Liver Diseases, 16*, 65–73.

Gonzalez, M. E., & Carrasco, L. (2003). Viroporins. *FEBS Letters, 552*, 28–34.

Gottwein, J. M., Scheel, T. K., Hoegh, A. M., Lademann, J. B., Eugen-Olsen, J., et al. (2007). Robust hepatitis C genotype 3a cell culture releasing adapted intergenotypic 3a/2a (S52/JFH1) viruses. *Gastroenterology, 133*, 1614–1626.

Gottwein, J. M., Scheel, T. K., Jensen, T. B., Lademann, J. B., Prentoe, J. C., et al. (2009). Development and characterization of hepatitis C virus genotype 1-7 cell culture systems: Role of CD81 and scavenger receptor class B type I and effect of antiviral drugs. *Hepatology, 49*, 364–377.

Grove, J., Huby, T., Stamataki, Z., Vanwolleghem, T., Meuleman, P., et al. (2007). Scavenger receptor BI and BII expression levels modulate hepatitis C virus infectivity. *Journal of Virology, 81*, 3162–3169.

Grove, J., Nielsen, S., Zhong, J., Bassendine, M. F., Drummer, H. E., et al. (2008). Identification of a residue in hepatitis C virus E2 glycoprotein that determines scavenger receptor BI and CD81 receptor dependency and sensitivity to neutralizing antibodies. *Journal of Virology, 82*, 12020–12029.

Guttman, J. A., & Finlay, B. B. (2009). Tight junctions as targets of infectious agents. *Biochimica et Biophysica Acta, 1788*, 832–841.

Harris, H. J., Davis, C., Mullins, J. G. L., Hu, K., Goodall, M., et al. (2010). Claudin association with CD81 defines hepatitis C virus entry. Journal of Biological Chemistry, in press.

Harris, H. J., Farquhar, M. J., Mee, C. J., Davis, C., Reynolds, G. M., et al. (2008). CD81 and claudin 1 coreceptor association: Role in hepatitis C virus entry. *Journal of Virology, 82*, 5007–5020.

Haskins, J., Gu, L., Wittchen, E. S., Hibbard, J., & Stevenson, B. R. (1998). ZO-3, a novel member of the MAGUK protein family found at the tight junction, interacts with ZO-1 and occludin. *Journal of Cell Biology, 141*, 199–208.

Hemler, M. E. (2005). Tetraspanin functions and associated microdomains. *Nature Reviews. Molecular Cell Biology, 6*, 801–811.

Hoofnagle, J. H. (2002). Course and outcome of hepatitis C. *Hepatology, 36*, S21–S29.

Hsu, M., Zhang, J., Flint, M., Logvinoff, C., Cheng-Mayer, C., et al. (2003). Hepatitis C virus glycoproteins mediate pH-dependent cell entry of pseudotyped retroviral particles. *Proceedings of the National Academy of Sciences of the United States of America, 100*, 7271–7276.

Kitadokoro, K., Bordo, D., Galli, G., Petracca, R., Falugi, F., et al. (2001). CD81 extracellular domain 3D structure: Insight into the tetraspanin superfamily structural motifs. *EMBO Journal, 20*, 12–18.

Kovalenko, O. V., Yang, X. H., & Hemler, M. E. (2007). A novel cysteine cross-linking method reveals a direct association between claudin-1 and tetraspanin CD9. *Molecular and Cellular Proteomics, 6*, 1855–1867.

Krause, G., Winkler, L., Piehl, C., Blasig, I., Piontek, J., et al. (2009). Structure and function of extracellular claudin domains. *Annals of the New York Academy of Sciences, 1165*, 34–43.

Krey, T., d'Alayer, J., Kikuti, C. M., Saulnier, A., Damier-Piolle, L., et al. (2010). The disulfide bonds in glycoprotein e2 of hepatitis C virus reveal the tertiary organization of the molecule. *PLoS Pathogens, 6*, e1000762.

Krieger, M. (1997). The other side of scavenger receptors: Pattern recognition for host defense. *Current Opinion in Lipidology, 8*, 275–280.

Krieger, M. (1999). Charting the fate of the "good cholesterol": Identification and characterization of the high-density lipoprotein receptor SR-BI. *Annual Review of Biochemistry, 68*, 523–558.

Krieger, S. E., Zeisel, M. B., Davis, C., Thumann, C., Harris, H. J., et al. (2010). Inhibition of hepatitis C virus infection by anti-claudin-1 antibodies is mediated by neutralization of E2-CD81-Claudin-1 associations. *Hepatology, 51*(4), 1144–1157.

Lavie, M., Voisset, C., Vu-Dac, N., Zurawski, V., Duverlie, G., et al. (2006). Serum amyloid A has antiviral activity against hepatitis C virus by inhibiting virus entry in a cell culture system. *Hepatology, 44*, 1626–1634.

Lavillette, D., Tarr, A. W., Voisset, C., Donot, P., Bartosch, B., et al. (2005). Characterization of host-range and cell entry properties of the major genotypes and subtypes of hepatitis C virus. *Hepatology, 41*, 265–274.

Lindenbach, B. D., Evans, M. J., Syder, A. J., Wolk, B., Tellinghuisen, T. L., et al. (2005). Complete replication of hepatitis C virus in cell culture. *Science, 309*, 623–626.

Lindenbach, B. D., Meuleman, P., Ploss, A., Vanwolleghem, T., Syder, A. J., et al. (2006). Cell culture-grown hepatitis C virus is infectious in vivo and can be recultured in vitro. *Proceedings of the National Academy of Sciences of the United States of America, 103*, 3805–3809.

Liu, S., Yang, W., Shen, L., Turner, J. R., Coyne, C. B., et al. (2009). Tight junction proteins claudin-1 and occludin control hepatitis C virus entry and are downregulated during infection to prevent superinfection. *Journal of Virology, 83*, 2011–2014.

Lozach, P. Y., Lortat-Jacob, H., de Lacroix de Lavalette, A., Staropoli, I., Foung, S., et al. (2003). DC-SIGN and L-SIGN are high affinity binding receptors for hepatitis C virus glycoprotein E2. *The Journal of Biological Chemistry, 278*, 20358–20366.

Manns, M. P., Wedemeyer, H., & Cornberg, M. (2006). Treating viral hepatitis C: Efficacy, side effects, and complications. *Gut, 55*, 1350–1359.

Matsuda, M., Kubo, A., Furuse, M., & Tsukita, S. (2004). A peculiar internalization of claudins, tight junction-specific adhesion molecules, during the intercellular movement of epithelial cells. *The Journal of Cell Science, 117*, 1247–1257.

McKeating, J. A., Zhang, L. Q., Logvinoff, C., Flint, M., Zhang, J., et al. (2004). Diverse hepatitis C virus glycoproteins mediate viral infection in a CD81-dependent manner. *Journal of Virology, 78*, 8496–8505.

Mee, C. J., Farquhar, M. J., Harris, H. J., Hu, K., Ramma, W., et al. (2010). Hepatitis C virus infection reduces hepatocellular polarity in a vascular endothelial growth factor-dependent manner. *Gastroenterology, 138*(3), 1134–1142.

Mee, C. J., Grove, J., Harris, H. J., Hu, K., Balfe, P., et al. (2008). Effect of cell polarization on hepatitis C virus entry. *Journal of Virology, 82*, 461–470.

Mee, C. J., Harris, H. J., Farquhar, M. J., Wilson, G., Reynolds, G., et al. (2009). Polarization restricts hepatitis C virus entry into HepG2 hepatoma cells. *Journal of Virology, 83*, 6211–6221.

Meertens, L., Bertaux, C., Cukierman, L., Cormier, E., Lavillette, D., et al. (2008). The tight junction proteins claudin-1, -6, and -9 are entry cofactors for hepatitis C virus. *Journal of Virology, 82*, 3555–3560.

Molina, S., Castet, V., Fournier-Wirth, C., Pichard-Garcia, L., Avner, R., et al. (2007). The low-density lipoprotein receptor plays a role in the infection of primary human hepatocytes by hepatitis C virus. *Journal of Hepatology, 46*, 411–419.

Molina, S., Castet, V., Pichard-Garcia, L., Wychowski, C., Meurs, E., et al. (2008). Serum-derived hepatitis C virus infection of primary human hepatocytes is tetraspanin CD81 dependent. *Journal of Virology, 82*, 569–574.

Nakajima, H., Cocquerel, L., Kiyokawa, N., Fujimoto, J., & Levy, S. (2005). Kinetics of HCV envelope proteins' interaction with CD81 large extracellular loop. *Biochemical and Biophysical Research Communications, 328*, 1091–1100.

Nielsen, S. U., Bassendine, M. F., Burt, A. D., Bevitt, D. J., & Toms, G. L. (2004). Characterization of the genome and structural proteins of hepatitis C virus resolved from infected human liver. *The Journal of General Virology, 85*, 1497–1507.

Op De Beeck, A., Voisset, C., Bartosch, B., Ciczora, Y., Cocquerel, L., et al. (2004). Characterization of functional hepatitis C virus envelope glycoproteins. *Journal of Virology, 78*, 2994–3002.

Owen, D. M., Huang, H., Ye, J., & Gale, M. Jr. (2009). Apolipoprotein E on hepatitis C virion facilitates infection through interaction with low-density lipoprotein receptor. *Virology, 394*, 99–108.

Paris, L., Tonutti, L., Vannini, C., & Bazzoni, G. (2008). Structural organization of the tight junctions. *Biochimica et Biophysica Acta, 1778*, 646–659.

Patel, K. P., Coyne, C. B., & Bergelson, J. M. (2009). Dynamin- and lipid raft-dependent entry of decay-accelerating factor (DAF)-binding and non-DAF-binding coxsackieviruses into non-polarized cells. *Journal of Virology, 83*, 11064–11077.

Pestova, T. V., Shatsky, I. N., Fletcher, S. P., Jackson, R. J., & Hellen, C. U. (1998). A prokaryotic-like mode of cytoplasmic eukaryotic ribosome binding to the initiation codon during internal translation initiation of hepatitis C and classical swine fever virus RNAs. *Genes and Development, 12*, 67–83.

Pileri, P., Uematsu, Y., Campagnoli, S., Galli, G., Falugi, F., et al. (1998). Binding of hepatitis C virus to CD81. *Science, 282*, 938–941.

Ploss, A., Evans, M. J., Gaysinskaya, V. A., Panis, M., You, H., et al. (2009). Human occludin is a hepatitis C virus entry factor required for infection of mouse cells. *Nature, 457*, 882–886.

Ploss, A., Khetani, S. R., Jones, C. T., Syder, A. J., Trehan, K., et al. (2010). Persistent hepatitis C virus infection in microscale primary human hepatocyte cultures. *Proceedings of the National Academy of Sciences of the United States of America, 107*(7), 3141–3145.

Pohlmann, S., Zhang, J., Baribaud, F., Chen, Z., Leslie, G. J., et al. (2003). Hepatitis C virus glycoproteins interact with DC-SIGN and DC-SIGNR. *Journal of Virology, 77*, 4070–4080.

Popescu, C. I., & Dubuisson, J. (2009). Role of lipid metabolism in hepatitis C virus assembly and entry. *Biology of the Cell, 102*, 63–74.

Reynolds, G. M., Harris, H. J., Jennings, A., Hu, K., Grove, J., et al. (2008). Hepatitis C virus receptor expression in normal and diseased liver tissue. *Hepatology, 47*, 418–427.

Rusconi, S., Scozzafava, A., Mastrolorenzo, A., & Supuran, C. T. (2007). An update in the development of HIV entry inhibitors. *Current Topics in Medicinal Chemistry, 7*, 1273–1289.

Saitou, M., Furuse, M., Sasaki, H., Schulzke, J. D., Fromm, M., et al. (2000). Complex phenotype of mice lacking occludin, a component of tight junction strands. *Molecular Biology of the Cell, 11*, 4131–4142.

Scarselli, E., Ansuini, H., Cerino, R., Roccasecca, R. M., Acali, S., et al. (2002). The human scavenger receptor class B type I is a novel candidate receptor for the hepatitis C virus. *EMBO Journal, 21*, 5017–5025.

Scheel, T. K., Gottwein, J. M., Jensen, T. B., Prentoe, J. C., Hoegh, A. M., et al. (2008). Development of JFH1-based cell culture systems for hepatitis C virus genotype 4a and evidence for cross-genotype neutralization. *Proceedings of the National Academy of Sciences of the United States of America, 105*, 997–1002.

Selby, M. J., Glazer, E., Masiarz, F., & Houghton, M. (1994). Complex processing and protein: protein interactions in the E2:NS2 region of HCV. *Virology, 204*, 114–122.

Shen, L., Weber, C. R., & Turner, J. R. (2008). The tight junction protein complex undergoes rapid and continuous molecular remodeling at steady state. *Journal of Cell Biology, 181*, 683–695.

Shepard, C. W., Finelli, L., & Alter, M. J. (2005). Global epidemiology of hepatitis C virus infection. *The Lancet Infectious Diseases, 5*, 558–567.

Simmonds, P., Bukh, J., Combet, C., Deleage, G., Enomoto, N., et al. (2005). Consensus proposals for a unified system of nomenclature of hepatitis C virus genotypes. *Hepatology, 42*, 962–973.

Stamatovic, S. M., Keep, R. F., Wang, M. M., Jankovic, I., & Andjelkovic, A. V. (2009). Caveolae-mediated internalization of occludin and claudin-5 during CCL2-induced tight junction remodeling in brain endothelial cells. *The Journal of Biological Chemistry, 284*, 19053–19066.

Steinmann, E., Penin, F., Kallis, S., Patel, A. H., Bartenschlager, R., et al. (2007). Hepatitis C virus p7 protein is crucial for assembly and release of infectious virions. *PLoS Pathogens, 3*, e103.

StGelais, C., Foster, T. L., Verow, M., Atkins, E., Fishwick, C. W., et al. (2009). Determinants of hepatitis C virus p7 ion channel function and drug sensitivity identified in vitro. *Journal of Virology, 83*, 7970–7981.

Thomssen, R., Bonk, S., & Thiele, A. (1993). Density heterogeneities of hepatitis C virus in human sera due to the binding of beta-lipoproteins and immunoglobulins. *Medical Microbiology and Immunology, 182*, 329–334.

Tscherne, D. M., Jones, C. T., Evans, M. J., Lindenbach, B. D., McKeating, J. A., et al. (2006). Time- and temperature-dependent activation of hepatitis C virus for low-pH-triggered entry. *Journal of Virology, 80*, 1734–1741.

Van Itallie, C. M., Betts, L., Smedley, J. G., III., McClane, B. A., & Anderson, J. M. (2008). Structure of the claudin-binding domain of Clostridium perfringens enterotoxin. *The Journal of Biological Chemistry, 283*, 268–274.

Voisset, C., Callens, N., Blanchard, E., Op De Beeck, A., Dubuisson, J., et al. (2005). High density lipoproteins facilitate hepatitis C virus entry through the scavenger receptor class B type I. *The Journal of Biological Chemistry, 280*, 7793–7799.

von Hahn, T., Lindenbach, B. D., Boullier, A., Quehenberger, O., Paulson, M., et al. (2006). Oxidized low-density lipoprotein inhibits hepatitis C virus cell entry in human hepatoma cells. *Hepatology, 43*, 932–942.

Wakita, T., Pietschmann, T., Kato, T., Date, T., Miyamoto, M., et al. (2005). Production of infectious hepatitis C virus in tissue culture from a cloned viral genome. *Nature Medicine, 11*, 791–796.

Whidby, J., Mateu, G., Scarborough, H., Demeler, B., Grakoui, A., et al. (2009). Blocking hepatitis C virus infection with recombinant form of envelope protein 2 ectodomain. *Journal of Virology, 83*, 11078–11089.

Yang, W., Qiu, C., Biswas, N., Jin, J., Watkins, S. C., et al. (2008). Correlation of the tight junction-like distribution of Claudin-1 to the cellular tropism of hepatitis C virus. *The Journal of Biological Chemistry, 283*, 8643–8653.

Zhang, J., Randall, G., Higginbottom, A., Monk, P., Rice, C. M., et al. (2004). CD81 is required for hepatitis C virus glycoprotein-mediated viral infection. *Journal of Virology, 78*, 1448–1455.

Zheng, A., Yuan, F., Li, Y., Zhu, F., Hou, P., et al. (2007). Claudin-6 and claudin-9 function as additional coreceptors for hepatitis C virus. *Journal of Virology, 81*, 12465–12471.

Zhong, J., Gastaminza, P., Chung, J., Stamataki, Z., Isogawa, M., et al. (2006). Persistent hepatitis C virus infection in vitro: Coevolution of virus and host. *Journal of Virology, 80*, 11082–11093.

CHAPTER 13

Claudins in Cancer Biology

Blanca L. Valle* and Patrice J. Morin*,†

*Laboratory of Cellular and Molecular Biology, National Institute on Aging, NIH, Biomedical Research Center, Baltimore, Maryland, USA

†Department of Pathology, Johns Hopkins Medical Institutions, Baltimore, Maryland, USA

I. OVERVIEW

Claudins are a family of 24 integral membrane proteins that have crucial roles in tight junction (TJ) formation and function (Krause et al., 2008; Lal-Nag & Morin, 2009; Tsukita & Furuse, 2000). The cellular TJ, located at the most apical area of the basolateral membrane of epithelial and endothelial cells, is a structure involved in maintenance of cell polarity and paracellular ion transport. Different tissues have different transport requirements, and it is believed that the different combinations of claudins in a given

1063-5823/10 $35.00
DOI: 10.1016/S1063-5823(10)65013-2

cell determines the tightness and selectivity of the TJ (Krause et al.; Van Itallie & Anderson, 2006), explaining the heterogeneity in claudin expression among different tissues. A large number of studies have now demonstrated that the expression of claudin proteins is frequently altered in several human malignancies. In this chapter, we review the changes in claudin expression observed in various cancers, their mechanisms of regulation, as well as the functional studies that may suggest a role for these proteins in cancer. Finally, we summarize the extensive literature suggesting that claudins may become useful markers for detection and diagnosis, as well as possible targets for therapy.

II. INTRODUCTION

The claudin genes encode transmembrane proteins that belong to the PMP22/EMP/MP20 superfamily of proteins (Lal-Nag & Morin, 2009; Van Itallie & Anderson, 2006) that have essential roles in TJ formation and function. TJ represent the apical-most intercellular junction in epithelial and endothelial cells and are crucial in regulating paracellular transport as well as cell polarity. The integral membrane claudin proteins, which form the backbone of TJs, are relatively small proteins, with sizes ranging between 22 and 24 kDa, although some members are larger (up to 34 kDa) due to larger N- or C-terminal tails (Hewitt, Agarwal, & Morin, 2006; Lal-Nag & Morin). Based on hydropathy plots, as well as several experimental lines of evidence, it has been shown that claudin proteins have four transmembrane helices, with their N- and C-terminal regions extending into the cytoplasm. This configuration gives rise to two extracellular loops, as well as one short intracellular loop between transmembrane helices 2 and 3. The first extracellular loop is believed to play crucial roles in paracellular charge selectivity of the TJ (Colegio, Van Itallie, McCrea, Rahner, & Anderson, 2002). The second extracellular loop is believed to be important in lateral TJ strand interactions between claudins located on opposing cells, and is therefore essential in forming the paired TJ strands (Piontek et al., 2008). The region that exhibits the most size and sequence heterogeneity is the C-terminus. This region contains sequences that have been shown to be important in signaling, including phosphorylation and palmitoylation sites, as well as a PDZ domain-binding motif at the extreme C-terminus. The PDZ domain-binding motif has been implicated in the binding of claudin proteins to several cytoplasmic scaffolding proteins such as the TJ-associated proteins MUPP1 (Hamazaki, Itoh, Sasaki, Furuse, & Tsukita, 2002), PATJ (Roh, Liu, Laurinec, & Margolis, 2002), ZO-1, ZO-2, and ZO-3, and MAGUKs (Itoh et al., 1999).

III. ABERRANT EXPRESSION OF CLAUDINS IN CANCER

Claudins are crucial components of TJs, and as such, are expressed in all organs and tissues lined by or containing epithelial or endothelial cells. The physiological requirement for TJ tightness is highly variable among various tissues, and for this reason the pattern of claudin expression is highly tissue-specific (Krause et al., 2008). For example, claudin-7, -8, -12, -13, and -15 are found expressed in various parts of the small and large intestines, while claudin-6, -9, -10, -11, -14, -16, -18, and -19 are not observed in this tissue (Fujita et al., 2006). Interestingly, expression levels of the various claudins exhibit significant heterogeneity along the intestinal tract or even along the crypt–surface axis, suggesting that this tissue can fine-tune its transport properties in various areas by spatially altering claudin composition (Fujita et al.). While the exact claudin composition varies, similar complex patterns of claudin expression have been reported for other tissues such as the kidney (Kiuchi-Saishin et al., 2002; Li, Huey, & Yu, 2004; Reyes et al., 2002), the inner ear (Kitajiri et al., 2004), the prostate (Sakai et al., 2007), and others (Krause et al.).

The expression of claudin proteins is altered in several neoplastic tissues compared to their normal counterparts (Morin, 2005). Claudins have been found both elevated and decreased in tumors, depending on the exact claudin and cancer studied. Table I summarizes the changes in claudin expression that have been reported to date in cancer, and shows that 14 different claudin family members have been found aberrantly expressed in various malignancies. Claudin-3 and claudin-4 are among the most frequently deregulated claudins in cancer and are typically found highly expressed in ovarian cancer, breast cancer, prostate cancer, and pancreatic cancer (Table I). Other tumors which express claudin-3 and -4 include those in bladder, thyroid, stomach, colon, and uterus.

Claudin-1 and -7 have also been found frequently deregulated in various malignancies but their expression patterns are more variable. Claudin-1, for example, is found to be downregulated in breast cancer, colon cancer, lung cancer, and other malignancies (Table I). Other studies, however, have found claudin-1 to be upregulated in these cancers, as well as other neoplasms. The reasons for the discrepancy are unclear, but may include differences in sample collection, sample handling, staining, and/or the exact stages and grades of the cancers collected. Claudin-7 is downregulated in head and neck carcinomas, squamous cell carcinoma of the esophagus, breast carcinomas, and others; however, it was found to be elevated in cervical cancer, colorectal carcinomas, as well as other malignancies (Table I).

TABLE I
Patterns of Claudin Expression in Human Cancer

Claudin	Up/ Down	Cancer site	References
CLDN1	Up	Breast	Blanchard et al. (2009)
		Cervical	Lee, Lee, et al. (2005), Sobel et al. (2005), Szabo, Kiss, Schaff, & Sobel (2009), Vazquez-Ortiz et al. (2005)
		Colorectal	de Oliveira, de Oliveira, De Souza, & Morgado-Diaz (2005), Dhawan et al. (2005), Grone et al. (2007), Huo et al. (2009), Kinugasa et al. (2007), Mees et al. (2009), Oshima, Miwa, et al. (2008)
		Esophageal	Takala, Saarnio, Wiik, & Soini (2007)
		Gastric	Resnick, Gavilanez, et al. (2005), Wu, Zhang, Wang, & Chen (2008)
		IBD	Weber, Nalle, Tretiakova, Rubin, & Turner (2008)
		Lung	Liu et al. (2007)
		Melanoma	Leotlela et al. (2007)
		Oral	Dos Reis et al. (2008)
		Ovarian	Kleinberg et al. (2008), Soini & Talvensaari-Mattila (2006)
		RCC	Fritzsche et al. (2008)
		Sarcoma	Thway, Fisher, Debiec-Rychter, & Calonje (2009)
		Thyroid	Fluge, Bruland, Akslen, Lillehaug, & Varhaug (2006)
		Tongue	Bello et al. (2008)
		Urothelial	Nakanishi et al. (2008)
	Down	Breast	Kramer, White, Kubbies, Swisshelm, & Weber (2000), Martins et al. (2009), Morohashi et al. (2007), Tokes et al. (2005)
		Esophageal	Miyamoto et al. (2008)
		HCC	Higashi et al. (2007)
		Lung	Chao et al. (2009), Paschoud, Bongiovanni, Pache, & Citi (2007)
		Melanoma	Cohn et al. (2005)
		Prostate	Sheehan et al. (2007), Szasz et al. (2009)
		Thyroid	Tzelepi, Tsamandas, Vlotinou, Vagianos, & Scopa (2008)
	Varied	Cervical	Sobel et al. (2005)
		Colorectal	Resnick, Konkin, et al. (2005)
		Gastric	Soini, Tommola, Helin, & Martikainen (2006)
		Lung	Moldvay et al. (2007)

(continued)

TABLE I (*continued*)

Claudin	Up/ Down	Cancer site	References
		Pancreatic	Tsukahara et al. (2005)
		Sarcoma	Billings et al. (2004)
		Skin	Morita, Tsukita, & Miyachi (2004)
CLDN2	Up	Colorectal	Aung et al. (2006), Kinugasa et al. (2007)
		Endometrial	Szabo et al. (2009)
		Gastric	Song et al. (2008)
		IBD	Weber et al. (2008)
		Rectal	Ishida, Kushima, & Okabe (2009)
	Down	Breast	Kim et al. (2008)
		Prostate	Szasz et al. (2009), Vare et al. (2008)
	Varied	Breast	Soini (2004)
		Cervical	Sobel, Paska, et al. (2005)
		Lung	Moldvay et al. (2007)
CLDN3	Up	Breast	Blanchard et al. (2009), Kominsky et al. (2004)
		Colorectal	de Oliveira et al. (2005), Mees et al., (2009), Oshima, Kunisaki, et al. (2008)
		Endometrial	Konecny et al. (2008), Pan et al. (2007), Szabo et al. (2009)
		Esophageal	Montgomery et al. (2006), Takala et al. (2007)
		Gastric	Resnick, Gavilanez, et al. (2005)
		Ovarian	Bignotti et al. (2006), Choi, Kim, Kwon, et al. (2007), Davidson et al. (2006), Heinzelmann-Schwarz et al. (2004), Hough et al. (2000), Kleinberg et al. (2008), Lu et al. (2004), Rangel et al. (2003), Rosen et al. (2005), Santin et al. (2004), Szabo et al. (2009), Zhu, Brannstrom, Janson, & Sundfeldt (2006)
		Prostate	Long et al. (2001), Sheehan et al. (2007), Szasz et al. (2009), Vare et al. (2008)
		Rectal	Ishida et al. (2009)
		RCC	Lechpammer et al. (2008)
		Urothelial	Nakanishi et al. (2008)
		Uterine	Santin, Bellone, Marizzoni, et al. (2007), Santin, Bellone, Siegel, et al. (2007), Santin, Zhan, et al. (2005)
	Varied	Breast	Soini (2004)
		Gastric	Matsuda et al. (2007), Satake et al. (2008), Soini, Tommola, et al. (2006)
		Lung	Moldvay et al. (2007)

(*continued*)

TABLE I (*continued*)

Claudin	Up/ Down	Cancer site	References
CLDN4	Up	Breast	Blanchard et al. (2009), Kominsky et al., (2004), Kulka et al. (2009)
		Cervical	Lee, Lee, et al. (2005), Sobel, Szabo, et al. (2005)
		Cholangio-carcinoma	Nishino et al. (2008)
		Colorectal	de Oliveira et al. (2005), Mees et al. (2009), Oshima, Kunisaki, et al. (2008), Resnick, Konkin, et al. (2005)
		Endometrial	Konecny et al. (2008), Pan et al. (2007), Szabo et al. (2009)
		Esophageal	Montgomery et al. (2006), Takala et al. (2007)
		Gastric	Cunningham et al. (2006), Resnick, Gavilanez, et al. (2005)
		Lung	Hanada et al. (2008), Jung et al. (2009), Paschoud et al. (2007)
		Ovarian	Bignotti et al. (2006), Choi, Kim, Kwon, et al. (2007), Davidson et al. (2006), Hibbs et al. (2004), Hough et al. (2000), Kleinberg et al. (2008), Litkouhi et al. (2007), Rangel, Agarwal, et al. (2003), Santin et al. (2004), Soini & Talvensaari-Mattila (2006, Stewart et al. (2006), Szabo et al. (2009), Zhu et al. (2006)
		Pancreatic	Gress et al. (1996), Michl et al. (2003), Nichols, Ashfaq & Iacobuzio-Donahue (2004), Sato et al. (2004), Terris et al. (2002)
		Prostate	Landers et al. (2008), Long et al. (2001), Sheehan et al. (2007), Szasz et al. (2009), Vare et al. (2008)
		Rectal	Ishida et al. (2009)
		RCC	Lechpammer et al. (2008)
		Urothelial	Nakanishi et al. (2008)
		Uterine	Santin, Bellone, Marizzoni, et al., 2007; Santin, Bellone, Siegel, et al., 2007; Santin, Zhan, et al. (2005)
	Down	Breast	Tokes et al. (2005)
		Colorectal	Ueda et al. (2007)
		Gastric	Lee, Moon, et al. (2005)
		Thyroid	Tzelepi et al. (2008)
	Varied	Breast	Soini (2004)
		Cervical	Sobel, Paska, et al. (2005)
		Gastric	Matsuda et al. (2007), Satake et al. (2008), Soini, Tommola, et al. (2006), Wu et al. (2006)

(*continued*)

TABLE I (*continued*)

Claudin	Up/ Down	Cancer site	References
		Lung	Moldvay et al. (2007)
		Skin	Morita et al. (2004)
CLDN5	Up	Esophageal	Takala et al. (2007)
		Lung	Jung et al. (2009), Paschoud et al. (2007)
	Down	Prostate	Vare et al. (2008)
	Varied	Breast	Soini (2004)
		Gastric	Soini, Tommola, et al. (2006)
CLDN7	Up	Cervical	Lee, Lee, et al. (2005), Sobel, Szabo, et al. (2005), Szabo et al. (2009)
		Colorectal	Darido et al. (2008)
		Esophageal	Montgomery et al. (2006), Takala et al. (2007)
		Gastric	Johnson et al. (2005), Park et al. (2007),
		Ovarian	Banz et al. (2009), Kleinberg et al. (2008), Santin et al. (2004), Soini & Talvensaari-Mattila (2006), Tassi et al. (2008)
		Renal	Choi, Kim, Ryu, et al. (2007), Schuetz et al. (2005)
		Tongue	Bello et al. (2008)
		Urothelial	Nakanishi et al. (2008)
	Down	Breast	Kominsky et al. (2003), Park, Karesen, Axcrona, Noren, & Sauer (2007)
		Colorectal	Nakayama et al. (2008)
		Head/neck	Al Moustafa et al. (2002)
		Esophageal	Usami et al. (2006)
		Prostate	Sheehan et al. (2007)
		Thyroid	Tzelepi et al. (2008)
	Varied	Cervical	Sobel, Paska, et al. (2005)
		Lung	Moldvay et al. (2007)
		Renal	Li et al. (2008), Osunkoya et al. (2009)
CLDN8	Down	Colorectal	Grone et al. (2007)
	Varied	Renal	Osunkoya et al. (2009)
CLDN10	Up	Thyroid	Aldred et al. (2004)
	Down	Biliary tract	Nemeth et al. (2009)
CLDN11	Down	Gastric	Agarwal et al. (2009)
CLDN12	Up	Melanoma	Morita et al. (2008)
	Up	Colorectal	Grone et al. (2007)

(*continued*)

TABLE I (*continued*)

Claudin	Up/ Down	Cancer site	References
CLDN15	Up	Mesothelioma	Davidson et al. (2006)
CLDN16	Up	Ovarian	Rangel et al. (2003)
		Thyroid	Fluge et al. (2006)
	Down	Breast	Martin et al. (2008)
CLDN18	Up	Pancreatic	Karanjawala et al. (2008)
	Down	Gastric	Sanada et al. (2006), Yasui, Oue, Sentani, Sakamoto, & Motoshita (2009)
	Varied	Gastric	Matsuda et al. (2007), Satake et al. (2008)
		Signet ring cell	Sentani et al. (2008)
CLDN23	Down	Gastric	Katoh & Katoh (2003)

IV. REGULATION OF CLAUDIN EXPRESSION

Multiple stimuli, including growth factors, cytokines, and certain drugs, have been shown to affect claudin expression. The exact mechanisms by which the claudin levels are affected are not completely understood, but claudin expression can be regulated at the transcriptional, posttranscriptional, and posttranslational levels. Epigenetic mechanisms have also been shown to influence claudin expression in cancer. In this section, we will discuss the pathways and mechanisms that affect claudin expression in cancer.

A. Pathways Implicated in Claudin Regulation

Numerous studies have implicated EGF in the regulation of claudin expression in normal and transformed cells. In the MDCK canine kidney cells, claudin-4 expression was induced by EGF and appeared to be mediated through SP1 activation (Ikari et al., 2009). The EGF-induced expression of claudin-4 was inhibited by the MEK inhibitor U0126, suggesting involvement of this pathway in the regulation of claudin-4 expression. However, the basal expression of claudin-4 was not inhibited by U0126, suggesting that additional pathways regulate claudin-4 expression. In a previous study, EGF treatment of MDCK cells led to an increase in claudin-1, -3, and -4 and a decrease in claudin-2 expression, with a corresponding increase in TJ strength (Singh & Harris, 2004). In alveolar epithelial cells, EGF treatment was found to lead to increases in claudin-4 and -7 and decreases in claudin-3 and -5 (Chen et al., 2005), demonstrating that different claudins might be differentially regulated by EGF. In the non-small cell lung cancer cell line A549, EGF treatment increased the expression of claudin-2, resulting in reduced permeability and increased cell colonization (Peter, Comellas, Levantini, Ingenito, & Shapiro, 2009). Furthermore, claudin-7 was among the genes overexpressed in EGF-induced liver tumors in EGF2B transgenic mice (Borlak, Meier, Halter, Spanel, & Spanel-Borowski, 2005), suggesting that EGF appears to also regulate expression of claudins *in vivo*. Hepatocyte growth factor (HGF) can also regulate the expression of claudins. Indeed, the expression of claudin-2 was found to be reduced in MDCK cells following treatment with HGF (Lipschutz, Li, Arisco, & Balkovetz, 2005). HGF treatment decreased expression of claudin-2 and increased transepithelial resistance and these effects were mediated through the ERK MAPK pathway, as demonstrated through the use of the MEK inhibitor U0126.

Proinflammatory cytokines can also affect claudin expression. For instance, interleukin-17 (IL-17) treatment induced expression of claudin-1 and -2 in colon adenocarcinoma T84 cells, leading to TJ formation (Kinugasa, Sakaguchi, Gu, & Reinecker, 2000). Treatment with the MEK inhibitor PD98059 decreased the expression of claudin-2 and TJ formation, again implicating the ERK pathway in regulation of claudin expression and function. However, the expression of claudin-1 was not modulated by the MEK inhibitor, suggesting that other pathways may also be involved. Expression of claudin-1 was upregulated in the colon cell line Caco-2 in response to IFN, tumor necrosis factor (TNF), and IL-1β treatment (Han, Fink, & Delude, 2003). Increased claudin-1 expression along with decreased expression of occludin and zonula occludens 1 (ZO-1) resulted in decreased barrier function. Claudin-1, occludin, and ZO-1 were mislocalized in Caco-2 cells upon treatment. In another study, IL-1β induced the expression of claudin-2 in rat hepatocytes (Yamamoto et al., 2004). Increase in claudin-2 expression by IL-1β was mediated by the PI3K and p38 pathways.

Nonsteroidal anti-inflammatory drugs (NSAIDs) also have been shown to affect claudin expression. The NSAIDs indomethacin, celecoxib, and diclofenac induced the expression of claudin-4 in the gastric carcinoma cells AGS, probably through an increase in intracellular Ca^{2+} (Mima et al., 2005). Claudin expression in nonneoplastic cells could also be affected by NSAIDs, as NSAID treatment inhibited claudin-2 in gastric epithelial cells, and led to a reduction in migration and invasion (Mima et al., 2008). In another gastric epithelial cell line, MKN28, claudin-7 expression was downregulated by treatment with aspirin, resulting in increased permeability to dextran and decreased TJ strength (Oshima, Miwa, & Joh, 2008). The p38 inhibitor SB203580, blocked the effects of aspirin on permeability and claudin-7 expression, suggesting that the p38 pathway regulates the expression of claudin-7. Another anti-inflammatory drug, troglitazone, increased the expression of claudin-4 and E-cadherin in the pancreatic cancer cell line PK-1 (Kumei, Motomura, Yoshizaki, Takakusaki, & Okumura, 2009) through inhibition of the ERK/MAPK pathway. 5-Fluorouracil, a commonly used chemotherapeutic agent, induced the expression of claudin-1 in nasopharyngeal carcinoma cells, leading to a reduction in apoptosis (Lee et al., 2010). Reexpression of E-cadherin decreased claudin-1 expression and decreased the antiapoptotic effect of claudin-1.

B. Epigenetic Regulation

Promoter hypermethylation has been implicated in the regulation of expression of the CLDN3, 4, 6, and 7 genes (Honda, Pazin, D'Souza, Ji, & Morin, 2007; Honda, Pazin, Ji, Wernyj, & Morin, 2006; Kominsky et al., 2003; Osanai, Murata, Chiba, Kojima, & Sawada, 2007; Roth et al., 2006). Loss of CLDN7 expression in the breast cancer cells MDA-MB-231, MDA-MB-435, and HS578T correlated with promoter hypermethylation in these cells as compared to breast cancer cells that express this gene (Kominsky et al.). In addition, treatment of these cells with a demethylating agent, 5-aza-dC, produced an increase in expression of CLDN7, supporting a role for promoter methylation in CLDN7 silencing in breast cancer cells. However, invasive breast carcinomas that lack CLDN7 expression did not exhibit promoter hypermethylation suggesting that other mechanisms are responsible for CLDN7 downregulation in invasive breast tumors (Kominsky et al.).

Extensive work has shown that the expression of the CLDN3 and CLDN4 genes can be regulated by promoter hypermethylation and histone acetylation (Boireau et al., 2007; Honda et al., 2006, 2007; Litkouhi et al., 2007). Ovarian cancer cell lines that do not express claudin-3 and claudin-4, exhibited hypermethylation of these promoters while the inverse was true for cell lines that expressed high levels of these genes (Honda et al., 2006, 2007). The demethylating agent 5-aza-dC, together with the histone deacetylase inhibitor TSA, could induce expression of CLDN3 and CLDN4 in ovarian cancer cells with low or no expression of these genes (Honda et al., 2006, 2007). In high-grade bladder carcinomas, CLDN4 expression was decreased and correlated with hypermethylation in the coding sequence of the CLDN4 gene (Boireau et al.). Again, the demethylating agent 5-aza-CdR increased the expression of CLDN4 in primary cells derived from the high-grade bladder carcinomas. In transgenic mice overexpressing DNA methyltransferase 3a, the expression of Cldn4 was undetectable in bladder umbrella cells, suggesting that *in vivo* Cldn4 expression might also be regulated by hypermethylation. CLDN3 gene methylation was also observed in esophageal squamous cell carcinoma and its precursor lesions (Roth et al., 2006).

In the breast adenocarcinoma cell line MCF7, the promoter region of the CLDN6 gene was found hypermethylated. Expression of CLDN6 could be induced by treatment with the demethylating agent 5-aza-dC and with the histone deacetylase inhibitor TSA (Osanai et al., 2007), suggesting that expression of CLDN6 can also be regulated, at least in part, by epigenetic mechanisms.

C. Transcriptional Regulation

Various transcription factors induce or repress expression of claudins. Snail, a transcriptional repressor known to promote epithelial to mesenchymal transition (EMT), represses the expression of several claudins. In the canine kidney MDCK cells, expression of Snail decreased the expression of claudin-2, -4, and -7 and this correlated with decreased transepithelial electrical resistance (TER), a measure of TJ strength (Carrozzino et al., 2005). Expression of Snail and Slug, another member of the Snail family, also decreases the expression of claudin-1 in MDCK cells, resulting in decreased TER (Martinez-Estrada et al., 2006). Snail and Slug were shown to bind the E-box motifs (CACCTG) in the claudin-1 promoter. In various breast cancer cell lines, claudin-1 and Snail/Slug expression were inversely correlated, as well as in a panel of invasive breast tumors (Martinez-Estrada et al.). Expression of claudin-1 and claudin-7 was also found inversely correlated with Snail expression in esophageal squamous cell carcinomas (Usami et al., 2008). In another study, the expression of claudin-3 was repressed by a complex between SNAIL1 and SMAD3 and 4 (Vincent et al., 2009). TGF-β-induced EMT in breast epithelial cells while expression of occludin, E-cadherin, Coxsackie adenovirus receptor (CAR), and claudin-3 was repressed by the Snail1–SMAD3/4 complex (Vincent et al.). SNAIL1 and SMAD4 interacted at adjacent E-boxes and Smad-binding elements (SBEs) in the CAR promoter. Claudin-3, occludin, and E-cadherin have SBEs close to E-boxes, suggesting their expression could be repressed in similar fashion by the SNAIL/SMAD complex. In addition, in breast cancer cells, forced expression of SNAI1P, a SNAIL homolog, repressed the expression of claudin-7 (Mittal, Myers, Bailey, Misra, & Chaudhuri, 2010). SNAI1P was recruited to the CLDN7 promoter together with HDAC1 and the corepressor CtBP1, and the deacetylase inhibitor trichostin A decreased the SNAI1P-induced repression.

The expression levels of Smad4 and claudin-1 were inversely correlated in colon cancer samples, and overexpression of Smad4 repressed CLDN1 promoter activity in colon cancer cell lines, probably though modulation of the TCF/β-catenin transcription complex (Shiou et al., 2007). In addition, the Cdx2 transcription factor was shown to regulate claudin-3 and claudin-4 expression during intestinal differentiation of gastric carcinoma (Satake et al., 2008). In gastric cancer, CLDN1 was shown to be a direct transcriptional target of the gastric tumor suppressor RUNX3 (Chang et al., 2010). Interestingly, claudin-1 expression was found to repress tumorigenicity of cells lacking RUNX3 expression, suggesting a significant role for claudin-1 as a downstream effector of RUNX3 (Chang et al.). In mice, Cldn1 was shown to be a target of the p63 transcription factor, and this pathway was suggested to be involved in ectodermal dysplasia (Lopardo et al., 2008).

Finally, SP1 sites are present in several claudin genes such as CLDN3/4 (Honda et al., 2006, 2007; Ikari et al., 2009), CLDN5 (Burek & Forster, 2009), and CLDN1 (Dufresne & Cyr, 2007). In several of these cases, the SP1 sites were shown to be functionally involved in the expression of the CLDN genes in cancer (Honda et al., 2006, 2007; Ikari et al.).

D. Posttranslational Regulation

Claudin localization and function is known to be influenced by diverse posttranslational modifications and by interaction with other proteins. Conversely, interaction of claudins with other proteins influences the functions of these proteins and the cellular outcome. Phosphorylation of claudins by diverse kinases has been reported by several groups (Aono & Hirai, 2008; Banan et al., 2005; French et al., 2009; Fujibe et al., 2004; Ikari et al., 2006, 2008; Ishizaki et al., 2003; Nunbhakdi-Craig et al., 2002; Soma et al., 2004; Tanaka, Kamata, & Sakai, 2005; Tatum et al., 2007; Yamamoto et al., 2008), and in most of these cases, phosphorylation had functional consequences on claudin function and/or localization. The exact roles of claudin phosphorylation events in cancer are not completely clear, but may be involved in the reduction in TJ function. For example, in ovarian cancer cells, claudin-4 was found to be phosphorylated by protein kinase C and this phosphorylation was found to lead to translocation of claudin-4 from the membrane to the cytoplasm and a decrease in TJ function (D'Souza, Indig, & Morin, 2007). Similarly, claudin-3 is phosphorylated in ovarian cells by cAMP-dependent protein kinase (PKA), and phosphorylation of claudin-3 leads to a decrease in TJ strength (D'Souza, Agarwal, & Morin, 2005). Another study found that claudin-1, -2, -3, and -4 were phosphorylated by the WNK4 kinase and that phosphorylation was an important event in claudin and TJ regulation

(Yamauchi et al., 2004). Claudin proteins are also known to be regulated by palmitoylation, as palmitoylation of claudin-14 was shown to be crucial for its proper localization (Van Itallie, Gambling, Carson, & Anderson, 2005).

E. Localization of Claudins

Localization of claudins is frequently altered in claudin overexpressing cancer cells. Indeed, claudins, which are typically found in the membrane, are sometimes found mislocalized to the cytoplasm or to the nucleus. A very good example is a recent report of a gradual localization shift from the membrane to the cytoplasm of several different claudin proteins during DMBA-induced mouse tumorigenesis (Arabzadeh, Troy, & Turksen, 2007). While membrane localization is typically associated with their function in TJ, little is known about the significance and role of claudins located to the cytoplasm or nucleus. However, increased expression of claudins in several cancers generally leads to an increase in survival and invasion, suggesting that cytoplasmic or nuclear claudins might be important for tumorigenesis and/or progression of cancer. The mislocalization of claudins might cause disruption of the TJ structure, promoting a reduction in adhesion and increased motility. Furthermore, overexpressed cytoplasmic or nuclear claudins might be involved in signal transduction leading to cancer progression. Interestingly, in addition to being mislocalized in cancer, claudins can also be localized to the cytoplasm in normal cells as well (Blackman, Russell, Nordeen, Medina, & Neville, 2005), again suggesting possible cytoplasmic functions for these proteins.

The first report suggesting that claudin may become mislocalized in certain circumstances showed that claudin-1 could be localized to the cytoplasm of Ras-transformed MDCK cells (Chen, Lu, Schneeberger, & Goodenough, 2000). The use of the MEK inhibitor PD98059 could promote the relocalization of claudin-1 to the cell membrane, and the phenotype went from fibroblast to epithelial. Consequently, the ERK MAP kinase pathway appears to mediate the localization of claudin-1 to the cytoplasm in Ras-transformed MDCK cells. In melanoma tissue samples and cell lines expressing claudin-1, this protein was localized mainly to the cytoplasm (French et al., 2009; Leotlela et al., 2007). In particular, metastatic lesions had higher cytoplasmic staining than primary cancers or nevi. In nevi, claudin-1 was almost exclusively localized to the nucleus, and in normal epidermis, claudin-1 could be detected at the cell junctions (Leotlela et al.). Interestingly, when melanoma cells were manipulated to express either nuclear or cytoplasmic claudin-1, cytoplasmic but not nuclear claudin-1 promoted invasion, suggesting that cytoplasmic claudin-1 might play an important role in enhancing the migration and metastatic potential in melanoma cells. In primary and metastatic colorectal carcinomas, claudin-1 was principally found in the cytoplasm and nucleus, as compared to normal colon tissue, which had claudin-1 at the membrane (Dhawan et al., 2005). In particular, liver and lymph node metastases had nuclear claudin-1, whereas normal colon cells did not. In ApcMin/+ mice, which express mutant APC, the intestinal adenomas also had claudin-1 localized to the nucleus and cytoplasm, suggesting that loss of the tumor suppressor function of APC might be associated in the regulation/localization of claudin-1 (Dhawan et al.). Claudin-7 has also been found mislocalized to the cytoplasm in esophageal squamous cell carcinoma (Lioni et al., 2007).

There is ample evidence that the localization of claudins may be controlled, at least in part, by phosphorylation. In ovarian cancer cell lines, claudin-3 and claudin-4 localizations were affected by PKA and PKC phosphorylation, respectively (D'Souza et al., 2005, 2007). In fact, expression of a mutant mimicking the phosphorylated state of claudin-3 or treatment with forskolin, a PKA activator produced a decrease in TJ function and diffuse claudin-3 staining at cell to cell contacts, suggesting a relocalization of claudin-3 upon phosphorylation (D'Souza et al., 2005). Similarly, claudin-4 in OVCA433 cells is localized to the cytoplasm after TPA treatment (D'Souza et al., 2007). In addition, TPA reduced TJ function and PKC inhibitors prevented this effect, suggesting a role for PKC in the localization of claudin-4 and TJ function.

V. ROLES OF CLAUDINS IN TUMORIGENESIS

The obvious question resulting from the observation of deregulated claudins in cancer is whether these changes have functional consequences in cancer progression. It has been suggested that the loss of claudin expression could be important in the disruption of TJ structure often observed in cancer and possibly lead to decreased cell–cell adhesion. In this scenario, claudins could be envisioned as having a tumor-suppressive role due to their roles in TJ formation and cell adhesion. Additionally, disruption of TJs could lead to unrestricted passage of growth factors and nutrients needed for tumor growth and progression. However, it appears that claudins are most often found elevated in various cancers, and that they may act to promote tumorigenesis. The exact roles of these upregulated claudins have not been completely elucidated, but patterns have begun to emerge. In this section, we will review some of the functional studies that have enhanced our understanding of the roles of claudins in tumorigenesis.

As mentioned earlier, some claudins are downregulated in cancer (Table I), which has led to the hypothesis that they may be involved in suppressing tumorigenesis and progression. In breast cancer, claudin-7 is decreased in *in situ* and invasive ductal carcinomas of the breast and its expression is inversely correlated with histological grade, where significant loss of claudin-7 expression occurs in high-grade lesions (Kominsky et al., 2003). In addition, claudin-7 is also lost in the majority of lobular *in situ* carcinomas. Similar to the high-grade invasive ductal carcinomas, lobular *in situ* carcinomas are characterized by loss of cell to cell adhesion, suggesting that loss of claudin-7 expression might correlate with loss of cellular adhesion which ultimately might increase metastasis. Similarly, the siRNA-mediated knockdown of claudin-7 expression in the TE esophageal squamous cell carcinoma line leads to an increase in cellular growth and invasion, associated with decreased expression of E-cadherin (Lioni et al., 2007). On the other hand, reexpression of claudin-7 in these cells reverses the phenotype, leading to more-adhesive and less-invasive cells. This correlated with an increased E-cadherin expression, further supporting a role for claudin-7 in cell to cell adhesion (Lioni et al.). Similarly, several lines of evidence have suggested a tumor suppressor role for claudin-1 in cancer. Knockdown of claudin-1 expression was found to increase the tumorigenicity of human gastric cancer cells (Chang et al., 2010), while overexpression of claudin-1 inhibited cancer cell dissociation and suppressed cancer cell migration, invasion, and metastasis in lung cancer (Chao et al., 2009). Reexpression of claudin-1 has also been found to induce apoptosis in breast tumor spheroids (Hoevel, Macek, Swisshelm, & Kubbies, 2004). The mechanisms of claudin-1 downregulation are unclear, but it was suggested that RON, a cell membrane protein tyrosine kinase, may be important in downregulating claudin-1 in breast cancer cells and that claudin-1 downregulation may be important in mediating RON-induced increases in tumorigenicity (Zhang, Yao, & Wang, 2008). In addition, HOXB7, which mediates increases in invasion and tumorigenicity in breast cancer cells, has also been suggested to be an upstream negative modulator of claudin-1 (Wu et al., 2006).

Claudin-16 expression is reduced in breast cancer and forced expression of claudin-16 in the breast adenocarcinoma cell line MDA-MB-231 decreased the motility and invasion of these cells (Martin, Harrison, Watkins, & Jiang, 2008). *In vivo*, tumors induced by claudin-16 overexpressing cells in nude mice grew slower and were smaller in volume, as compared to the control, suggesting a possible tumor-suppressive role for claudin-16 in breast cancer. Claudin-6 expression is also lower in breast cancer cells and tissue (Osanai et al., 2007; Quan & Lu, 2003). In cells derived from Copenhagen rats, which are resistant to mammary tumor formation by carcinogens, claudin-6 expression was higher, as compared to expression in cells derived from

Buffalo rats susceptible to carcinogens (Quan & Lu). In the breast adenocarcinoma cell line MCF-7, claudin-6 expression is reduced due to promoter methylation, further inhibition of claudin-6 expression by siRNA produced cells resistant to apoptosis and increased colony formation (Osanai et al.). Induced expression of claudin-6, by a demethylating agent and TSA, sensitized MCF-7 cells to apoptosis and reduced tumorigenicity. Together, this suggests that claudin-6 might have a tumor-suppressive role in breast cancer. Claudin-11 has also been found downregulated in gastric cancer, and this downregulation appeared relevant for invasiveness of these cells (Agarwal et al., 2009).

On the other hand, claudins are often found upregulated in multiple cancers (Morin, 2005) (Table I). For example, claudin-7 has been found elevated in multiple cancers, and may have important roles in promoting cancer progression. Indeed, an interaction between claudin-7 and epithelial cell adhesion molecule (EpCAM) was found in tumors of the gastrointestinal tract, pancreatic and colon cancer cell lines, as well as in normal gastrointestinal tissue (Ladwein et al., 2005). Claudin-7 forms a complex with EpCAM and is recruited to tetraspanin-enriched membrane microdomains (TEM) and associate with CO-029 and CD44v6 in the human embryonic kidney cells HEK 293T, and in the colon cancer cell lines SW948 and SW480 (Kuhn et al., 2007). In the absence of claudin-7, the association of EpCAM with CO-029 and CD44v6 as well as the recruitment to TEMs, is reduced, suggesting that claudin-7 is necessary for complex formation and that it can modulate the effects of EpCAM (Kuhn et al.). The colon cancer cell lines HT-29 and SW948, which express high levels of the four proteins that form the complex, had increased apoptosis resistance when treated with cisplatin, as compared to Lovo, SW707, and SW480 cells, which express low levels of CD44v6 or CO-029. In AS cells coexpressing EpCAM and claudin-7, there was an increase in ERK and AKT phosphorylation, increase in expression of the antiapoptotic proteins bcl-2 and bcl-xl, inhibition of the proapoptotic protein bad and resistance to cisplatin (Nubel et al., 2009). *In vivo*, injection of cancer cells coexpressing EpCAM and claudin-7 into nude mice led to increased tumorigenicity, increased tumor growth, and ascites production (Nubel et al.). In addition, expression of claudin-7, EpCAM, CO-029, and CD44v6 was upregulated in colon tumors and liver metastases, and coexpression was inversely correlated with disease-free survival (Kuhn et al.), implying that the association of claudin-7 and EpCAM is important for colon cancer progression.

There have been numerous studies investigating the roles of claudin-3 and claudin-4 in cancer. In the human ovarian surface epithelial (HOSE) cells, the overexpression of claudin-3 and -4 led to an increase in migration, invasion, and survival (Agarwal, D'Souza, & Morin, 2005). In addition, increased matrix metalloproteinase-2 (MMP-2) activity was detected in these cells overexpressing claudin-3 and -4, providing a possible mechanism for the increase in invasion. Similarly, overexpression of claudin-4 has been shown to stimulate the invasive activity of the colon cancer cell line Caco-2, likely through activation of MMP-2 and MMP-9 (Takehara, Nishimura, Mima, Hoshino, & Mizushima, 2009). Experiments performed in kidney 293T cells showed that the activation of MMP by claudin proteins was shown to occur through direct interaction between claudins and MMPs, and that several claudins could induce activation of pro-MMP2, including claudin-1,-2,-3, and -5 (Miyamori et al., 2001). Yet another study suggested that claudin-4 may be important in MMP-9 expression in gastric cancer and may also determine the development of intestinal-type over diffuse gastric cancer (Lee et al., 2008). In hepatocellular carcinoma cells, claudin-10 expression promoted cell survival, motility, and invasiveness, while MMP2 was upregulated (Ip, Cheung, Lee, Ho, & Fan, 2007).

HOSE cells stably transfected with claudin-4 preferentially expressed genes known to be involved in survival, migration, and angiogenesis (Li, Chigurupati, et al., 2009). In particular, genes encoding CXCL8 (IL-8), CXCL1 (Gro-α), CXCL2 (Gro-β), IL1B, and POSTN, all known to be involved in angiogenesis, were upregulated. Importantly, claudin-4 overexpressing cells promoted angiogenesis in *in vitro* and *in vivo* models, suggesting that claudin-4 expression may indeed lead to the secretion of factors that promote angiogenesis. In another study, injection of claudin-3 siRNA into three different ovarian cancer mouse models decreased tumor growth, increased apoptosis, and decreased proliferation (Huang et al., 2009). The production of ascites was decreased in the siRNA-injected mice, consistent with inhibition of metastasis. Claudin-4 is also highly overexpressed in pancreatic cancer cells (Michl et al., 2003); however, in contrast to ovarian cancer, its overexpression was associated with decreased invasion. The overexpressed claudin-4, in this case, was found primarily at cell to cell contacts and there was an increase in TJ formation in these cells. In colorectal cancer, claudin-4 expression is decreased particularly in invasive and metastatic lesions (Ueda et al., 2007). Reduced claudin-4 expression correlated significantly with depth of invasion and metastasis. In addition, siRNA silencing of claudin-4 in the colorectal cancer cells SW480 increased cell motility.

Several studies have implicated claudins in important oncogenic pathways. For example, claudin-1 has been suggested to be involved in TNF-α-dependent growth signals and proliferation in pancreatic cancer cells (Kondo et al., 2008). Claudin-1 and claudin-4 interact with the Eph ligand Ephrin-B1 in COS1 cells coexpressing these proteins (Tanaka et al., 2005), and interaction between claudins and Ephrin-B1 can induce phosphorylation of Ephrin-B1, which then leads to decreased cell to cell adhesion. It has been suggested that HGF can disrupt TJs in human breast cancer cells through changes in the expression of TJ components, including claudins (Martin, Watkins, Mansel, & Jiang, 2004). The simian virus 40 small tumor antigen induces deregulation of TJs in kidney epithelial cells through changes in several TJ molecules, including claudin-1 (Nunbhakdi-Craig, Craig, Machleidt, & Sontag, 2003). Claudin-1 has also been implicated in the regulation of E-cadherin as well as the β-catenin/Tcf signaling pathway in human colorectal cancers (Dhawan et al., 2005; Miwa et al., 2001). Indeed, claudin-1 is overexpressed in colon cancer and metastatic lesions, as compared to normal colon epithelium and the overexpressed claudin-1 was primarily mislocalized to the cytoplasm and nucleus (Dhawan et al.). Because of the patterns of expression of claudin-1 and E-cadherin in colon cells, a role for claudin-1 in the regulation of events that lead to an EMT in these cells was suggested. Activity of MMP-2 and MMP-9 is greater in the colon carcinoma cells SW480 stably transfected with claudin-1 and these cells exhibit increased tumorigenicity and metastatic behavior in nude mice. Claudin-1 is also overexpressed in melanoma cells, and similar to colon cancer, it was found primarily mislocalized to the cytoplasm in a melanoma tissue array and in the melanoma cell line M93-047 (Leotlela et al., 2007). The overexpression of claudin-1 in 1205LU melanoma cells increased motility as well as the activity of MMP-2. These experiments suggested a role for claudin-1 in migration and metastatic potential of melanoma cells (Leotlela et al.), although another study found claudin-1 expression was decreased in metastatic melanomas (Cohn et al., 2005).

Claudin-1 is overexpressed in oral squamous cell carcinoma, and its overexpression is associated with increased invasion (Dos Reis et al., 2008; Oku, Sasabe, Ueta, Yamamoto, & Osaki, 2006). Claudin-1 expression was higher in the highly invasive OSC-4 and NOS-2 cell lines and lower in the less-invasive OSC-7 cell line (Oku et al.). The invasive potential of the oral squamous cell carcinoma OSC-4 cells decreased when claudin-1 expression was knocked down with siRNA. In addition, the expression of MT1-MMP, the activation of MMP-2, and cleavage of laminin-5 $\gamma 2$ were reduced in OSC-4 cells treated with CLDN-1 siRNA. It was therefore suggested that claudin-1 expression may promote invasion of oral squamous cell carcinoma cells through increased expression, and activation of MT1-MMP and MMP-2 leading to the cleavage of laminin-5 $\gamma 2$ and induction of EGFR signaling (Oku et al.). Indeed, consistent with this view, the use of EGFR antibody or EGFR inhibitors decreased invasion. In an additional study, the overexpression of claudin-1 increased the invasion potential and, conversely, siRNA-mediated inhibition of claudin-1 expression decreased the invasive potential of several oral squamous cell carcinoma cell lines (Dos Reis et al.). In addition, in tissues, claudin-1 expression correlated with angiolymphatic and perineural invasion, further supporting a role for claudin-1 in invasion in this and other cancers (Dos Reis et al.).

Little is known about the roles of other claudins in cancer progression. In hepatocellular carcinoma, claudin-10 expression was associated with recurrence after hepatectomy (Cheung et al., 2005). Forced expression of claudin-10 in the hepatocellular carcinoma Hep3B cells increased the survival, migration, and invasion of these cells, whereas cell proliferation was unchanged. In addition, active MMP-2 was increased in these claudin-10 overexpressing cells and the expression of MT1-MMP increased at the mRNA and protein level. Interestingly, overexpression of claudin-10 could enhance the expression of claudin-1, claudin-2, and claudin-4. Consistent with these studies, hepatocellular carcinoma invasion could be suppressed through claudin-10 inhibition in HLE cells (Ip et al., 2007).

In conclusion, the expression, localization, and effects of claudins appear to be highly cancer- and claudin-specific. However, while they differ in their expression patterns, several common features are apparent. Aberrant expression of claudins appears to affect, positively or negatively, cell motility and invasion. In addition, in many instances, an increase in invasion appears to be correlated with an increase in expression and/or activation of metalloproteinases, in particular MMP-2 and MT1-MMP.

VI. CLINICAL IMPLICATIONS OF CLAUDIN OVEREXPRESSION IN CANCER

A. Detection and Diagnosis

Differential claudin expression in tumors as compared to normal tissues suggests that they might have utility as possible biomarkers for early detection, differential diagnosis, prognosis, or response to therapy. Table II summarizes the reports suggesting the possible use of claudins as clinical biomarkers. Because claudins are membrane proteins, not believed to be secreted, these proteins are not typically thought of as serum detection markers. However, a recent study shows that claudin proteins can be detected in the plasma of women with ovarian cancer (Li, Sherman-Baust, et al. 2009), suggesting they might have relevance as detection biomarkers. Indeed, claudin-4 was released, within small lipid particles named exosomes, into the plasma of ovarian cancer patients. Claudin-4 was detected in half of the ovarian cancer patient blood samples, but only very rarely in the blood of healthy individuals. While the claudin-4 test did not yield sensitivity and specificity values that were superior to the commonly used ovarian cancer biomarker CA125, claudin-4 was detected by immunoblotting and the development of better detection strategies may improve the value of claudins as detection biomarkers. Interestingly, expression of claudin-3 was detected in

TABLE II

Findings with Potential Prognostic and Diagnostic Value

Cancer	Claudin	Findings	References
Breast	Claudin-1,4	Expression of claudins-1, -4 associated with the basal-like subtype	Blanchard et al. (2009)
	Claudin-1	Decreased expression of claudin-1 correlated with short disease-free interval	Morohashi et al. (2007)
	Claudin-16	Low claudin-16 expression associated with node positive tumors, poor prognosis	Martin et al. (2008)
	Claudin-4	Claudin-4 expression associated with poor prognosis and high tumor grade	Lanigan et al. (2009)
	Claudin-4	Claudin-4 expression significantly higher in the basal-like tumors	Kulka et al. (2009)
	Claudin-7	Low claudin-7 correlated with higher tumor grade, metastatic disease	Sauer, Pedersen, Ebeltoft, & Naess (2005)
	Claudin-7	Loss of claudin-7 correlated with histological grade	Kominsky et al. (2003)
Cervical	Claudin-1	Claudin-1 RNA suggested in detection of cervical cancer	Steinau et al. (2007)
Colorectal	Claudin-7	Reduced claudin-7 correlates with venous invasion and liver metastasis	Oshima, Kunisaki, et al. (2008)
Endometrial	Claudin-1	Claudin-1 differentiates endometrioid from serous papillary subtypes	Sobel et al. (2006)
	Claudin-3,4	Expression of claudin-3 and -4 associated with poor clinical outcome	Konecny et al. (2008)
Esophageal	Claudin-1	Low claudin-1 correlated with recurrence status, short disease-free survival	Miyamoto et al. (2008)

(continued)

TABLE II (*continued*)

Cancer	Claudin	Findings	References
	Claudin-3,4	Loss of claudin-3 and -4 expression is associated with metastatic behavior	Takala et al. (2007)
	Claudin-5	Claudin-5 associated with lymph node metastasis	Chiba et al. (2009)
	Claudin-7	Low claudin-7 associated with, stage, lymphatic vessel invasion, metastasis	Usami et al. (2006)
Gastric	Claudin-3	Claudin-3 expression correlated with better prognosis	Soini, Tommola, et al. (2006)
	Claudin-3,4	The expression associated with histopathological type	Kamata et al. (2009)
	Claudin-3,4	High claudins, higher risk of synchronous and metachronous secondary tumors	Semba, Hasuo, Satake, Nakayama, & Yokozaki (2008)
	Claudin-4	Claudin-4 staining associated with decreased survival	Resnick, Gavilanez, et al. (2005)
	Claudin-4	Low claudin-4 correlated with tumor aggressiveness and survival	Ohtani et al. (2009)
	Claudin-4	Claudin-4 used to distinguish intestinal from diffuse gastric cancer	Wu, Lee, et al. (2006)
	Claudin-4	Low claudin-4 correlated with poor differentiation	Lee, Moon, et al. (2005)
HCC	Claudin-1	Low claudin-1 associated with dedifferentiation and portal invasion	Higashi et al. (2007)
	Claudin-4	To distinguish biliary tract carcinomas from hepatocellular carcinomas	Lodi et al. (2006)
	Claudin-10	Claudin-10 expression associated with recurrence	Cheung et al. (2005)

(*continued*)

TABLE II (*continued*)

Cancer	Claudin	Findings	References
Lung	Claudin-1	Low claudin-1 expression correlated with shorter overall survival	Chao et al. (2009)
	Claudin-1,4,5	To distinguish squamous cell carcinomas and adenocarcinomas.	Jung et al. (2009), Paschoud et al. (2007)
Meningioma	Claudin-1	To distinguish anaplastic meningiomas versus meningeal hemangiopericytomas	Hahn, Bundock, & Hornick (2006), Rajaram, Brat, & Perry (2004)
Mesothelioma	Claudin-1,3,4,5,7	Claudins 1, -3, -4, -5, and -7 less positive in mesothelioma	Soini, Kinnula, Kahlos, & Paakko (2006)
	Claudin-4	Claudin-4 to rule out the diagnosis of mesothelioma	Facchetti et al. (2007)
Oral	Claudin-1	High claudin-1 associated with increased invasiveness	Dos Reis et al. (2008)
Ovarian	Claudin-1,3,7	High claudin in ovarian carcinoma effusions associated with poor survival	Kleinberg et al. (2008)
	Claudin-3	High claudin-3 associated with shorter survival	Choi, Kim, Kwon, et al. (2007)
	Claudin-4	Claudin-4 as a possible detection biomarker	Li, Sherman-Baust, et al. (2009)
Pancreatic	Claudin-5,7	To distinguish solid-pseudopapillary from other tumors	Comper et al. (2009)
Perineurioma	Claudin-1	To distinguish soft tissue perineurioma from potential mimics	Folpe et al. (2002)
Prostate	Claudin-1	Low claudin-1 high tumor grade and recurrence	Sheehan et al. (2007)
	Claudin-1,4,5	To distinguish between patients with or without metastases	Szasz et al. (2009)
	Claudin-1,5	More strongly expressed in tumors with a lower Gleason score	Vare et al. (2008)

(*continued*)

TABLE II (*continued*)

Cancer	Claudin	Findings	References
	Claudin-3	High claudin-3 associated with advanced stage and recurrence	Sheehan et al. (2007)
	Claudin-4	High claudin-4 associated with advanced stage	Sheehan et al. (2007)
	Claudin-7	Low claudin-1 associated with high tumor grade	Sheehan et al. (2007)
RCC	Claudin-1	Claudin-1 associated with short disease-specific survival	Fritzsche et al. (2008)
	Claudin-3,4	Claudins-3 and -4 in prognosis	Lechpammer et al. (2008)
	Claudin-7,8	To distinguish chromophobe renal cell carcinomas from oncocytomas	Choi, Kim, Ryu, et al. (2007), Hornsby et al. (2007), Kim et al. (2009), Lechpammer et al. (2008) Osunkoya et al. (2009), Rohan et al. (2006)
Thyroid	Claudin-1	Loss of claudin-1 expression associated with worse disease-free survival	Tzelepi et al. (2008)
Tongue	Claudin-7	Low/high claudin-7 associated with decreased survival	Bello et al. (2008)
Urothelial	Claudin-1,4	Claudins-1 and -4 expression was significantly associated with stage	Nakanishi et al. (2008)
	Claudin-3	Claudin-3 expression was significantly associated with stage, grade, growth pattern	Nakanishi et al. (2008)

all epithelial ovarian cancers that express little or no CA125 (Rosen et al., 2005), suggesting this claudin may complement CA125 if a blood test could indeed be optimized.

Because of their very specific patterns of expression, claudins have been suggested to be useful for differential diagnosis (Table II). Distinguishing between histological subtypes of cancers, or tissue of origin, may have significant clinical implications for therapy and prognosis. For example, claudin-1, -3, and-7 were found to be useful in distinguishing ovarian cancer from cancers of other origins by examining expression in peritoneal effusions (Kleinberg et al., 2007). In biliary tract carcinomas claudin-4 was highly expressed, whereas it was absent in hepatocellular carcinomas, making claudin-4 a potential marker for distinguishing between these cancers (Lodi et al., 2006). Claudin-3 was differentially expressed in serous, endometrioid, and clear cell carcinomas as compared to mucinous tumors and normal epithelium (Lu et al., 2004). In lung cancer, claudin-4 and claudin-5 are strongly expressed in adenocarcinomas, whereas claudin-1 is upregulated in squamous cell carcinomas compared to adenocarcinomas, suggesting that together they could be useful in distinguishing between squamous cell carcinomas and adenocarcinomas in lung (Jung et al., 2009). In renal cell carcinomas, the expression of claudin-7 and claudin-8 were useful in distinguishing chromophobe renal cell carcinomas from oncocytomas, respectively (Choi, Kim, Ryu, et al., 2007; Hornsby et al., 2007; Kim et al., 2009; Lechpammer et al., 2008; Osunkoya et al., 2009; Rohan et al., 2006). In another renal cell carcinomas study, claudin-1 was absent in the majority of clear cell renal cell carcinomas, whereas papillary renal cell carcinomas almost always expressed claudin-1 (Fritzsche et al., 2008). Interestingly, claudin-4 appears to be present in a large number of epithelial neoplasms that can metastasize to serous membranes, and may therefore represent a specific pancarcinoma marker with extremely high sensitivity and specificity. Claudin-4 may therefore be used as an immunohistochemical reagent to rule out the diagnosis of mesothelioma (Facchetti et al., 2007).

B. Prognosis

Claudins have been suggested as potential markers of prognosis for several cancers (Table II). Generally high expression of claudins correlates with a poorer prognosis. Claudin-3 and -4 have often been shown to have prognosis ability. For example, overexpression of claudin-3 and/or claudin-4 has been associated with worse survival in endometrial cancer (Konecny et al., 2008), renal cell carcinoma (Lechpammer et al., 2008), breast cancer (Lanigan et al., 2009), ovarian cancer (Choi, Kim, Kwon, et al., 2007; Kleinberg, Holth, Trope, Reich, & Davidson, 2008), and gastric adenocarcinomas (Resnick, Gavilanez, et al., 2005). In lung adenocarcinomas, renal cell carcinoma, and stage II colon carcinomas, lower claudin-1 expression has been associated with reduced overall survival (Chao et al., 2009; Fritzsche et al., 2008; Resnick, Konkin, Routhier, Sabo, & Pricolo, 2005). In prostate adenocarcinomas, low expression of a combination of several claudins correlated with a high Gleason score and poor prognosis (Vare, Loikkanen, Hirvikoski, Vaarala, & Soini, 2008).

Claudins might be useful to predict patient outcome following therapy. Decreased expression of claudin-1, for instance, correlated with recurrence and short disease-free interval amongst breast cancer patients after simple mastectomy and axillary lymph node dissection (Morohashi et al., 2007), and claudin-4 expression predicted a poorer outcome in breast cancer patients that received adjuvant tamoxifen therapy (Lanigan et al., 2009). Claudin-10 expression in HCC appeared to be a marker of disease recurrence after curative hepatectomy (Cheung et al., 2005). In esophageal squamous cell carcinoma, decreased expression of claudin-1 also correlated with recurrence in patients that underwent esophagectomy and lymph node dissection (Miyamoto et al., 2008).

C. Therapy

The overexpression of claudins in various cancers and their localization at the plasma membrane makes them attractive candidates for targeted therapy. Several groups have generated antibodies targeting claudin-3 and -4 for possible therapy (Offner et al., 2005; Romani et al., 2009; Suzuki et al., 2009). In one of these studies, peptides derived from the two extracellular loops of claudin-1, claudin-3, and claudin-4 were used to generate antibodies that could bind to the surface of multiple cancer cell lines (Offner et al.). In another study, a humanized antibody against the second extracellular loop of claudin-3 was generated in a single-chain fragment format (scFv) (Romani et al.). The anti-claudin-3 antibody scFv H6 bound to multiple ovarian and endometrial cancer cells. An antibody to claudin-4 (KM3900) was also generated by immunizing BXSB mice with the pancreatic cell line Capan-2 and was shown to bind to the second extracellular loop of claudin-4 (Suzuki et al.). Importantly, a chimeric mouse–human antibody KM3934 suppressed the growth of tumors in MCAS or CFPAC-1 xenografts in SCID mice, implying that claudin-4 could be a relevant target for therapy. Anti-claudin-4 has been suggested as a possible radiotherapeutic agent for human pancreatic cancers (Foss et al., 2007). However, it is unclear at this juncture whether any of these antibodies will be useful therapeutically.

Claudin-3 and claudin-4 are receptors for *Clostridium perfringens* enterotoxin (CPE) (Katahira, Inoue, Horiguchi, Matsuda, & Sugimoto, 1997; Katahira et al., 1997), a 35 kDa polypeptide that is the cause of *C. perfringens* food poisoning in humans (McClane, Hanna, & Wnek, 1988). Upon binding to its receptors, CPE induces pore formation, leading to altered membrane permeability and lysis of epithelial cells (Kominsky et al., 2004; Long, Crean, Lee, Cummings, & Gabig, 2001; McClane, 2000). CPE has therefore been suggested as a potential therapy for tumors that overexpress claudin-3 and claudin-4 (Kominsky, 2006; Morin, 2005, 2007). Starting with early studies in prostate and pancreatic cancers, several papers have now shown that CPE therapy may be a viable therapeutic possibility for breast cancer (Kominsky et al., 2004), breast cancer brain metastasis (Kominsky et al., 2007), uterine serous papillary carcinoma (Santin et al., 2007), chemoresistant ovarian cancers (Santin, Cane, et al., 2005), and uterine carcinosarcomas (Santin et al., 2007). However, since some normal cells also express claudin-3 and -4, toxicity is an issue to consider making systemic delivery potentially problematic, although experiments in mice suggest that toxicity may be overcome. Certainly, the possibility of local CPE treatment for certain cancers may be an attractive alternative to systemic treatment. The claudin binding domain of CPE has also been suggested as a possible delivery agent of toxic compounds to tumors expressing claudin-3 and claudin-4. Indeed, a fragment of CPE, which has no cytolytic activity but maintains binding affinity to claudin-3 and claudin-4, was used to deliver TNF into ovarian cancer cells (Yuan et al., 2009). In these cells, TNF fused to the C-terminus of CPE ($CPE_{290-319}$–TNF) was more effective than TNF alone in killing ovarian cancer cells. In addition shRNA-mediated knockdown of claudin-3 or claudin-4 reduced the effects of $CPE_{290-319}$–TNF, indicating that its effects were specific. In another study, the protein synthesis inhibitory factor (PSIF) from *Pseudomonas aeruginosa* was fused to the C-terminus of CPE (C-CPE–PSIF) (Ebihara et al., 2006; Saeki et al., 2009). C-CPE–PSIF caused cell death in multiple cell lines expressing claudin-4. The intratumoral injection of C-CPE–PSIF decreased tumor volume in BALB/c mice with tumor xenografts, suggesting that C-CPE–PSIF or similar strategies might have potential therapeutic use *in vivo*.

While claudin proteins represent promising potential targets for therapy, several issues should be taken into consideration in the claudin-targeted therapies: (1) claudins are sometimes expressed in the cytoplasm or nucleus of cancer cells, making them unavailable as targets for CPE or antibody therapy. (2) Although they are overexpressed in most cancers, they are also expressed in certain normal tissues, which may cause toxicity. These toxic effects may be lessened by the fact that claudins are typically parts of TJs in normal tissues and potentially less available than claudins on cancer cells, which are not thought to form strong TJs. In addition, local delivery of CPE or antibodies may also provide an alternative approach with much reduced toxicity.

VII. CONCLUDING REMARKS

Claudin proteins are now clearly recognized as important players in the tumorigenesis of multiple tumors. Many studies have shown that various claudins may have diagnosis and prognosis utility. In addition, claudins may become useful as specific targets in cancer therapy. Future research in the field will likely concentrate on clarifying the exact roles for these proteins in cancer development and identifying possible uses for these proteins in the detection, diagnosis, and therapy of human cancer.

References

Agarwal, R., D'Souza, T., & Morin, P. J. (2005). Claudin-3 and claudin-4 expression in ovarian epithelial cells enhances invasion and is associated with increased matrix metalloproteinase-2 activity. *Cancer Research, 65*(16), 7378–7385.

Agarwal, R., Mori, Y., Cheng, Y., Jin, Z., Olaru, A. V., Hamilton, J. P., et al. (2009). Silencing of claudin-11 is associated with increased invasiveness of gastric cancer cells. *PLoS ONE, 4*(11), e8002.

Aldred, M. A., Huang, Y., Liyanarachchi, S., Pellegata, N. S., Gimm, O., Jhiang, S., et al. (2004). Papillary and follicular thyroid carcinomas show distinctly different microarray expression profiles and can be distinguished by a minimum of five genes. *Journal of Clinical Oncology, 22*(17), 3531–3539.

Al Moustafa, A. E., Alaoui-Jamali, M. A., Batist, G., Hernandez-Perez, M., Serruya, C., Alpert, L., et al. (2002). Identification of genes associated with head and neck carcinogenesis by cDNA microarray comparison between matched primary normal epithelial and squamous carcinoma cells. *Oncogene, 21*(17), 2634–2640.

Aono, S., & Hirai, Y. (2008). Phosphorylation of claudin-4 is required for tight junction formation in a human keratinocyte cell line. *Experimental Cell Research, 314*(18), 3326–3339.

Arabzadeh, A., Troy, T. C., & Turksen, K. (2007). Changes in the distribution pattern of Claudin tight junction proteins during the progression of mouse skin tumorigenesis. *BMC Cancer, 7*(196).

Aung, P. P., Mitani, Y., Sanada, Y., Nakayama, H., Matsusaki, K., & Yasui, W. (2006). Differential expression of claudin-2 in normal human tissues and gastrointestinal carcinomas. *Virchows Archiv, 448*(4), 428–434.

Banan, A., Zhang, L. J., Shaikh, M., Fields, J. Z., Choudhary, S., Forsyth, C. B., et al. (2005). Theta Isoform of protein kinase C alters barrier function in intestinal epithelium through modulation of distinct claudin isotypes: A novel mechanism for regulation of permeability. *The Journal of Pharmacology and Experimental Therapeutics, 313*(3), 962–982.

Banz, C., Ungethuem, U., Kuban, R. J., Diedrich, K., Lengyel, E., & Hornung, D. (2009). The molecular signature of endometriosis-associated endometrioid ovarian cancer differs significantly from endometriosis-independent endometrioid ovarian cancer. *Fertility and Sterility* [Epub ahead of print].

Bello, I. O., Vilen, S. T., Niinimaa, A., Kantola, S., Soini, Y., & Salo, T. (2008). Expression of claudins 1, 4, 5, and 7 and occludin, and relationship with prognosis in squamous cell carcinoma of the tongue. *Human Pathology, 39*(8), 1212–1220.

Bignotti, E., Tassi, R. A., Calza, S., Ravaggi, A., Romani, C., Rossi, E., et al. (2006). Differential gene expression profiles between tumor biopsies and short-term primary cultures of ovarian serous carcinomas: Identification of novel molecular biomarkers for early diagnosis and therapy. *Gynecologic Oncology, 103*(2), 405–416.

Billings, S. D., Walsh, S. V., Fisher, C., Nusrat, A., Weiss, S. W., & Folpe, A. L. (2004). Aberrant expression of tight junction-related proteins ZO-1, claudin-1 and occludin in synovial sarcoma: An immunohistochemical study with ultrastructural correlation. *Modern Pathology, 17*(2), 141–149.

Blackman, B., Russell, T., Nordeen, S. K., Medina, D., & Neville, M. C. (2005). Claudin 7 expression and localization in the normal murine mammary gland and murine mammary tumors. *Breast Cancer Research, 7*(2), R248–R255.

Blanchard, A. A., Skliris, G. P., Watson, P. H., Murphy, L. C., Penner, C., Tomes, L., et al. (2009). Claudins 1, 3, and 4 protein expression in ER negative breast cancer correlates with markers of the basal phenotype. *Virchows Archiv, 454*(6), 647–656.

Boireau, S., Buchert, M., Samuel, M. S., Pannequin, J., Ryan, J. L., Choquet, A., et al. (2007). DNA-methylation-dependent alterations of claudin-4 expression in human bladder carcinoma. *Carcinogenesis, 28*(2), 246–258.

Borlak, J., Meier, T., Halter, R., Spanel, R., & Spanel-Borowski, K. (2005). Epidermal growth factor-induced hepatocellular carcinoma: Gene expression profiles in precursor lesions, early stage and solitary tumours. *Oncogene, 24*(11), 1809–1819.

Burek, M., & Forster, C. Y. (2009). Cloning and characterization of the murine claudin-5 promoter. *Molecular and Cellular Endocrinology, 298*(1–2), 19–24.

Carrozzino, F., Soulie, P., Huber, D., Mensi, N., Orci, L., Cano, A., et al. (2005). Inducible expression of Snail selectively increases paracellular ion permeability and differentially modulates tight junction proteins. *American Journal of Physiology Cell Physiology, 289*(4), C1002–C1014.

Chang, T. L., Ito, K., Ko, T. K., Liu, Q., Salto-Tellez, M., Yeoh, K. G., et al. (2010). Claudin-1 has tumor suppressive activity and is a direct target of RUNX3 in gastric epithelial cells. *Gastroenterology, 138*(1), 255–265. e1-3.

Chao, Y. C., Pan, S. H., Yang, S. C., Yu, S. L., Che, T. F., Lin, C. W., et al. (2009). Claudin-1 is a metastasis suppressor and correlates with clinical outcome in lung adenocarcinoma. *American Journal of Respiratory and Critical Care Medicine, 179*(2), 123–133.

Chen, Y., Lu, Q., Schneeberger, E. E., & Goodenough, D. A. (2000). Restoration of tight junction structure and barrier function by down-regulation of the mitogen-activated protein kinase pathway in ras-transformed Madin-Darby canine kidney cells. *Molecular Biology of the Cell, 11*(3), 849–862.

Chen, S. P., Zhou, B., Willis, B. C., Sandoval, A. J., Liebler, J. M., Kim, K. J., et al. (2005). Effects of transdifferentiation and EGF on claudin isoform expression in alveolar epithelial cells. *Journal of Applied Physiology, 98*(1), 322–328.

Cheung, S. T., Leung, K. L., Ip, Y. C., Chen, X., Fong, D. Y., Ng, I. O., et al. (2005). Claudin-10 expression level is associated with recurrence of primary hepatocellular carcinoma. *Clinical Cancer Research, 11*(2 Pt 1), 551–556.

Chiba, T., Kawachi, H., Kawano, T., Kumagai, J., Kitagaki, K., Sekine, M., et al. (2009). Independent histological risk factors for lymph node metastasis of superficial esophageal squamous cell carcinoma; implication of claudin-5 immunohistochemistry for expanding the indications of endoscopic resection. *Diseases of the Esophagus* [Epub ahead of print].

Choi, Y. L., Kim, J., Kwon, M. J., Choi, J. S., Kim, T. J., Bae, D. S., et al. (2007). Expression profile of tight junction protein claudin 3 and claudin 4 in ovarian serous adenocarcinoma with prognostic correlation. *Histology and Histopathology, 22*(11), 1185–1195.

Choi, Y. D., Kim, K. S., Ryu, S., Park, Y., Cho, N. H., Rha, S. H., et al. (2007). Claudin-7 is highly expressed in chromophobe renal cell carcinoma and renal oncocytoma. *Journal of Korean Medical Science, 22*(2), 305–310.

Cohn, M. L., Goncharuk, V. N., Diwan, A. H., Zhang, P. S., Shen, S. S., & Prieto, V. G. (2005). Loss of claudin-1 expression in tumor-associated vessels correlates with acquisition of metastatic phenotype in melanocytic neoplasms. *Journal of Cutaneous Pathology, 32*(8), 533–536.

Colegio, O. R., Van Itallie, C. M., McCrea, H. J., Rahner, C., & Anderson, J. M. (2002). Claudins create charge-selective channels in the paracellular pathway between epithelial cells. *American Journal of Physiology Cell Physiology, 283*(1), C142–C147.

Comper, F., Antonello, D., Beghelli, S., Gobbo, S., Montagna, L., Pederzoli, P., et al. (2009). Expression pattern of claudins 5 and 7 distinguishes solid-pseudopapillary from pancreatoblastoma, acinar cell and endocrine tumors of the pancreas. *American Journal of Surgical Pathology, 33*(5), 768–774.

Cunningham, S. C., Kamangar, F., Kim, M. P., Hammoud, S., Haque, R., Iacobuzio-Donahue, C. A., et al. (2006). Claudin-4, mitogen-activated protein kinase kinase 4, and stratifin are markers of gastric adenocarcinoma precursor lesions. *Cancer Epidemiology, Biomarkers and Prevention, 15*(2), 281–287.

Darido, C., Buchert, M., Pannequin, J., Bastide, P., Zalzali, H., Mantamadiotis, T., et al. (2008). Defective claudin-7 regulation by Tcf-4 and Sox-9 disrupts the polarity and increases the tumorigenicity of colorectal cancer cells. *Cancer Research, 68*(11), 4258–4268.

Davidson, B., Zhang, Z., Kleinberg, L., Li, M., Florenes, V. A., Wang, T. L., et al. (2006). Gene expression signatures differentiate ovarian/peritoneal serous carcinoma from diffuse malignant peritoneal mesothelioma. *Clinical Cancer Research, 12*(20 Pt 1), 5944–5950.

de Oliveira, S. S., de Oliveira, I. M., De Souza, W., & Morgado-Diaz, J. A. (2005). Claudins upregulation in human colorectal cancer. *FEBS Letters, 579*(27), 6179–6185.

Dhawan, P., Singh, A. B., Deane, N. G., No, Y., Shiou, S. R., Schmidt, C., et al. (2005). Claudin-1 regulates cellular transformation and metastatic behavior in colon cancer. *The Journal of Clinical Investigation, 115*(7), 1765–1776.

Dos Reis, P. P., Bharadwaj, R. R., Machado, J., Macmillan, C., Pintilie, M., Sukhai, M. A., et al. (2008). Claudin 1 overexpression increases invasion and is associated with aggressive histological features in oral squamous cell carcinoma. *Cancer, 113*(11), 3169–3180.

D'Souza, T., Agarwal, R., & Morin, P. J. (2005). Phosphorylation of claudin-3 at threonine 192 by cAMP-dependent protein kinase regulates tight junction barrier function in ovarian cancer cells. *The Journal of Biological Chemistry, 280*(28), 26233–26240.

D'Souza, T., Indig, F. E., & Morin, P. J. (2007). Phosphorylation of claudin-4 by PKCepsilon regulates tight junction barrier function in ovarian cancer cells. *Experimental Cell Research, 313*(15), 3364–3375.

Dufresne, J., & Cyr, D. G. (2007). Activation of an SP binding site is crucial for the expression of claudin 1 in rat epididymal principal cells. *Biology of Reproduction, 76*(5), 825–832.

Ebihara, C., Kondoh, M., Hasuike, N., Harada, M., Mizuguchi, H., Horiguchi, Y., et al. (2006). Preparation of a claudin-targeting molecule using a C-terminal fragment of Clostridium perfringens enterotoxin. *The Journal of Pharmacology and Experimental Therapeutics, 316*(1), 255–260.

Facchetti, F., Lonardi, S., Gentili, F., Bercich, L., Falchetti, M., Tardanico, R., et al. (2007). Claudin 4 identifies a wide spectrum of epithelial neoplasms and represents a very useful marker for carcinoma versus mesothelioma diagnosis in pleural and peritoneal biopsies and effusions. *Virchows Archiv, 451*(3), 669–680.

Fluge, O., Bruland, O., Akslen, L. A., Lillehaug, J. R., & Varhaug, J. E. (2006). Gene expression in poorly differentiated papillary thyroid carcinomas. *Thyroid, 16*(2), 161–175.

Folpe, A. L., Billings, S. D., McKenney, J. K., Walsh, S. V., Nusrat, A., & Weiss, S. W. (2002). Expression of claudin-1, a recently described tight junction-associated protein, distinguishes soft tissue perineurioma from potential mimics. *American Journal of Surgical Pathology, 26*(12), 1620–1626.

Foss, C. A., Fox, J. J., Feldmann, G., Maitra, A., Iacobuzio-Donohue, C., Kern, S. E., et al. (2007). Radiolabeled anti-claudin 4 and anti-prostate stem cell antigen: Initial imaging in experimental models of pancreatic cancer. *Molecular Imaging, 6*(2), 131–139.

French, A. D., Fiori, J. L., Camilli, T. C., Leotlela, P. D., O'Connell, M. P., Frank, B. P., et al. (2009). PKC and PKA phosphorylation affect the subcellular localization of claudin-1 in melanoma cells. *International Journal of Medical Sciences, 6*(2), 93–101.

Fritzsche, F. R., Oelrich, B., Johannsen, M., Kristiansen, I., Moch, H., Jung, K., et al. (2008). Claudin-1 protein expression is a prognostic marker of patient survival in renal cell carcinomas. *Clinical Cancer Research, 14*(21), 7035–7042.

Fujibe, M., Chiba, H., Kojima, T., Soma, T., Wada, T., Yamashita, T., et al. (2004). Thr203 of claudin-1, a putative phosphorylation site for MAP kinase, is required to promote the barrier function of tight junctions. *Experimental Cell Research, 295*(1), 36–47.

Fujita, H., Chiba, H., Yokozaki, H., Sakai, N., Sugimoto, K., Wada, T., et al. (2006). Differential expression and subcellular localization of claudin-7, -8, -12, -13, and -15 along the mouse intestine. *The Journal of Histochemistry and Cytochemistry, 54*(8), 933–944.

Gress, T. M., Muller-Pillasch, F., Geng, M., Zimmerhackl, F., Zehetner, G., Friess, H., et al. (1996). A pancreatic cancer-specific expression profile. *Oncogene, 13*(8), 1819–1830.

Grone, J., Weber, B., Staub, E., Heinze, M., Klaman, I., Pilarsky, C., et al. (2007). Differential expression of genes encoding tight junction proteins in colorectal cancer: Frequent dysregulation of claudin-1, -8 and -12. *International Journal of Colorectal Disease, 22*(6), 651–659.

Hahn, H. P., Bundock, E. A., & Hornick, J. L. (2006). Immunohistochemical staining for claudin-1 can help distinguish meningiomas from histologic mimics. *American Journal of Clinical Pathology, 125*(2), 203–208.

Hamazaki, Y., Itoh, M., Sasaki, H., Furuse, M., & Tsukita, S. (2002). Multi-PDZ domain protein 1 (MUPP1) is concentrated at tight junctions through its possible interaction with claudin-1 and junctional adhesion molecule. *The Journal of Biological Chemistry, 277*(1), 455–461.

Han, X., Fink, M. P., & Delude, R. L. (2003). Proinflammatory cytokines cause NO*-dependent and -independent changes in expression and localization of tight junction proteins in intestinal epithelial cells. *Shock, 19*(3), 229–237.

Hanada, S., Maeshima, A., Matsuno, Y., Ohta, T., Ohki, M., Yoshida, T., et al. (2008). Expression profile of early lung adenocarcinoma: Identification of MRP3 as a molecular marker for early progression. *The Journal of Pathology, 216*(1), 75–82.

Heinzelmann-Schwarz, V. A., Gardiner-Garden, M., Henshall, S. M., Scurry, J., Scolyer, R. A., Davies, M. J., et al. (2004). Overexpression of the cell adhesion molecules DDR1, Claudin 3, and Ep-CAM in metaplastic ovarian epithelium and ovarian cancer. *Clinical Cancer Research, 10*(13), 4427–4436.

Hewitt, K. J., Agarwal, R., & Morin, P. J. (2006). The claudin gene family: Expression in normal and neoplastic tissues. *BMC Cancer, 6*, 186.

Hibbs, K., Skubitz, K. M., Pambuccian, S. E., Casey, R. C., Burleson, K. M., Oegema, T. R. Jr., et al. (2004). Differential gene expression in ovarian carcinoma: Identification of potential biomarkers. *American Journal of Pathology, 165*(2), 397–414.

Higashi, Y., Suzuki, S., Sakaguchi, T., Nakamura, T., Baba, S., Reinecker, H. C., et al. (2007). Loss of claudin-1 expression correlates with malignancy of hepatocellular carcinoma. *The Journal of Surgical Research, 139*(1), 68–76.

Hoevel, T., Macek, R., Swisshelm, K., & Kubbies, M. (2004). Reexpression of the TJ protein CLDN1 induces apoptosis in breast tumor spheroids. *International Journal of Cancer, 108*(3), 374–383.

Honda, H., Pazin, M. J., D'Souza, T., Ji, H., & Morin, P. J. (2007). Regulation of the CLDN3 gene in ovarian cancer cells. *Cancer Biology & Therapy, 6*(11), 1733–1742.

Honda, H., Pazin, M. J., Ji, H., Wernyj, R. P., & Morin, P. J. (2006). Crucial roles of Sp1 and epigenetic modifications in the regulation of the CLDN4 promoter in ovarian cancer cells. *The Journal of Biological Chemistry, 281*(30), 21433–21444.

Hornsby, C. D., Cohen, C., Amin, M. B., Picken, M. M., Lawson, D., Yin-Goen, Q., et al. (2007). Claudin-7 immunohistochemistry in renal tumors: A candidate marker for chromophobe renal cell carcinoma identified by gene expression profiling. *Archives of Pathology and Laboratory Medicine, 131*(10), 1541–1546.

Hough, C. D., Sherman-Baust, C. A., Pizer, E. S., Montz, F. J., Im, D. D., Rosenshein, N. B., et al. (2000). Large-scale serial analysis of gene expression reveals genes differentially expressed in ovarian cancer. *Cancer Research, 60*, 6281–6287.

Huang, Y. H., Bao, Y., Peng, W., Goldberg, M., Love, K., Bumcrot, D. A., et al. (2009). Claudin-3 gene silencing with siRNA suppresses ovarian tumor growth and metastasis. *Proceedings of the National Academy of Sciences of the United States of America, 106*(9), 3426–3430.

Huo, Q., Kinugasa, T., Wang, L., Huang, J., Zhao, J., Shibaguchi, H., et al. (2009). Claudin-1 protein is a major factor involved in the tumorigenesis of colorectal cancer. *Anticancer Research, 29*(3), 851–857.

Ikari, A., Atomi, K., Takiguchi, A., Yamazaki, Y., Miwa, M., & Sugatani, J. (2009). Epidermal growth factor increases claudin-4 expression mediated by Sp1 elevation in MDCK cells. *Biochemical and Biophysical Research Communications, 384*(3), 306–310.

Ikari, A., Ito, M., Okude, C., Sawada, H., Harada, H., Degawa, M., et al. (2008). Claudin-16 is directly phosphorylated by protein kinase A independently of a vasodilator-stimulated phosphoprotein-mediated pathway. *Journal of Cellular Physiology, 214*(1), 221–229.

Ikari, A., Matsumoto, S., Harada, H., Takagi, K., Hayashi, H., Suzuki, Y., et al. (2006). Phosphorylation of paracellin-1 at Ser217 by protein kinase A is essential for localization in tight junctions. *Journal of Cell Science, 119*(Pt 9), 1781–1789.

Ip, Y. C., Cheung, S. T., Lee, Y. T., Ho, J. C., & Fan, S. T. (2007). Inhibition of hepatocellular carcinoma invasion by suppression of claudin-10 in HLE cells. *Molecular Cancer Therapeutics, 6*(11), 2858–2867.

Ishida, M., Kushima, R., & Okabe, H. (2009). Claudin expression in rectal well-differentiated endocrine neoplasms (carcinoid tumors). *Oncology Reports, 21*(1), 113–117.

Ishizaki, T., Chiba, H., Kojima, T., Fujibe, M., Soma, T., Miyajima, H., et al. (2003). Cyclic AMP induces phosphorylation of claudin-5 immunoprecipitates and expression of claudin-5 gene in blood-brain-barrier endothelial cells via protein kinase A-dependent and -independent pathways. *Experimental Cell Research, 290*(2), 275–288.

Itoh, M., Furuse, M., Morita, K., Kubota, K., Saitou, M., & Tsukita, S. (1999). Direct binding of three tight junction-associated MAGUKs, ZO-1, ZO-2, and ZO-3, with the COOH termini of claudins. *The Journal of Cell Biology, 147*(6), 1351–1363.

Johnson, A. H., Frierson, H. F., Zaika, A., Powell, S. M., Roche, J., Crowe, S., et al. (2005). Expression of tight-junction protein claudin-7 is an early event in gastric tumorigenesis. *American Journal of Pathology, 167*(2), 577–584.

Jung, J. H., Jung, C. K., Choi, H. J., Jun, K. H., Yoo, J., Kang, S. J., et al. (2009). Diagnostic utility of expression of claudins in non-small cell lung cancer: Different expression profiles in squamous cell carcinomas and adenocarcinomas. *Pathology, Research and Practice, 205*(6), 409–416.

Kamata, I., Ishikawa, Y., Akishima-Fukasawa, Y., Ito, K., Akasaka, Y., Uzuki, M., et al. (2009). Significance of lymphatic invasion and cancer invasion-related proteins on lymph node metastasis in gastric cancer. *Journal of Gastroenterology and Hepatology, 24*(9), 1527–1533.

Karanjawala, Z. E., Illei, P. B., Ashfaq, R., Infante, J. R., Murphy, K., Pandey, A., et al. (2008). New markers of pancreatic cancer identified through differential gene expression analyses: Claudin 18 and annexin A8. *American Journal of Surgical Pathology, 32*(2), 188–196.

Katahira, J., Inoue, N., Horiguchi, Y., Matsuda, M., & Sugimoto, N. (1997). Molecular cloning and functional characterization of the receptor for Clostridium perfringens enterotoxin. *The Journal of Cell Biology, 136*(6), 1239–1247.

Katahira, J., Sugiyama, H., Inoue, N., Horiguchi, Y., Matsuda, M., & Sugimoto, N. (1997). Clostridium perfringens enterotoxin utilizes two structurally related membrane proteins as functional receptors in vivo. *The Journal of Biological Chemistry, 272*(42), 26652–26658.

Katoh, M., & Katoh, M. (2003). CLDN23 gene, frequently down-regulated in intestinal-type gastric cancer, is a novel member of CLAUDIN gene family. *International Journal of Molecular Medicine, 11*(6), 683–689.

Kim, S. S., Choi, Y. D., Jin, X. M., Cho, Y. M., Jang, J. J., Juhng, S. W., et al. (2009). Immunohistochemical stain for cytokeratin 7, S100A1 and claudin 8 is valuable in differential diagnosis of chromophobe renal cell carcinoma from renal oncocytoma. *Histopathology, 54*(5), 633–635.

Kim, T. H., Huh, J. H., Lee, S., Kang, H., Kim, G. I., & An, H. J. (2008). Down-regulation of claudin-2 in breast carcinomas is associated with advanced disease. *Histopathology, 53*(1), 48–55.

Kinugasa, T., Huo, Q., Higashi, D., Shibaguchi, H., Kuroki, M., Tanaka, T., et al. (2007). Selective up-regulation of claudin-1 and claudin-2 in colorectal cancer. *Anticancer Research, 27*(6A), 3729–3734.

Kinugasa, T., Sakaguchi, T., Gu, X., & Reinecker, H. C. (2000). Claudins regulate the intestinal barrier in response to immune mediators. *Gastroenterology, 118*(6), 1001–1011.

Kitajiri, S. I., Furuse, M., Morita, K., Saishin-Kiuchi, Y., Kido, H., Ito, J., et al. (2004). Expression patterns of claudins, tight junction adhesion molecules, in the inner ear. *Hearing Research, 187*(1–2), 25–34.

Kiuchi-Saishin, Y., Gotoh, S., Furuse, M., Takasuga, A., Tano, Y., & Tsukita, S. (2002). Differential expression patterns of claudins, tight junction membrane proteins, in mouse nephron segments. *Journal of the American Society of Nephrology, 13*(4), 875–886.

Kleinberg, L., Holth, A., Fridman, E., Schwartz, I., Shih Ie, M., & Davidson, B. (2007). The diagnostic role of claudins in serous effusions. *American Journal of Clinical Pathology, 127*(6), 928–937.

Kleinberg, L., Holth, A., Trope, C. G., Reich, R., & Davidson, B. (2008). Claudin upregulation in ovarian carcinoma effusions is associated with poor survival. *Human Pathology, 39*(5), 747–757.

Kominsky, S. L. (2006). Claudins: Emerging targets for cancer therapy. *Expert Reviews in Molecular Medicine, 8*(18), 1–11.

Kominsky, S. L., Argani, P., Korz, D., Evron, E., Raman, V., Garrett, E., et al. (2003). Loss of the tight junction protein claudin-7 correlates with histological grade in both ductal carcinoma in situ and invasive ductal carcinoma of the breast. *Oncogene, 22*(13), 2021–2033.

Kominsky, S. L., Tyler, B., Sosnowski, J., Brady, K., Doucet, M., Nell, D., et al. (2007). Clostridium perfringens enterotoxin as a novel-targeted therapeutic for brain metastasis. *Cancer Research, 67*(17), 7977–7982.

Kominsky, S. L., Vali, M., Korz, D., Gabig, T. G., Weitzman, S. A., Argani, P., et al. (2004). Clostridium perfringens enterotoxin elicits rapid and specific cytolysis of breast carcinoma cells mediated through tight junction proteins claudin 3 and 4. *American Journal of Pathology, 164*(5), 1627–1633.

Kondo, J., Sato, F., Kusumi, T., Liu, Y., Motonari, O., Sato, T., et al. (2008). Claudin-1 expression is induced by tumor necrosis factor-alpha in human pancreatic cancer cells. *International Journal of Mol Med, 22*(5), 645–649.

Konecny, G. E., Agarwal, R., Keeney, G. A., Winterhoff, B., Jones, M. B., Mariani, A., et al. (2008). Claudin-3 and claudin-4 expression in serous papillary, clear-cell, and endometrioid endometrial cancer. *Gynecologic Oncology, 109*(2), 263–269.

Kramer, F., White, K., Kubbies, M., Swisshelm, K., & Weber, B. H. (2000). Genomic organization of claudin-1 and its assessment in hereditary and sporadic breast cancer. *Human Genetics, 107*(3), 249–256.

Krause, G., Winkler, L., Mueller, S. L., Haseloff, R. F., Piontek, J., & Blasig, I. E. (2008). Structure and function of claudins. *Biochimica et Biophysica Acta, 1778*(3), 631–645.

Kuhn, S., Koch, M., Nubel, T., Ladwein, M., Antolovic, D., Klingbeil, P., et al. (2007). A complex of EpCAM, claudin-7, CD44 variant isoforms, and tetraspanins promotes colorectal cancer progression. *Molecular Cancer Research, 5*(6), 553–567.

Kulka, J., Szasz, A. M., Nemeth, Z., Madaras, L., Schaff, Z., Molnar, I. A., et al. (2009). Expression of tight junction protein claudin-4 in basal-like breast carcinomas. *Pathology and Oncology Research, 15*(1), 59–64.

Kumei, S., Motomura, W., Yoshizaki, T., Takakusaki, K., & Okumura, T. (2009). Troglitazone increases expression of E-cadherin and claudin 4 in human pancreatic cancer cells. *Biochemical and Biophysical Research Communications, 380*(3), 614–619.

Ladwein, M., Pape, U. F., Schmidt, D. S., Schnolzer, M., Fiedler, S., Langbein, L., et al. (2005). The cell-cell adhesion molecule EpCAM interacts directly with the tight junction protein claudin-7. *Experimental Cell Research, 309*(2), 345–357.

Lal-Nag, M., & Morin, P. J. (2009). The claudins. *Genome Biology, 10*(8), 235.

Landers, K. A., Samaratunga, H., Teng, L., Buck, M., Burger, M. J., Scells, B., et al. (2008). Identification of claudin-4 as a marker highly overexpressed in both primary and metastatic prostate cancer. *British Journal of Cancer, 99*(3), 491–501.

Lanigan, F., McKiernan, E., Brennan, D. J., Hegarty, S., Millikan, R. C., McBryan, J., et al. (2009). Increased claudin-4 expression is associated with poor prognosis and high tumour grade in breast cancer. *International Journal of Cancer, 124*(9), 2088–2097.

Lechpammer, M., Resnick, M. B., Sabo, E., Yakirevich, E., Greaves, W. O., Sciandra, K. T., et al. (2008). The diagnostic and prognostic utility of claudin expression in renal cell neoplasms. *Modern Pathology, 21*(11), 1320–1329.

Lee, J. W., Hsiao, W. T., Chen, H. Y., Hsu, L. P., Chen, P. R., Lin, M. D., et al. (2010). Up-regulated claudin-1 expression confers resistance to cell death of nasopharyngeal carcinoma cells. *International Journal of Cancer, 126*(6), 1353–1366.

Lee, J. W., Lee, S. J., Seo, J., Song, S. Y., Ahn, G., Park, C. S., et al. (2005). Increased expressions of claudin-1 and claudin-7 during the progression of cervical neoplasia. *Gynecologic Oncology, 97*(1), 53–59.

Lee, S. K., Moon, J., Park, S. W., Song, S. Y., Chung, J. B., & Kang, J. K. (2005). Loss of the tight junction protein claudin 4 correlates with histological growth-pattern and differentiation in advanced gastric adenocarcinoma. *Oncol Rep, 13*(2), 193–199.

Lee, L. Y., Wu, C. M., Wang, C. C., Yu, J. S., Liang, Y., Huang, K. H., et al. (2008). Expression of matrix metalloproteinases MMP-2 and MMP-9 in gastric cancer and their relation to claudin-4 expression. *Histology and Histopathology, 23*(5), 515–521.

Leotlela, P. D., Wade, M. S., Duray, P. H., Rhode, M. J., Brown, H. F., Rosenthal, D. T., et al. (2007). Claudin-1 overexpression in melanoma is regulated by PKC and contributes to melanoma cell motility. *Oncogene, 26*(26), 3846–3856.

Li, J., Chigurupati, S., Agarwal, R., Mughal, M. R., Mattson, M. P., Becker, K. G., et al. (2009). Possible angiogenic roles for claudin-4 in ovarian cancer. *Cancer Biology & Therapy, 8*(19) 1806–1814.

Li, W. Y., Huey, C. L., & Yu, A. S. (2004). Expression of claudin-7 and -8 along the mouse nephron. *American Journal of Physical-Renal Physiology, 286*(6), F1063–F1071.

Li, J., Sherman-Baust, C. A., Tsai-Turton, M., Bristow, R. E., Roden, R. B., & Morin, P. J. (2009). Claudin-containing exosomes in the peripheral circulation of women with ovarian cancer. *BMC Cancer, 9*, 244.

Li, L., Yao, J. L., di Sant'agnese, P. A., Bourne, P. A., Picken, M. M., Young, A. N., et al. (2008). Expression of claudin-7 in benign kidney and kidney tumors. *International Journal of Clinical and Experimental Pathology, 1*(1), 57–64.

Lioni, M., Brafford, P., Andl, C., Rustgi, A., El-Deiry, W., Herlyn, M., et al. (2007). Dysregulation of claudin-7 leads to loss of E-cadherin expression and the increased invasion of esophageal squamous cell carcinoma cells. *American Journal of Pathology, 170*(2), 709–721.

Lipschutz, J. H., Li, S., Arisco, A., & Balkovetz, D. F. (2005). Extracellular signal-regulated kinases 1/2 control claudin-2 expression in Madin-Darby canine kidney strain I and II cells. *The Journal of Biological Chemistry, 280*(5), 3780–3788.

Litkouhi, B., Kwong, J., Lo, C. M., Smedley, III, J.G., McClane, B. A., Aponte, M., et al. (2007). Claudin-4 overexpression in epithelial ovarian cancer is associated with hypomethylation and is a potential target for modulation of tight junction barrier function using a C-terminal fragment of Clostridium perfringens enterotoxin. *Neoplasia, 9*(4), 304–314.

Liu, Y., Sun, W., Zhang, K., Zheng, H., Ma, Y., Lin, D., et al. (2007). Identification of genes differentially expressed in human primary lung squamous cell carcinoma. *Lung Cancer, 56*(3), 307–317.

Lodi, C., Szabo, E., Holczbauer, A., Batmunkh, E., Szijarto, A., Kupcsulik, P., et al. (2006). Claudin-4 differentiates biliary tract cancers from hepatocellular carcinomas. *Modern Pathology, 19*(3), 460–469.

Long, H., Crean, C. D., Lee, W. H., Cummings, O. W., & Gabig, T. G. (2001). Expression of Clostridium perfringens enterotoxin receptors claudin-3 and claudin-4 in prostate cancer epithelium. *Cancer Research, 61*(21), 7878–7881.

Lopardo, T., Lo Iacono, N., Marinari, B., Giustizieri, M. L., Cyr, D. G., Merlo, G., et al. (2008). Claudin-1 is a p63 target gene with a crucial role in epithelial development. *PLoS ONE, 3*(7), e2715.

Lu, K. H., Patterson, A. P., Wang, L., Marquez, R. T., Atkinson, E. N., Baggerly, K. A., et al. (2004). Selection of potential markers for epithelial ovarian cancer with gene expression arrays and recursive descent partition analysis. *Clinical Cancer Research, 10*(10), 3291–3300.

Martin, T. A., Harrison, G. M., Watkins, G., & Jiang, W. G. (2008). Claudin-16 reduces the aggressive behavior of human breast cancer cells. *Journal of Cellular Biochemistry 105*(1), 41–52.

Martin, T. A., Watkins, G., Mansel, R. E., & Jiang, W. G. (2004). Hepatocyte growth factor disrupts tight junctions in human breast cancer cells. *Cell Biology International, 28*(5), 361–371.

Martinez-Estrada, O. M., Culleres, A., Soriano, F. X., Peinado, H., Bolos, V., Martinez, F. O., et al. (2006). The transcription factors Slug and Snail act as repressors of Claudin-1 expression in epithelial cells. *Biochemical Journal, 394*(Pt 2), 449–457.

Martins, F. C., Teixeira, F., Reis, I., Geraldes, N., Cabrita, A. M., & Dias, M. F. (2009). Increased transglutaminase 2 and GLUT-1 expression in breast tumors not susceptible to chemoprevention with antioxidants. *Tumori, 95*(2), 227–232.

Matsuda, Y., Semba, S., Ueda, J., Fuku, T., Hasuo, T., Chiba, H., et al. (2007). Gastric and intestinal claudin expression at the invasive front of gastric carcinoma. *Cancer Science, 98*(7), 1014–1019.

McClane, B. A. (2000). Clostridium perfringens enterotoxin and intestinal tight junctions. *Trends in Microbiology, 8*(4), 145–146.

McClane, B. A., Hanna, P. C., & Wnek, A. P. (1988). Clostridium perfringens enterotoxin. *Microbial Pathogenesis, 4*(5), 317–323.

Mees, S. T., Mennigen, R., Spieker, T., Rijcken, E., Senninger, N., Haier, J., et al. (2009). Expression of tight and adherens junction proteins in ulcerative colitis associated colorectal carcinoma: Upregulation of claudin-1, claudin-3, claudin-4, and beta-catenin. *International Journal of Colorectal Disease, 24*(4), 361–368.

Michl, P., Barth, C., Buchholz, M., Lerch, M. M., Rolke, M., Holzmann, K. H., et al. (2003). Claudin-4 expression decreases invasiveness and metastatic potential of pancreatic cancer. *Cancer Research, 63*(19), 6265–6271.

Mima, S., Tsutsumi, S., Ushijima, H., Takeda, M., Fukuda, I., Yokomizo, K., et al. (2005). Induction of claudin-4 by nonsteroidal anti-inflammatory drugs and its contribution to their chemopreventive effect. *Cancer Research, 65*(5), 1868–1876.

Mima, S., Takehara, M., Takada, H., Nishimura, T., Hoshino, T., & Mizushima, T. (2008). NSAIDs suppress the expression of claudin-2 to promote invasion activity of cancer cells. *Carcinogenesis, 29*(10), 1994–2000.

Mittal, M. K., Myers, J. N., Bailey, C. K., Misra, S., & Chaudhuri, G. (2010). Mode of action of the retrogene product SNAI1P, a SNAIL homolog, in human breast cancer cells. *Molecular Biology Reports, 37*(3), 1221–1227.

Miwa, N., Furuse, M., Tsukita, S., Niikawa, N., Nakamura, Y., & Furukawa, Y. (2001). Involvement of claudin-1 in the beta-catenin/Tcf signaling pathway and its frequent upregulation in human colorectal cancers. *Oncology Research, 12*(11–12), 469–476.

Miyamori, H., Takino, T., Kobayashi, Y., Tokai, H., Itoh, Y., Seiki, M., et al. (2001). Claudin promotes activation of pro-matrix metalloproteinase-2 mediated by membrane-type matrix metalloproteinases. *The Journal of Biological Chemistry, 276*(30), 28204–28211.

Miyamoto, K., Kusumi, T., Sato, F., Kawasaki, H., Shibata, S., Ohashi, M., et al. (2008). Decreased expression of claudin-1 is correlated with recurrence status in esophageal squamous cell carcinoma. *Biomedical Research, 29*(2), 71–76.

Moldvay, J., Jackel, M., Paska, C., Soltesz, I., Schaff, Z., & Kiss, A. (2007). Distinct claudin expression profile in histologic subtypes of lung cancer. *Lung Cancer, 57*(2), 159–167.

Montgomery, E., Mamelak, A. J., Gibson, M., Maitra, A., Sheikh, S., Amr, S. S., et al. (2006). Overexpression of claudin proteins in esophageal adenocarcinoma and its precursor lesions. *Applied Immunohistochemistry & Molecular Morphology, 14*(1), 24–30.

Morin, P. J. (2005). Claudin proteins in human cancer: Promising new targets for diagnosis and therapy. *Cancer Research, 65*(21), 9603–9606.

Morin, P. J. (2007). Claudin proteins in ovarian cancer. *Disease Markers, 23*(5–6), 453–457.

Morita, K., Morita, N. I., Nemoto, K., Nakamura, Y., Miyachi, Y., & Muto, M. (2008). Expression of claudin in melanoma cells. *The Journal of Dermatology, 35*(1), 36–38.

Morita, K., Tsukita, S., & Miyachi, Y. (2004). Tight junction-associated proteins (occludin, ZO-1, claudin-1, claudin-4) in squamous cell carcinoma and Bowen's disease. *British Journal of Dermatology, 151*(2), 328–334.

Morohashi, S., Kusumi, T., Sato, F., Odagiri, H., Chiba, H., Yoshihara, S., et al. (2007). Decreased expression of claudin-1 correlates with recurrence status in breast cancer. *International Journal of Molecular Medicine, 20*(2), 139–143.

Nakanishi, K., Ogata, S., Hiroi, S., Tominaga, S., Aida, S., & Kawai, T. (2008). Expression of occludin and claudins 1, 3, 4, and 7 in urothelial carcinoma of the upper urinary tract. *American Journal of Clinical Pathology, 130*(1), 43–49.

Nakayama, F., Semba, S., Usami, Y., Chiba, H., Sawada, N., & Yokozaki, H. (2008). Hypermethylation-modulated downregulation of claudin-7 expression promotes the progression of colorectal carcinoma. *Pathobiology, 75*(3), 177–185.

Nemeth, Z., Szasz, A. M., Tatrai, P., Nemeth, J., Gyorffy, H., Somoracz, A., et al. (2009). Claudin-1, -2, -3, -4, -7, -8, and -10 protein expression in biliary tract cancers. *The Journal of Histochemistry and Cytochemistry, 57*(2), 113–121.

Nichols, L. S., Ashfaq, R., & Iacobuzio-Donahue, C. A. (2004). Claudin 4 protein expression in primary and metastatic pancreatic cancer: Support for use as a therapeutic target. *American Journal of Clinical Pathology, 121*(2), 226–230.

Nishino, R., Honda, M., Yamashita, T., Takatori, H., Minato, H., Zen, Y., et al. (2008). Identification of novel candidate tumour marker genes for intrahepatic cholangiocarcinoma. *Journal of Hepatology, 49*(2), 207–216.

Nubel, T., Preobraschenski, J., Tuncay, H., Weiss, T., Kuhn, S., Ladwein, M., et al. (2009). Claudin-7 regulates EpCAM-mediated functions in tumor progression. *Molecular Cancer Research, 7*(3), 285–299.

Nunbhakdi-Craig, V., Craig, L., Machleidt, T., & Sontag, E. (2003). Simian virus 40 small tumor antigen induces deregulation of the actin cytoskeleton and tight junctions in kidney epithelial cells. *Journal of Virology, 77*(5), 2807–2818.

Nunbhakdi-Craig, V., Machleidt, T., Ogris, E., Bellotto, D., White, III, C.L., Sontag, E. (2002). Protein phosphatase 2A associates with and regulates atypical PKC and the epithelial tight junction complex. *The Journal of Cell Biology, 158*(5), 967–978.

Offner, S., Hekele, A., Teichmann, U., Weinberger, S., Gross, S., Kufer, P., et al. (2005). Epithelial tight junction proteins as potential antibody targets for pancarcinoma therapy. *Cancer Immunology, Immunotherapy, 54*(5), 431–445.

Ohtani, S., Terashima, M., Satoh, J., Soeta, N., Saze, Z., Kashimura, S., et al. (2009). Expression of tight-junction-associated proteins in human gastric cancer: Downregulation of claudin-4 correlates with tumor aggressiveness and survival. *Gastric Cancer, 12*(1), 43–51.

Oku, N., Sasabe, E., Ueta, E., Yamamoto, T., & Osaki, T. (2006). Tight junction protein claudin-1 enhances the invasive activity of oral squamous cell carcinoma cells by promoting cleavage of laminin-5 gamma2 chain via matrix metalloproteinase (MMP)-2 and membrane-type MMP-1. *Cancer Research, 66*(10), 5251–5257.

Osanai, M., Murata, M., Chiba, H., Kojima, T., & Sawada, N. (2007). Epigenetic silencing of claudin-6 promotes anchorage-independent growth of breast carcinoma cells. *Cancer Science, 98*(10), 1557–1562.

Oshima, T., Kunisaki, C., Yoshihara, K., Yamada, R., Yamamoto, N., Sato, T., et al. (2008). Reduced expression of the claudin-7 gene correlates with venous invasion and liver metastasis in colorectal cancer. *Oncology Reports, 19*(4), 953–959.

Oshima, T., Miwa, H., & Joh, T. (2008). Aspirin induces gastric epithelial barrier dysfunction by activating p38 MAPK via claudin-7. *American Journal of Physiology Cell Physiology, 295*(3), C800–C806.

Osunkoya, A. O., Cohen, C., Lawson, D., Picken, M. M., Amin, M. B., & Young, A. N. (2009). Claudin-7 and claudin-8: Immunohistochemical markers for the differential diagnosis of chromophobe renal cell carcinoma and renal oncocytoma. *Human Pathology, 40*(2), 206–210.

Pan, X. Y., Wang, B., Che, Y. C., Weng, Z. P., Dai, H. Y., & Peng, W. (2007). Expression of claudin-3 and claudin-4 in normal, hyperplastic, and malignant endometrial tissue. *International Journal of Gynecological Cancer, 17*(1), 233–241.

Park, D., Karesen, R., Axcrona, U., Noren, T., & Sauer, T. (2007). Expression pattern of adhesion molecules (E-cadherin, alpha-, beta-, gamma-catenin and claudin-7), their influence on survival in primary breast carcinoma, and their corresponding axillary lymph node metastasis. *Apmis, 115*(1), 52–65.

Park, J. Y., Park, K. H., Oh, T. Y., Hong, S. P., Jeon, T. J., Kim, C. H., et al. (2007). Up-regulated claudin 7 expression in intestinal-type gastric carcinoma. *Oncology Reports, 18*(2), 377–382.

Paschoud, S., Bongiovanni, M., Pache, J. C., & Citi, S. (2007). Claudin-1 and claudin-5 expression patterns differentiate lung squamous cell carcinomas from adenocarcinomas. *Modern Pathology, 20*(9), 947–954.

Peter, Y., Comellas, A., Levantini, E., Ingenito, E. P., & Shapiro, S. D. (2009). Epidermal growth factor receptor and claudin-2 participate in A549 permeability and remodeling: Implications for non-small cell lung cancer tumor colonization. *Molecular Carcinogenesis, 48*(6), 488–497.

Piontek, J., Winkler, L., Wolburg, H., Muller, S. L., Zuleger, N., Piehl, C., et al. (2008). Formation of tight junction: Determinants of homophilic interaction between classic claudins. *FASEB Journal, 22*(1), 146–158.

Quan, C., & Lu, S. J. (2003). Identification of genes preferentially expressed in mammary epithelial cells of Copenhagen rat using subtractive hybridization and microarrays. *Carcinogenesis, 24*(10), 1593–1599.

Rajaram, V., Brat, D. J., & Perry, A. (2004). Anaplastic meningioma versus meningeal hemangiopericytoma: Immunohistochemical and genetic markers. *Human Pathology, 35*(11), 1413–1418.

Rangel, L. B. A., Agarwal, R., D'Souza, T., Pizer, E. S., Alò, P. L., Lancaster, W. D., et al. (2003). Tight junction proteins Claudin-3 and Claudin-4 are frequently overexpressed in ovarian cancer but not in ovarian cystadenomas. *Clinical Cancer Research, 9*, 2567–2575.

Rangel, L. B. A., Sherman-Baust, C. A., Wernyj, R. P., Schwartz, D. R., Cho, K. R., & Morin, P. J. (2003). Characterization of novel human ovarian cancer-specific transcripts (HOSTs) identified by serial analysis of gene expression. *Oncogene, 22*, 7225–7232.

Resnick, M. B., Gavilanez, M., Newton, E., Konkin, T., Bhattacharya, B., Britt, D. E., et al. (2005). Claudin expression in gastric adenocarcinomas: A tissue microarray study with prognostic correlation. *Human Pathology, 36*(8), 886–892.

Resnick, M. B., Konkin, T., Routhier, J., Sabo, E., & Pricolo, V. E. (2005). Claudin-1 is a strong prognostic indicator in stage II colonic cancer: A tissue microarray study. *Modern Pathology, 18*(4), 511–518.

Reyes, J. L., Lamas, M., Martin, D., del Carmen Namorado, M., Islas, S., Luna, J., et al. (2002). The renal segmental distribution of claudins changes with development. *Kidney International, 62*(2), 476–487.

Roh, M. H., Liu, C. J., Laurinec, S., & Margolis, B. (2002). The carboxyl terminus of zona occludens-3 binds and recruits a mammalian homologue of discs lost to tight junctions. *The Journal of Biological Chemistry, 277*(30), 27501–27509.

Rohan, S., Tu, J. J., Kao, J., Mukherjee, P., Campagne, F., Zhou, X. K., et al. (2006). Gene expression profiling separates chromophobe renal cell carcinoma from oncocytoma and identifies vesicular transport and cell junction proteins as differentially expressed genes. *Clinical Cancer Research, 12*(23), 6937–6945.

Romani, C., Comper, F., Bandiera, E., Ravaggi, A., Bignotti, E., Tassi, R. A., et al. (2009). Development and characterization of a human single-chain antibody fragment against claudin-3: A novel therapeutic target in ovarian and uterine carcinomas. *American Journal of Obstetrics and Gynecology, 201*(1), 70e1, 70e9.

Rosen, D. G., Wang, L., Atkinson, J. N., Yu, Y., Lu, K. H., Diamandis, E. P., et al. (2005). Potential markers that complement expression of CA125 in epithelial ovarian cancer. *Gynecologic Oncology, 99*(2), 267–277.

Roth, M. J., Abnet, C. C., Hu, N., Wang, Q. H., Wei, W. Q., Green, L., et al. (2006). p16, MGMT, RARbeta2, CLDN3, CRBP and MT1G gene methylation in esophageal squamous cell carcinoma and its precursor lesions. *Oncology Reports, 15*(6), 1591–1597.

Saeki, R., Kondoh, M., Kakutani, H., Tsunoda, S., Mochizuki, Y., Hamakubo, T., et al. (2009). A novel tumor-targeted therapy using a claudin-4-targeting molecule. *Molecular Pharmacology, 76*(4), 918–926.

Sakai, N., Chiba, H., Fujita, H., Akashi, Y., Osanai, M., Kojima, T., et al. (2007). Expression patterns of claudin family of tight-junction proteins in the mouse prostate. *Histochemistry and Cell Biology, 127*(4), 457–462.

Sanada, Y., Oue, N., Mitani, Y., Yoshida, K., Nakayama, H., & Yasui, W. (2006). Down-regulation of the claudin-18 gene, identified through serial analysis of gene expression data analysis, in gastric cancer with an intestinal phenotype. *The Journal of Pathology, 208*(5), 633–642.

Santin, A. D., Bellone, S., Marizzoni, M., Palmieri, M., Siegel, E. R., McKenney, J. K., et al. (2007). Overexpression of claudin-3 and claudin-4 receptors in uterine serous papillary carcinoma: Novel targets for a type-specific therapy using Clostridium perfringens enterotoxin (CPE). *Cancer, 109*(7), 1312–1322.

Santin, A. D., Bellone, S., Siegel, E. R., McKenney, J. K., Thomas, M., Roman, J. J., et al. (2007). Overexpression of Clostridium perfringens enterotoxin receptors claudin-3 and claudin-4 in uterine carcinosarcomas. *Clinical Cancer Research, 13*(11), 3339–3346.

Santin, A. D., Cane, S., Bellone, S., Palmieri, M., Siegel, E. R., Thomas, M., et al. (2005). Treatment of chemotherapy-resistant human ovarian cancer xenografts in C.B-17/SCID mice by intraperitoneal administration of Clostridium perfringens enterotoxin. *Cancer Research, 65*(10), 4334–4342.

Santin, A. D., Zhan, F., Bellone, S., Palmieri, M., Cane, S., Bignotti, E., et al. (2004). Gene expression profiles in primary ovarian serous papillary tumors and normal ovarian epithelium: Identification of candidate molecular markers for ovarian cancer diagnosis and therapy. *International Journal of Cancer, 112*(1), 14–25.

Santin, A. D., Zhan, F., Cane, S., Bellone, S., Palmieri, M., Thomas, M., et al. (2005). Gene expression fingerprint of uterine serous papillary carcinoma: Identification of novel molecular markers for uterine serous cancer diagnosis and therapy. *British Journal of Cancer, 92*(8), 1561–1573.

Satake, S., Semba, S., Matsuda, Y., Usami, Y., Chiba, H., Sawada, N., et al. (2008). Cdx2 transcription factor regulates claudin-3 and claudin-4 expression during intestinal differentiation of gastric carcinoma. *Pathology International, 58*(3), 156–163.

Sato, N., Fukushima, N., Maitra, A., Iacobuzio-Donahue, C. A., van Heek, N. T., Cameron, J. L., et al. (2004). Gene expression profiling identifies genes associated with invasive intraductal papillary mucinous neoplasms of the pancreas. *American Journal of Pathology, 164*(3), 903–914.

Sauer, T., Pedersen, M. K., Ebeltoft, K., & Naess, O. (2005). Reduced expression of Claudin-7 in fine needle aspirates from breast carcinomas correlate with grading and metastatic disease. *Cytopathology, 16*(4), 193–198.

Schuetz, A. N., Yin-Goen, Q., Amin, M. B., Moreno, C. S., Cohen, C., Hornsby, C. D., et al. (2005). Molecular classification of renal tumors by gene expression profiling. *The Journal of Molecular Diagnostics, 7*(2), 206–218.

Semba, S., Hasuo, T., Satake, S., Nakayama, F., & Yokozaki, H. (2008). Prognostic significance of intestinal claudins in high-risk synchronous and metachronous multiple gastric epithelial neoplasias after initial endoscopic submucosal dissection. *Pathology International, 58*(6), 371–377.

Sentani, K., Oue, N., Tashiro, T., Sakamoto, N., Nishisaka, T., Fukuhara, T., et al. (2008). Immunohistochemical staining of Reg IV and claudin-18 is useful in the diagnosis of gastrointestinal signet ring cell carcinoma. *American Journal of Surgical Pathology, 32*(8), 1182–1189.

Sheehan, G. M., Kallakury, B. V., Sheehan, C. E., Fisher, H. A., Kaufman, R. P., Jr., Ross, J. S. (2007). Loss of claudins-1 and -7 and expression of claudins-3 and -4 correlate with prognostic variables in prostatic adenocarcinomas. *Human Pathology, 38*(4), 564–569.

Shiou, S. R., Singh, A. B., Moorthy, K., Datta, P. K., Washington, M. K., Beauchamp, R. D., et al. (2007). Smad4 regulates claudin-1 expression in a transforming growth factor-beta-independent manner in colon cancer cells. *Cancer Research, 67*(4), 1571–1579.

Singh, A. B., & Harris, R. C. (2004). Epidermal growth factor receptor activation differentially regulates claudin expression and enhances transepithelial resistance in Madin-Darby canine kidney cells. *The Journal of Biological Chemistry, 279*(5), 3543–3552.

Sobel, G., Nemeth, J., Kiss, A., Lotz, G., Szabo, I., Udvarhelyi, N., et al. (2006). Claudin 1 differentiates endometrioid and serous papillary endometrial adenocarcinoma. *Gynecologic Oncology, 103*(2), 591–598.

Sobel, G., Paska, C., Szabo, I., Kiss, A., Kadar, A., & Schaff, Z. (2005). Increased expression of claudins in cervical squamous intraepithelial neoplasia and invasive carcinoma. *Human Pathology, 36*(2), 162–169.

Sobel, G., Szabo, I., Paska, C., Kiss, A., Kovalszky, I., Kadar, A., et al. (2005). Changes of cell adhesion and extracellular matrix (ECM) components in cervical intraepithelial neoplasia. *Pathology Oncology Research, 11*(1), 26–31.

Soini, Y. (2004). Claudins 2, 3, 4, and 5 in Paget's disease and breast carcinoma. *Human Pathology, 35*(12), 1531–1536.

Soini, Y., & Talvensaari-Mattila, A. (2006). Expression of claudins 1, 4, 5, and 7 in ovarian tumors of diverse types. *International Journal of Gynecological Pathology, 25*(4), 330–335.

Soini, Y., Kinnula, V., Kahlos, K., & Paakko, P. (2006). Claudins in differential diagnosis between mesothelioma and metastatic adenocarcinoma of the pleura. *Journal of Clinical Pathology, 59*(3), 250–254.

Soini, Y., Tommola, S., Helin, H., & Martikainen, P. (2006). Claudins 1, 3, 4 and 5 in gastric carcinoma, loss of claudin expression associates with the diffuse subtype. *Virchows Archiv, 448*(1), 52–58.

Soma, T., Chiba, H., Kato-Mori, Y., Wada, T., Yamashita, T., Kojima, T., et al. (2004). Thr (207) of claudin-5 is involved in size-selective loosening of the endothelial barrier by cyclic AMP. *Experimental Cell Research, 300*(1), 202–212.

Song, X., Li, X., Tang, Y., Chen, H., Wong, B., Wang, J., et al. (2008). Expression of claudin-2 in the multistage process of gastric carcinogenesis. *Histology and Histopathology, 23*(6), 673–682.

Steinau, M., Rajeevan, M. S., Lee, D. R., Ruffin, M. T., Horowitz, I. R., Flowers, L. C., et al. (2007). Evaluation of RNA markers for early detection of cervical neoplasia in exfoliated cervical cells. *Cancer Epidemiology, Biomarkers and Prevention, 16*(2), 295–301.

Stewart, J. J., White, J. T., Yan, X., Collins, S., Drescher, C. W., Urban, N. D., et al. (2006). Proteins associated with Cisplatin resistance in ovarian cancer cells identified by quantitative proteomic technology and integrated with mRNA expression levels. *Molecular & Cellular Proteomics, 5*(3), 433–443.

Suzuki, M., Kato-Nakano, M., Kawamoto, S., Furuya, A., Abe, Y., Misaka, H., et al. (2009). Therapeutic antitumor efficacy of monoclonal antibody against Claudin-4 for pancreatic and ovarian cancers. *Cancer Science, 100*(9), 1623–1630.

Szabo, I., Kiss, A., Schaff, Z., & Sobel, G. (2009). Claudins as diagnostic and prognostic markers in gynecological cancer. *Histology and Histopathology, 24*(12), 1607–1615.

Szasz, A. M., Nyirady, P., Majoros, A., Szendroi, A., Szucs, M., Szekely, E., et al. (2009). Beta-catenin expression and claudin expression pattern as prognostic factors of prostatic cancer progression. *BJU International.*

Takala, H., Saarnio, J., Wiik, H., & Soini, Y. (2007). Claudins 1, 3, 4, 5 and 7 in esophageal cancer: Loss of claudin 3 and 4 expression is associated with metastatic behavior. *Apmis, 115*(7), 838–847.

Takehara, M., Nishimura, T., Mima, S., Hoshino, T., & Mizushima, T. (2009). Effect of claudin expression on paracellular permeability, migration and invasion of colonic cancer cells. *Biological and Pharmaceutical Bulletin, 32*(5), 825–831.

Tanaka, M., Kamata, R., & Sakai, R. (2005). EphA2 phosphorylates the cytoplasmic tail of Claudin-4 and mediates paracellular permeability. *The Journal of Biological Chemistry, 280*(51), 42375–42382.

Tassi, R. A., Bignotti, E., Falchetti, M., Ravanini, M., Calza, S., Ravaggi, A., et al. (2008). Claudin-7 expression in human epithelial ovarian cancer. *International Journal of Gynecological Cancer, 18*(6), 1262–1271.

Tatum, R., Zhang, Y., Lu, Q., Kim, K., Jeansonne, B. G., & Chen, Y. H. (2007). WNK4 phosphorylates ser(206) of claudin-7 and promotes paracellular Cl(-) permeability. *FEBS Letters, 581*(20), 3887–3891.

Terris, B., Blaveri, E., Crnogorac-Jurcevic, T., Jones, M., Missiaglia, E., Ruszniewski, P., et al. (2002). Characterization of gene expression profiles in intraductal papillary-mucinous tumors of the pancreas. *American Journal of Pathology, 160*(5), 1745–1754.

Thway, K., Fisher, C., Debiec-Rychter, M., & Calonje, E. (2009). Claudin-1 is expressed in perineurioma-like low-grade fibromyxoid sarcoma. *Human Pathology, 40*(11), 1586–1590.

Tokes, A. M., Kulka, J., Paku, S., Szik, A., Paska, C., Novak, P. K., et al. (2005). Claudin-1, -3 and -4 proteins and mRNA expression in benign and malignant breast lesions: A research study. *Breast Cancer Research, 7*(2), R296–R305.

Tsukahara, M., Nagai, H., Kamiakito, T., Kawata, H., Takayashiki, N., Saito, K., et al. (2005). Distinct expression patterns of claudin-1 and claudin-4 in intraductal papillary-mucinous tumors of the pancreas. *Pathology International, 55*(2), 63–69.

Tsukita, S., & Furuse, M. (2000). Pores in the wall: Claudins constitute tight junction strands containing aqueous pores. *The Journal of Cell Biology, 149*(1), 13–16.

Tzelepi, V. N., Tsamandas, A. C., Vlotinou, H. D., Vagianos, C. E., & Scopa, C. D. (2008). Tight junctions in thyroid carcinogenesis: Diverse expression of claudin-1, claudin-4, claudin-7 and occludin in thyroid neoplasms. *Modern Pathology, 21*(1), 22–30.

Ueda, J., Semba, S., Chiba, H., Sawada, N., Seo, Y., Kasuga, M., et al. (2007). Heterogeneous expression of claudin-4 in human colorectal cancer: Decreased claudin-4 expression at the invasive front correlates cancer invasion and metastasis. *Pathobiology, 74*(1), 32–41.

Usami, Y., Chiba, H., Nakayama, F., Ueda, J., Matsuda, Y., Sawada, N., et al. (2006). Reduced expression of claudin-7 correlates with invasion and metastasis in squamous cell carcinoma of the esophagus. *Human Pathology, 37*(5), 569–577.

Usami, Y., Satake, S., Nakayama, F., Matsumoto, M., Ohnuma, K., Komori, T., et al. (2008). Snail-associated epithelial-mesenchymal transition promotes oesophageal squamous cell carcinoma motility and progression. *The Journal of Pathology, 215*(3), 330–339.

Van Itallie, C. M., & Anderson, J. M. (2006). Claudins and epithelial paracellular transport. *Annual Review of Physiology, 68*, 403–429.

Van Itallie, C. M., Gambling, T. M., Carson, J. L., & Anderson, J. M. (2005). Palmitoylation of claudins is required for efficient tight-junction localization. *Journal of Cell Science, 118*(Pt 7), 1427–1436.

Vare, P., Loikkanen, I., Hirvikoski, P., Vaarala, M. H., & Soini, Y. (2008). Low claudin expression is associated with high Gleason grade in prostate adenocarcinoma. *Oncology Reports, 19*(1), 25–31.

Vazquez-Ortiz, G., Ciudad, C. J., Pina, P., Vazquez, K., Hidalgo, A., Alatorre, B., et al. (2005). Gene identification by cDNA arrays in HPV-positive cervical cancer. *Archives of Medical Research, 36*(5), 448–458.

Vincent, T., Neve, E. P., Johnson, J. R., Kukalev, A., Rojo, F., Albanell, J., et al. (2009). A SNAIL1-SMAD3/4 transcriptional repressor complex promotes TGF-beta mediated epithelial-mesenchymal transition. *Nature Cell Biology, 11*(8), 943–950.

Weber, C. R., Nalle, S. C., Tretiakova, M., Rubin, D. T., & Turner, J. R. (2008). Claudin-1 and claudin-2 expression is elevated in inflammatory bowel disease and may contribute to early neoplastic transformation. *Laboratory Investigation, 88*(10), 1110–1120.

Wu, X., Chen, H., Parker, B., Rubin, E., Zhu, T., Lee, J. S., et al. (2006). HOXB7, a homeodomain protein, is overexpressed in breast cancer and confers epithelial-mesenchymal transition. *Cancer Research, 66*(19), 9527–9534.

Wu, C. M., Lee, Y. S., Wang, T. H., Lee, L. Y., Kong, W. H., Chen, E. S., et al. (2006). Identification of differential gene expression between intestinal and diffuse gastric cancer using cDNA microarray. *Oncology Reports, 15*(1), 57–64.

Wu, Y. L., Zhang, S., Wang, G. R., & Chen, Y. P. (2008). Expression transformation of claudin-1 in the process of gastric adenocarcinoma invasion. *World Journal of Gastroenterology, 14* (31), 4943–4948.

Yamamoto, T., Kojima, T., Murata, M., Takano, K., Go, M., Chiba, H., et al. (2004). IL-1beta regulates expression of Cx32, occludin, and claudin-2 of rat hepatocytes via distinct signal transduction pathways. *Experimental Cell Research, 299*(2), 427–441.

Yamamoto, M., Ramirez, S. H., Sato, S., Kiyota, T., Cerny, R. L., Kaibuchi, K., et al. (2008). Phosphorylation of claudin-5 and occludin by rho kinase in brain endothelial cells. *American Journal of Pathology, 172*(2), 521–533.

Yamauchi, K., Rai, T., Kobayashi, K., Sohara, E., Suzuki, T., Itoh, T., et al. (2004). Disease-causing mutant WNK4 increases paracellular chloride permeability and phosphorylates claudins. *Proceedings of the National Academy of Sciences of the United States of America, 101*(13), 4690–4694.

Yasui, W., Oue, N., Sentani, K., Sakamoto, N., & Motoshita, J. (2009). Transcriptome dissection of gastric cancer: Identification of novel diagnostic and therapeutic targets from pathology specimens. *Pathology International, 59*(3), 121–136.

Yuan, X., Lin, X., Manorek, G., Kanatani, I., Cheung, L. H., Rosenblum, M. G., et al. (2009). Recombinant CPE fused to tumor necrosis factor targets human ovarian cancer cells expressing the claudin-3 and claudin-4 receptors. *Molecular Cancer Therapeutics, 8*(7), 1906–1915.

Zhang, K., Yao, H. P., & Wang, M. H. (2008). Activation of RON differentially regulates claudin expression and localization: Role of claudin-1 in RON-mediated epithelial cell motility. *Carcinogenesis, 29*(3), 552–559.

Zhu, Y., Brannstrom, M., Janson, P. O., & Sundfeldt, K. (2006). Differences in expression patterns of the tight junction proteins, claudin 1, 3, 4 and 5, in human ovarian surface epithelium as compared to epithelia in inclusion cysts and epithelial ovarian tumours. *International Journal of Cancer, 118*(8), 1884–1891.

Index

U

V

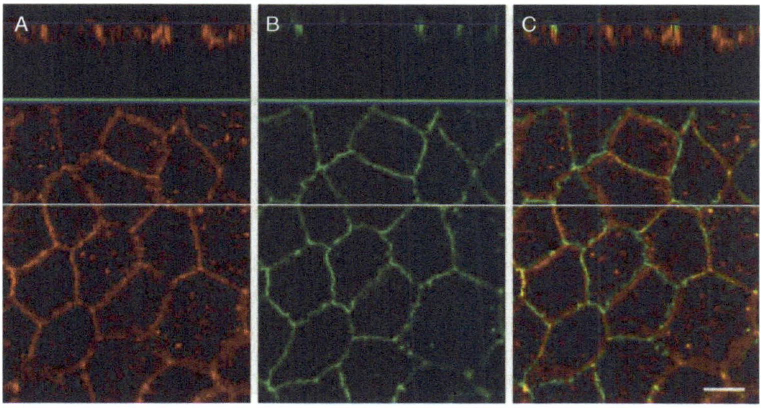

Please refer to Fig. 2 of Chapter 2 in the text for figure legend.

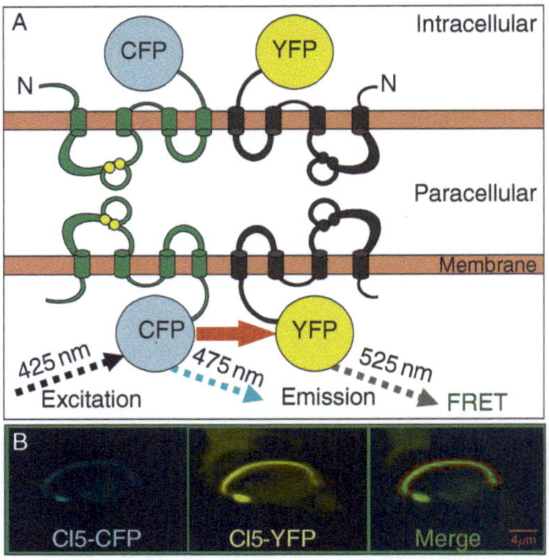

Please refer to Fig. 5 of Chapter 5 in the text for figure legend.

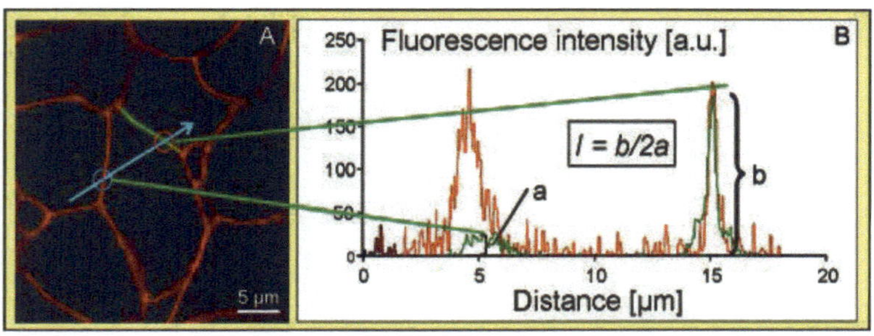

Please refer to Fig. 6 of Chapter 5 in the text for figure legend.

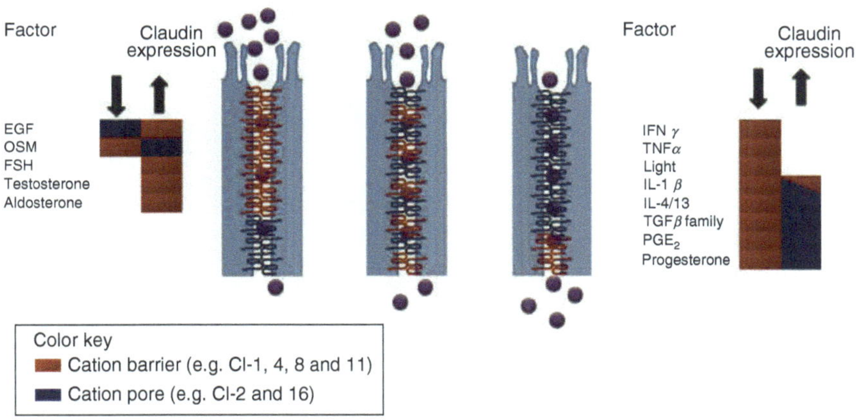

Please refer to Fig. 2 of Chapter 6 in the text for figure legend.

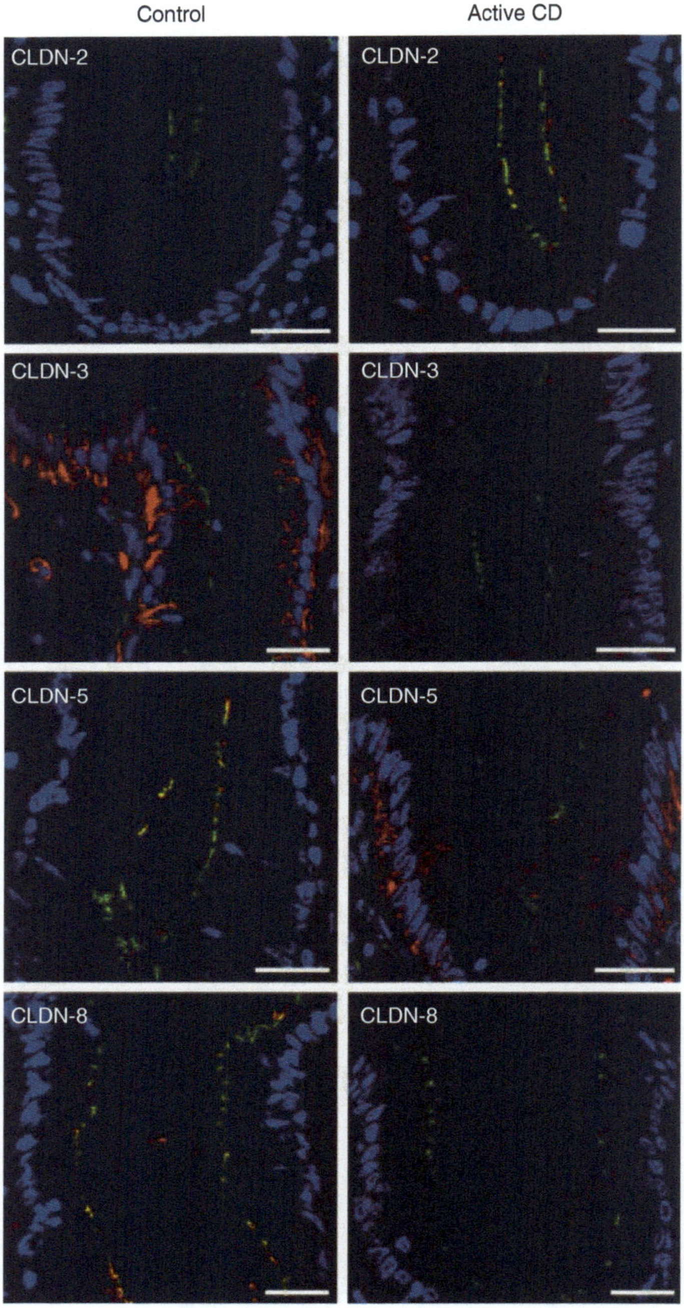

Please refer to Fig. 1 of Chapter 9 in the text for figure legend.

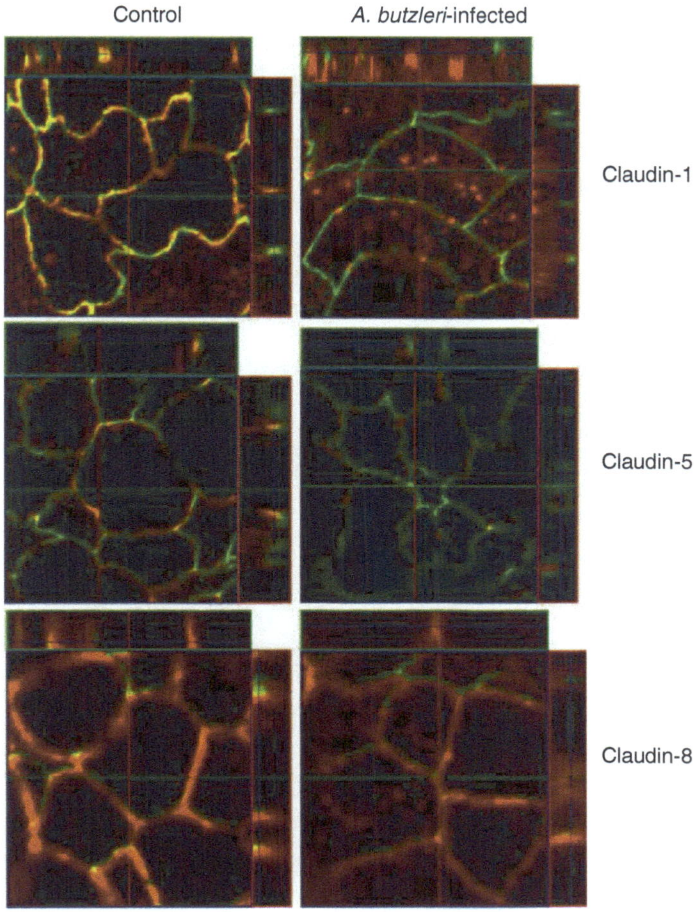

Please refer to Fig. 3 of Chapter 9 in the text for figure legend.

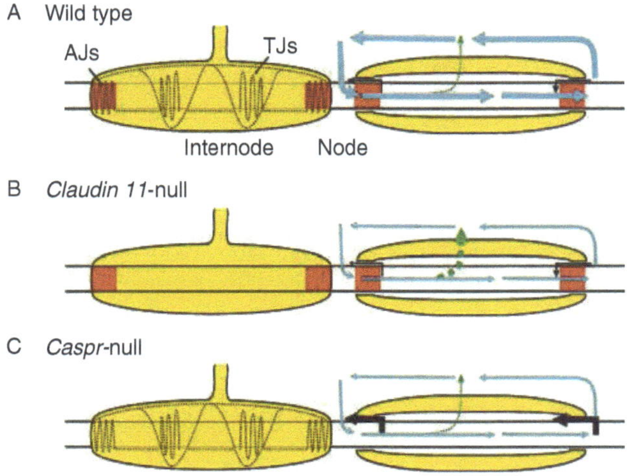

Please refer to Fig. 5 of Chapter 10 in the text for figure legend.

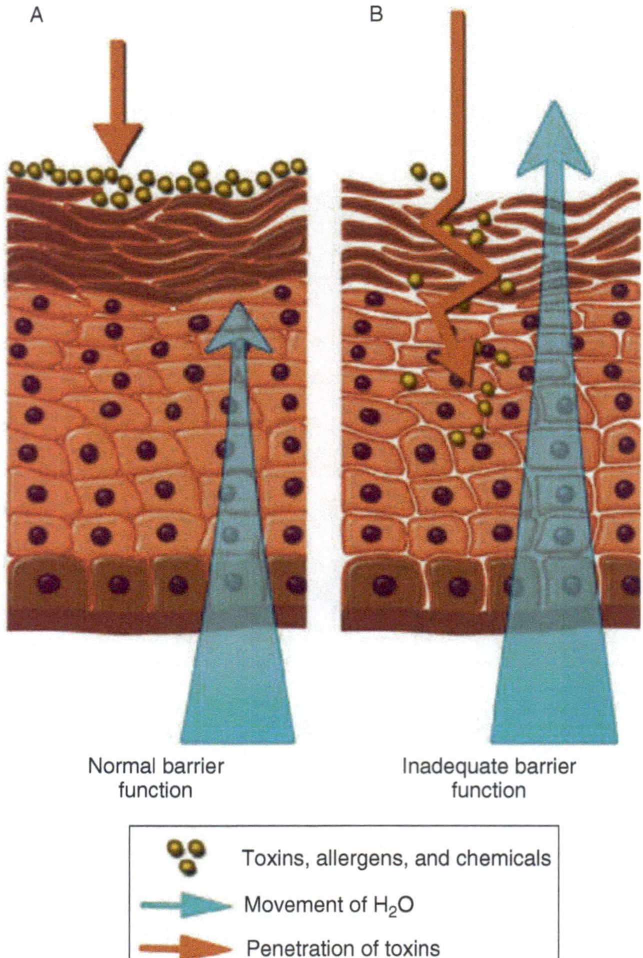

Please refer to Fig. 2 of Chapter 11 in the text for figure legend.

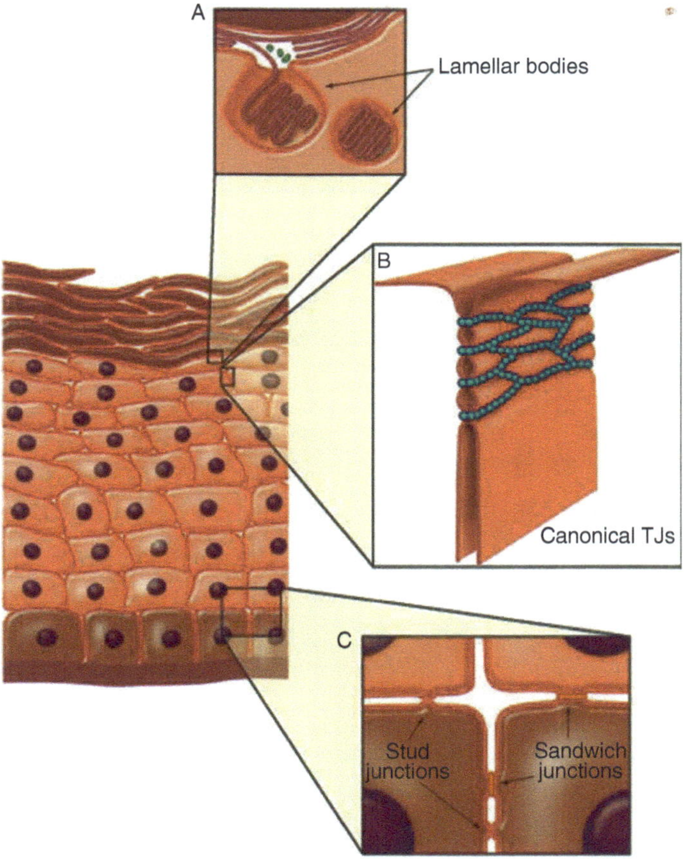

Please refer to Fig. 3 of Chapter 11 in the text for figure legend.

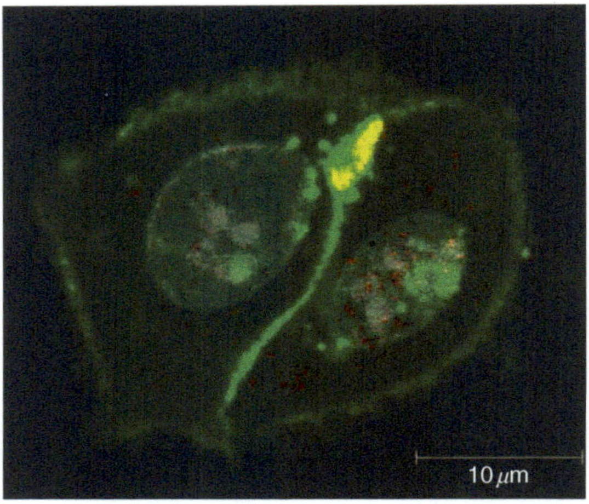

Please refer to Fig. 4 of Chapter 12 in the text for figure legend.

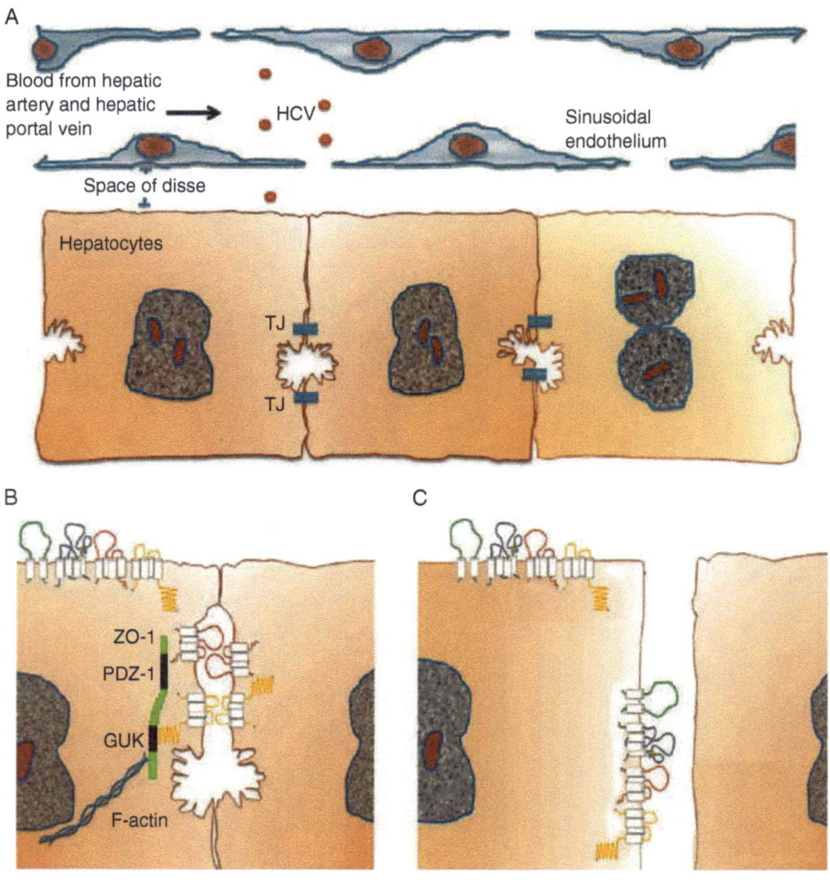

Please refer to Fig. 5 of Chapter 12 in the text for figure legend.

FSC
www.fsc.org
MIX
Papier | Fördert
gute Waldnutzung
FSC® C083411

Kolibri 360 GmbH
Ferdinand-Jühlke-Straße 7
99095 Erfurt, Deutschland
produktsicherheit@kolibri360.de